Eco-Pioneers

Practical Visionaries Solving Today's
Environmental Problems

Steve Lerner

The MIT Press
Cambridge, Massachusetts
London, England

This book was set in Sabon by Graphic Composition.

Printed and bound in the United States of America.

Library of Congress Cataloging-in-Publication Data
Lerner, Steve.
    Eco-pioneers : practical visionaries solving today's environmental problems / by Steve Lerner.
      p. cm.
    Includes bibliographical references and index.
    ISBN 0-262-12207-3
    1. Environmental degradation—United States—Case studies. 2. Sustainable development—United States—Case studies. 3. Environmental engineers—United States—Biography. 4. Environmentalists—United States—Biography.
I. Title.
GE150.L47 1997
363.7'0092'273—dc21
  [B]   97-8333
                                              CIP

Eco-Pioneers is dedicated to my mother and father, Edna and Max Lerner, who gave me a sense of adventure and a love of the wild.

Contents

Foreword

The great contribution of Steve Lerner's *Eco-Pioneers* is that it provides a rigorously documented and wide-ranging set of case studies of the men and women involved in creating sustainable forms of industry, agriculture, and human settlement. More important still, because Lerner describes cases selected from the United States, he sets out the parameters within which we can begin to look more closely at sustainable development in this country.

The book describes the efforts of eco-pioneers who are working on a range of fronts to sustainably harness the environment in a way that allows us to meet human needs and aspirations. The case studies show positive examples—transforming hazardous wastes into useful industrial materials; treating sewage with organic methods; building houses with sustainably harvested or recycled materials; raising crops with less use of pesticides and chemical fertilizers; minimizing the waste stream by diverting reusable goods; generating and saving energy by using renewable sources and energy conservation measures; providing inner-city jobs while cleaning up the urban environment; producing lumber by practicing a selective method of forestry; and teaching people to consume less while enjoying life more.

These real and encouraging stories reflect my own experience as co-chair of the President's Council on Sustainable Development. As we—the industry executives, environmentalist, labor, and civil rights leaders, and Cabinet members who sit on the council—looked around the country we found individuals, communities, and companies engaged in creative and effective strategies to protect the environment, create jobs, and improve lives.

I find the same to be true around the world. From Brazil to Scandinavia, I have seen anecdotal evidence of change—crucial demonstrations of what is possible when societies' will increases in the face of threats to the global commons as the connection between rich and poor, big and small, and present and future becomes clearer.

I believe that the uses of this book should be many. For the broad community of Americans concerned with the enormous challenge of achieving a sustainable future, *Eco-Pioneers* provides a new sense of hope that practical solutions to our environmental problems are already being discovered in a number of different fields.

For environmentalists, government officials, and those involved in the international movement for a sustainable future, *Eco-Pioneers* describes many of the state of the art efforts to achieve sustainable development in the United States.

For those teaching environment and sustainable development courses at colleges and universities, *Eco-Pioneers* should be required reading for any course that seeks to convey the issues and challenges that those committed to a sustainable future are wrestling with.

Perhaps most important, in an age in which young people are offered few authentic heroes, it would not be excessive to suggest that these "eco-pioneers" are among those whose contribution to humanity would deserve to be called heroic. For people young and old who wish to go beyond the descriptions of all that we have to overcome to achieve a sustainable future, Steve Lerner's *Eco-Pioneers* provides, if not a thousand, at least several dozen very real points of light.

JONATHAN LASH
President, World Resources Institute

Acknowledgments

This book would not have been written without the support of Commonweal, a nonprofit organization based in Bolinas, California, which focuses on environmental and health issues. Special thanks is due to my brother Michael Lerner, president of Commonweal, who provided invaluable advice and support as the manuscript took shape. Nor would the writing of the book have been possible without the help of my wife, Mary Ellin Barrett, who accompanied me on the initial trip around the country searching out eco-pioneers and helped with the editing.

Nine chapters of the book were previously published in somewhat altered form in two magazines. The Natural Resources Defense Council's *Amicus Journal* published the chapters on Chattanooga, Tennessee; Molten Metal Technology; eco-justice activists; Lorrie Otto; and Daniel Einstein and David Eagan. *Audubon Magazine* published the chapters on Daniel Knapp, Scott Bernstein, S. David Freeman, and the Reverend Jeffrey M. Golliher.

I am also indebted to my publisher Frank Urbanowski; my editors Madeline Sunley and Melissa Vaughn; Kate Mertes who compiled the index; and Philip Perlman, Michael Rafferty, and Andrew Sherman who helped solve computer-related problems.

Introduction

A new breed of environmental pioneer is emerging on the American landscape. With the inventive genius for which Americans are justifiably proud, these trailblazers are demonstrating that we can provide for the needs of citizens today without degrading the environment to the point where future generations are unable to meet their needs.

In both urban and rural settings, ecological innovators are modeling ways to log forests, grow food, raise livestock, manufacture goods, construct houses, build transportation systems, generate power, reuse materials, reduce waste, and design sustainable communities while minimizing damage to the web of life.

I call these men and women "eco-pioneers" because they are modern pathfinders who are mapping out a sustainable future for our nation. With little encouragement, they are working to reverse the accelerating pace of environmental degradation so that our children and grandchildren will not be forced to live in ecologically impoverished circumstances.

My search for American eco-pioneers began in June 1992, when I returned from the Earth Summit in Brazil. For the previous two years I had followed the tortuous negotiations at the United Nations Conference on Environment and Development (UNCED) that led up to the Earth Summit. And although the historical importance of developing a framework for international cooperation on environmental issues was not lost on me, I hungered for something more grounded than international conferences—some evidence that sustainable development was indeed possible.

Upon my return to the United States, I went up to the house I had built in Vermont in the early 1970s and immersed myself in some renovations, replacing the old single-pane windows with more energy-efficient ones and installing a composting toilet. These modest yet practical tasks were a refreshing counterpoint to the turgid international negotiations I had witnessed in Geneva and Rio. After a couple of months of this work, my wife Mary Ellin, our one-year-old son Benjamin, and I packed up our pickup truck and began a 10,000-mile trip meandering across the country looking for people who were involved in the nuts and bolts of sustainable development.

When I told people what I was doing, many of them laughed: "If you plan to write a book about people who solve environmental problems, I bet it will be a really short one," some wit observed. But as I began to dig into the subject, I found that a network of practitioners of sustainable development was already being formed, and as we traveled the country we were passed from one eco-pioneer to another.

Eco-pioneers are not just talking about the theory of sustainable development: they are practical visionaries and backyard mechanics who are going into the field to put sustainable development into practice. They are willing to get their hands dirty and experiment with new (often imperfect) approaches to solving problems. While none of them has invented a magic elixir or instant technological fix for the damage we are doing to the web of life, each has taken on a discrete environmental dilemma and immersed himself or herself in the mechanics of solving it.

John Todd is one such eco-pioneer. A biologist who co-founded the New Alchemy Institute and subsequently Ocean Arks International in Falmouth, Massachusetts, Todd invented a "living machine" for treating municipal sewage that uses the purifying power of nature to treat wastes.

On a typical day, Todd can be found in one of his greenhouses filling large tanks of water with a unique cocktail of microorganisms, algae, aquatic plants, snails, fish, and even some species of trees that don't mind getting their feet wet. This designer blend of marsh life feeds off the nutrients in the sewage, which, if released untreated, would cause algal blooms, which in turn use up dissolved oxygen in water and cause fish to suffocate and lakes and streams to become eutrophic.

In addition to consuming the nutrients in wastewater, the aerobic and anaerobic phases of Todd's living machines also break down toxic chemicals found in sewage, rendering them harmless. Certain aquatic plants and trees take up and sequester heavy metals found in the sewage. By the time the wastewater exits the greenhouse it has been treated to the point where it can be safely released.

To see Todd in one of his greenhouses pouring a bucket of swamp creatures into a living machine is to get a glimpse of where the movement to restore nature is headed. Watching Todd, one can discern a shift in the strategy of the environmental movement from an endless battle to regulate and clean up after environmentally destructive technologies to the search for more energy-efficient and less polluting methods of meeting human needs. With this shift, a tremendous amount of intellectual energy is unleashed as numerous avenues of inquiry open up.

For example, Chris Nagel and Bill Haney put their talents to work looking for a method to transform hazardous wastes into useful industrial materials. Nagel, a chemical engineer (who worked initially at U.S. Steel), devised a process that uses molten metal boiling at 3000° to break up complex toxic chemicals into their elemental components, thus permitting useful industrial materials to be siphoned off and reused. Haney, an eco-entrepreneur, turned this stroke of genius into Molten Metal Technology, Inc., a well-capitalized company headquartered near Boston that now recovers reusable elements in both radioactive and chemical wastes.

Eco-pioneers such as Todd, Nagel, and Haney can be distinguished by their uncanny ability to take a problem—such as that posed by the huge volumes of waste we generate—and craft a solution that provides both environmental benefits and jobs. "Conserving natural resources generally gives rise to more jobs than wasting them," writes Jonathan Rowe, who is working on an alternative to the Gross Domestic Product that will be an index of sustainable progress. "Investment in energy-saving technologies provides four times as many jobs as building new power plants. Recycling employs ten times as many people as dumping or incinerating the same amount of trash," he continues.[1]

This ability to extract value from what appears to be useless mimics the function of natural systems where the waste of one organism provides food for another. As Nancy Jack Todd and John Todd point out in their groundbreaking work *Bioshelters, Ocean Arks, City Farming: Ecology as the Basis of Design*:

There is a science to working with existing forms and structures. It is comprised of a peculiar mix of theory, research, and practicality—a science of "found objects." It does not attempt to build from scratch, but takes what exists and works to transform it to something useful or relevant. The French anthropologist Claude Lévi-Strauss has described it as *bricolage*. Practitioners are *bricoleurs*, which translates rather clumsily as 'enlightened tinkerers with what is at hand.' In an age of increasing scarcity, such a person is potentially a kind of hero, someone who can see with different eyes and utilize available resources.[2]

Paul and Julie Mankiewicz are two "bricoleurs" who want to transform the rooftops of Manhattan into urban farms. They believe that urban farming could result in a host of economic and environmental benefits such as reducing the need to transport fresh food over long distances while providing entry-level, inner-city jobs. But for years large-scale rooftop farming was impractical because normal soil (especially when watered) was so heavy that it could cause a roof to collapse unless it was reinforced at great expense.

In an inspired bit of tinkering, Mankiewicz developed a superlightweight soil made of an improbable blend of shredded, post-consumer Styrofoam and compost. To provide the compost for this lightweight soil, he conceived an ingenious system for collecting urban organic waste, which currently ends up in the landfill in most areas. Mankiewicz envisions sanitation workers driving small electric vehicles at night along deserted city sidewalks to collect urban kitchen scraps. This organic material would then be delivered to one of the highly efficient, fan-driven, odorless, compost units Mankiewicz designed, which can operate in the basements of apartment buildings. Together, these two technological breakthroughs—the lightweight soil and the urban composting units—make it possible to generate new jobs growing food in the city.

Other eco-pioneers are seeing how solutions to environmental problems can provide jobs and improve the quality of life in inner cities. For example, Scott Bernstein, president of the Center for Neighborhood Technology in Chicago, helps owners of small electroplating shops gain access to bank loans that allow them to buy equipment that reduces the volume of toxic waste they generate and lowers occupational exposure to hazardous chemicals. This permits electroplaters to meet increasingly stringent environmental standards, stay in business, and provide stable blue-collar jobs.

Bernstein relies on his community organizing skills to improve the environmental quality of urban life. He facilitates loans for the owners of small apartment buildings so they can retrofit their buildings to be more energy-efficient, thus improving their profit margin while keeping rents low enough to provide affordable housing. He also convinced officials at Metra, Chicago's commuter rail service, to change their schedule so that trains stop at inner-city stations in the morning and evening, allowing urban workers to ride mass transit to the jobs opening up in the suburbs.

In fact, once one begins to look around, practitioners of sustainable development can be found in all corners of American society. A remarkably heterogeneous group, eco-pioneers cross all geographic, political, racial, class, age, and gender lines.

Take Juana Beatríz Gutierrez, a resident in a low-income neighborhood in East Los Angeles, who saw how a water conservation program could provide jobs in her community where unemployment is high. Working with city officials, Gutierrez designed a program that distributes water-conserving toilets free to local residents in return for their old, water-guzzling, porcelain toilets, which are smashed to pieces and recycled as an underlayment for the streets of L.A. The money she earns from administering the program is used to hire young people who go door to door urging families to immunize their children and get them tested for lead poisoning.

Business acumen and entrepreneurial skills are also being used to solve environmental problems. Far from the inner cities, in Willapa Bay, Washington, a temperate rain forest that is one of the most productive ecosystems in North America, Alana Probst, a conservation-based development economist, is looking for ways to promote sustainable businesses that give local entrepreneurs a stake in protecting the natural resource base on which their enterprises depend. To this end she organized a bank that provides startup loans for sawmill operators, cranberry farmers, and fishermen who build ecologically sustainable businesses that provide local jobs while protecting the environment.

Eco-pioneers come in many guises. Some of them do not start out as environmentalists, nor are they poster material for environmental organizations. For instance, on first meeting Jack Turnell one would not immediately peg him as a practitioner of sustainable development. The manager of the Pitchfork Ranch in Meeteetse, Wyoming, Turnell hosted photo shoots for Marlboro Man cigarette ads on his ranch, featuring cowboys riding mustangs over the range.

Purists will argue that Turnell should not be included in a book about eco-pioneers because he runs cattle on arid western lands that are easily damaged by overgrazing. But I include Turnell (and some other controversial characters) because at a minimum they are improving environmental practices within their industry. They do not live in an ideal world where only pristine environmental practices are permitted; they live in the real world where compromises abound.

Turnell now employs a more sustainable method of ranching than the one he grew up with. He was sensitized to environmental issues and began looking for ways to improve his operation when an endangered species was found on his land. After a long search, he discovered a foreign breed of cattle that does less damage to the fragile ecosystem along his creeks than do his domestic cattle. He also reduced the amount of erosion his livestock cause as they climb to higher pastures during the summer by splitting his cattle into small units and herding them up separate watersheds.

Eco-pioneers may also be found among some of our oldest citizens who are living repositories of important information about how, in certain respects, people from previous generations lived in greater harmony with nature. In Franklin, North Carolina, eighty-five-year-old Walton Smith, a forester with fifty years' experience selectively cutting trees, points out that the U.S. Forestry Service practiced sustainable forestry in many areas up until the 1960s and that we could revive those methods. Smith, who had a long career with the Forest Service, is demonstrating that homeowners can improve the environmental quality and economic value of their land by practicing a sustainable method of forestry. A wood utilization expert by training, Smith cuts down mature trees and saws them up for lumber, but he insists that they be harvested in a fashion that makes only a limited hole in the forest canopy and that the felled logs be extracted with a rubber-tire tractor on well-designed logging roads to minimize soil erosion.

While many eco-pioneers, such as Smith, labor in relative obscurity for years, others are in great demand and are now actively recruited by communities that have decided to make sustainable development a central theme in their planning process. These communities are looking for help to jump-start zero-emissions industries that are energy and resource efficient. In Chattanooga, Tennessee, for example, Councilman David Crockett, a distant relative of the renowned frontiersman, is attempting to reshape his hometown as America's premier environmental city. Similarly, the residents of Pattonsburg, Missouri, a town that flooded frequently, decided to move to higher ground and rebuild along ecologically sustainable lines.

Among those called in to help with the urban renewal blueprint for Chattanooga was architect William McDonough, who seeks out sustainably harvested lumber for his buildings and becomes involved in redesign of skylights and carpets so that they will be more energy efficient and less toxic. In Pattonsburg, one of the experts imported to help with the design of the new town was Pliny Fisk III, an architect and inventor based in Austin, Texas, who searches out local building materials that can be used to replace unsustainably harvested lumber trucked in at great expense over long distances. One of the materials Fisk uses is called AshCrete, a substance he invented that is hard as cement yet made out of a waste product—the fly ash from coal-fired power plants and the bottom ash from aluminum smelters.

Other eco-pioneers are breaking new ground on the frontier of behavioral change by helping Americans waste less and overcome their addiction to a high-consumption lifestyle. Vicki Robin and Joe Dominguez, at the New Road Map Foundation in Seattle, are teaching people how to save money, free up more of their time to pursue their real interests, and preserve resources all at the same time. From the headquarters of the Global Action Plan in Connecticut, David Gershon offers families a four-month behavior modification program that enables them to radically reduce the volume of garbage they generate, their water and fuel consumption, and their output of carbon dioxide.

Similarly, in the spiritual arena, some members of the clergy and rabbis are beginning to question what spiritual and psychological changes will enable people to desist from despoiling nature and learn to cooperate in building a socially just and ecologically sustainable society. Among them is the Reverend Jeffrey M. Golliher, an Episcopal priest who travels from parish to parish in New England preaching a green gospel of reducing consumption and protecting God's creation.

These and other eco-pioneers constitute a new phase of the environmental movement. Until now, much of the energy of environmentalists has necessarily been devoted to stopping bad things from happening. Unfortunately, playing defense has meant they rarely have an opportunity to devise front-end strategies for reducing waste. Warning the public about environmental dangers has also left them with the unenviable job of being the harbingers of bad news.

My own work, writing for several decades about environmental threats to public health, earned me the nickname Grim Reaper. Those brave enough to invite me to dinner joked about my penchant for spicing mealtime conversations with diatribes about the pesticides in food or the toxic chemicals out-gassing from carpets. It took me years to learn to couch bad news such as this in the context of what could be done about it.

At least for the foreseeable future, environmentalists will continue to be obliged to play this role of public watchdog, warning us about specific threats to the environment and public health, organizing resistance to environmental degradation, and lobbying for legislative change. Keeping the public aware of the incremental damage we do to nature and public health is an important tradition, and alerting the public to the downside of emerging technologies will remain a necessary function.

Yet I cannot disguise my delight that eco-pioneers are now providing an upbeat note to a chorus that has been dominated for too long by the drumbeat of eco-dread. Environmentalists have known for some time now that they must move beyond the role of warning us that the sky is falling to the more demanding task of telling us what can be done about it. Thus, in addition to providing an astute critique of the unsustainable aspects of our society, environmentalists must also seek out and promote better ways of living and doing business that are less harmful to the environment.

This task requires a profound shift in mindset: something akin to teaching a professional wrestler the subtle art of aikido. Instead of attempting to dominate nature by brute force, today's environmental pioneers are experimenting with ways to redesign human activity so that it is less destructive to the environment and more compatible with natural cycles. Such ecological designs "adopt the wisdom and strategies of the natural world to human problems," observe the Todds.[3]

Unfortunately, there is no concerted federal effort to nurture this kind of activity, nor is there an official commitment to reorient our economy along ecologically sustainable lines. There is, in fact, no widely agreed-upon "green plan" for sustainable development in the United States—such as that advocated by Huey Johnson at the Resources Renewal Institute in San Francisco—because, as yet, there is no consensus that our economic system requires a fundamental overhaul.

The lack of a national green plan, the refusal to shift to green taxes, and the absence of a commitment to sustainable development are curious given that American diplomats are busy urging officials of developing nations to adopt sustainable development methods to avoid the pitfalls of environmental degradation.

I had a chance to witness this duplicity in action during the two years I spent following the Earth Summit negotiations. During that period I did a series of interviews with some of the leading thinkers and practitioners of sustainable development from around the world, which I compiled in two books entitled *Earth Summit: Conversations with Architects of an Ecologically Sustainable Future* and *Beyond the Earth Summit: Conversations with Advocates of Sustainable Development.*[4]

While doing the research for this project I was struck by the profound contradiction involved in U.S. diplomats exhorting officials of other nations, particularly poor ones, to practice sustainable development, while Americans continued to pursue a less admirable and less sustainable policy at home.

The message was clear: Do as we say and not as we do. Residents in developing nations were entreated to skip the unsustainable practices that had led Americans to great wealth and power in this country. Having cut down our ancient forests, planted our plains with monocrops dependent on large doses of pesticides and chemical fertilizers, permitted urban sprawl to pave prime agricultural land, dammed our rivers, digested unprecedented quantities of natural resources drawn from all over the world, and burned vast quantities of coal and oil, we seemed to have few qualms about lecturing others on the virtues of sustainable development.

While conciliatory and politically correct statements were made by U.S. representatives at the Earth Summit, beneath the diplomatic double talk the subtext was clearly that sustainable development was mostly applicable to developing nations and not to advanced industrialized countries such as our own. The absurdity of this proposition was not lost on representatives of developing nations who pointed out that the origins of the most pressing global environmental problems (ozone depletion, climate change, deforestation, extinction of species, toxic waste generation, and overconsumption of resources) were rooted in the unsustainable practices carried out on a massive scale by citizens of the so-called developed nations.

Americans and citizens of other industrialized nations not only use more than their fair share of energy and resources, they also appropriate more than their quota of "ecological space"—a concept coined to describe the finite capacity of the earth to produce the resources and absorb the byproducts of human populations. The "ecological footprint" of the average North American requires at least ten to twelve acres of productive land to support his or her consumer lifestyle, writes William E. Rees, director of the School of Community and Regional Planning at the University of British Columbia. "To bring just the present world population of almost 6 billion up to North American standards would require at least two additional earths, or else a three-fold improvement in efficiency of resource use and capacity for waste assimilation," he adds.[5] An oft-cited example of the limits of ecological space is the proposition that humans can only burn a finite amount of fossil fuels before they choke on the airborne pollutants and cook in the global warming that results from this combustion.

Representatives of developing nations now argue that heavily industrialized nations have used up more than their fair share of ecological space, leaving an inadequate supply for the world's poor to develop. Statistics to back up this contention are not hard to find. For instance, while the United States is responsible for 22 percent of all carbon dioxide emissions, India, with a population three times that of the United States, produces only 3% of CO_2 emissions.[6] By another measure, average annual per capita carbon emissions in the United States are 5.13 tons compared with 0.09 in India.[7]

In the negotiations between highly industrialized and developing nations over the apportionment of ecological space, the deal that is emerging will require that industrialized nations increase aid to developing nations, cut consumption, and use resources more efficiently, while Third World nations pursue development strategies that are ecologically sustainable. Already the United States and Europe are being pressed to reduce greenhouse gas emissions, but for this to occur, significant technological and behavioral change will be necessary.

It was while witnessing these glacial, stilted, and often depressing international negotiations that it became clear to me that unless Americans demonstrate the feasibility of sustainable practices at home (and indeed adopt some of the sustainable practices already implemented elsewhere), we will never have the moral authority to convince other nations to take this softer path to development.

While this will not be easy, there is clearly a need for Americans to join in a concerted effort to promote sustainable development both at home and abroad. Experts calculate that if the Chinese burn all their high-sulfur coal it will create air pollution that will cause both climate change and illness on a massive scale. President Clinton acknowledged this when he told China's President Jiang Zemin that the greatest threat that China posed to the United States is that the Chinese people will want to get rich exactly the same way Americans got rich. Unless improvements are made to triple automobile fuel efficiency and reduce greenhouse gas emissions, "we all won't be breathing very well," Clinton observed. "There are just so many more of you than there are of us, and if you behave exactly the same way we do, you will do irrevocable damage to the global environment. And it will be partly our fault because we got there first and we should be able to figure out how to help you solve this problem."[8]

While it is essential that we help the Chinese figure out ways to meet their needs without poisoning the environment, we Americans must begin to look within our own borders for ways to place our own economy under more sustainable management. We could begin this journey by identifying our own eco-pioneers and seeking ways to put their most successful ideas into practice.

From this starting point, Americans could initiate an honest dialogue with people from other nations about the need for a transition to a sustainable economy. We could admit that our economic system and much of our technology, although widely admired around the world, are deeply flawed from an environmental perspective. We could concede that our high-consumption lifestyle (while popular) is unsustainable. We could also acknowledge that Americans are only in the early stages of trying to reverse these trends and that entrenched economic interests that have invested heavily in unsustainable technologies are delaying our transformation to a more sustainable culture.

Having publicly recognized the challenges that face us, we could describe some of the best experiments going on in our country to solve these problems, while conceding that there are other areas where we remain stymied. We could solicit ideas from other nations wrestling with similar problems and pool our knowledge about what works.[9] Video documentaries of successful programs could be swapped and experts exchanged to increase the speed at which successful programs are replicated. International cooperation on sustainable development could become a source of pride for all of us. Monitoring of national and global progress toward a sustainable culture could become a focal point for political debate. And a multidisciplinary field of sustainable development could be established in academia and become a regular subject for mainstream media reports.

Early signs of this trajectory are already visible. The President's Council on Sustainable Development (PCSD), after two and a half years of deliberation, issued its report entitled *Sustainable America: A New Consensus for Prosperity, Opportunity, and a Healthy Environment for the Future.* Among its many recommendations, the PCSD called upon the President to appoint a commission to recommend changes in federal taxes and subsidies designed to promote sustainability. The report also encouraged life-cycle stewardship for manufactured products; management of resources on an ecosystem basis; respect for nature's carrying capacity; and headway toward a zero-waste society.

Progress is also being made on the diplomatic front: witness the fact that former Secretary of State Warren Christopher assigned a high priority to addressing international environmental issues. Further, since President Clinton's election, American diplomats assigned to the United Nations Commission on Sustainable Development have taken a less arrogant line than the edict issued during the Earth Summit by President Bush, who was widely quoted as saying that "the American lifestyle is not up for negotiation."

During his first term, President Clinton attempted to advance the sustainable development agenda through a low-key, piecemeal approach. He missed the opportunity to go on the air with a "fireside chat" telling Americans that the federal government was attempting to clean up its environmental act and challenge them to do their part. By downplaying the issue, Clinton failed to make the broad argument that in the long term it will pay to enact a progressive green plan for the United States (backed up by a shift to green taxes) because it will force industry to come up with products that are more energy efficient, less wasteful, longer lasting, and less polluting— all qualities that will sell well on the international market in the years to come.

However, if Clinton missed a chance to marshal the political will to make our economy more sustainable, to his credit he expended political capital on a number of initiatives that moved us in that direction. He explored the political feasibility of energy taxes, redesigned the White House along more environmentally sensible lines, directed federal agencies to purchase paper and other products with a higher percentage of recycled content, attempted to raise grazing fees, established a National Biological Survey, initiated a scheme to clean up and reuse abandoned and contaminated industrial sites, and resisted the Republican congressional effort to roll back environmental protections.

Yet the fact remains that in this country we rarely honor or reward our eco-pioneers and we resist changing our laws, taxes, and subsidies to encourage the replication of sustainable practices.

There is both good news and bad news about eco-pioneers. The good news is that they exist; the bad news is that the playing field they are engaged upon is uneven and does not favor their success.

First the bad news. Before we become euphoric about the ability of eco-pioneers to solve environmental problems, it is important to look at the scale on which these activities take place. Sadly, at present, the efforts of eco-pioneers are dwarfed by an industrial juggernaut that chews up resources and ecologically critical habitat at a fearsome rate, while wasting a huge percentage of this so-called throughput.

The invaluable series of *State of the World* reports published by the Worldwatch Institute have for years documented the way in which human demands exceed sustainable yields in sector after sector of our economy. "We have not yet learned how to stabilize our demands within the sustainable boundaries of earth's ecosystems," write Lester Brown and colleagues. During the period between 1950 and 1995 the world economy grew from $4 trillion to $20 trillion, the world population doubled, water use tripled, and we are using six times as much paper, four times as much fossil fuel, three times as much firewood, three times as much beef and mutton, and two times as much lumber as we did at midcentury. Not surprisingly, this crossing of environmental thresholds has led to water tables falling, forests shrinking, soils eroding, fisheries collapsing, species disappearing, and an ever increasing collection of chemical and radioactive pollutants in the biosphere.[10]

Former World Bank economist Herman Daly, now at the University of Maryland, and John Cobb define succinctly the dilemma we face: the economy is a subset of the ecosystem, so as the economy expands it places greater and greater strains on the parent system. In effect, the economy is like a parasite eating its host. The reason that traditional economic analysis fails to reflect the depreciation of natural capital is that nowhere do economists consider the carrying capacity of ecosystems despite the fact that all economic activity ultimately depends upon healthy ecosystems. This is an oversight that makes economic analysis increasingly divorced from biological reality. Thus the economy grows, huge fortunes are amassed, yet biologically the world becomes increasingly impoverished.

The best estimate of just how much ecological space humans appropriate is that we now use some 40 percent of land-based photosynthetic product. Considering that we are just one among many species, our voracious appetite leaves relatively little for the rest; nor does it leave us much time to develop environmentally sensible technologies, given that the human population is now doubling every 40 years, Daly and Cobb note.[11] Thus, while eco-pioneers are experimenting with ways to reorient human activities to be more environmentally sensitive, perhaps irreversible damage is being done to the natural world by an economic system that is spending down its natural capital.

The implications of the accelerating rate of environmental degradation that we witness around us are registered, either consciously or subconsciously, by all of us. While news commentators suggest that the modern "age of anxiety" is driven by downsizing and job insecurity, it also seems likely that much of this free-floating anxiety stems from our intuitive recognition that today's rapacious lifestyles and commercial practices are unsustainable and will eventually lead to environmental impoverishment and decimation of diversity, if not the outright extinction of our species.

Will humans survive their own compulsion to destroy the ecosystems on which they depend? There is no guarantee that a sustainable society will be established before man-made changes in the biosphere bring on a catastrophic event (or series of events) that causes our species to become extinct. Yet that does not mean that we are doomed. "Defining, and then achieving, the conditions for sustainability for the environment will be as tough a task as humanity has ever faced. But it is increasingly clear that the survival of what we call our civilization, and perhaps human life itself, demands nothing less," write economist Paul Ekins and colleagues.[12] Protestant theologian Robert Kinloch Massie Jr. also speaks of the need to "work together to avert the (ecological) tragedy to which our behavior has destined us."[13]

While it is not difficult to find thoughtful people with a dark view about the ultimate outcome of the contest between the forces of environmental destruction and those of environmental restoration, some of our best thinkers have considerable faith in the ability of humans to learn how to cooperate with the forces of nature.

In a lecture he gave in San Francisco shortly before he died, René Dubos, addressing an audience packed with ardent environmentalists, surprised many of them by taking a strikingly optimistic view. Dubos argued that it is only relatively recently that humans have begun to monitor their impact on the environment and that already the time lapse has been dramatically shortened between creating a problem, analyzing its environmental impact, and acting to solve it. An example of this is the discovery of the hole in the ozone layer and the subsequent Montreal Protocol banning the manufacture of most ozone-destroying chemicals.

The good news, then, is that while we do not have the answers to all our environmental problems, we have at least begun the process of building a sustainable culture in America. In his book *Building a Sustainable Society,* Lester Brown writes about the quietly growing impact of proenvironmental actions. "Each new hydroelectric generator, each new decline in the national birth rate, each new community garden, brings humanity closer to a sustainable society. Collectively, millions of small initiatives will bring forth a society that can endure. At first the changes are slow, but they are cumulative and they are accelerating. Mutually reinforcing trends may move us toward a sustainable society much more quickly than now seems likely." [14]

The book you have in your hands deliberately focuses on these modest initiatives that are moving us toward a sustainable culture. By showcasing the work of eco-pioneers who are providing practical solutions to environmental problems, I have intentionally chosen to accentuate the positive because I believe many Americans are so saturated with bad news about the environment that they feel practical solutions are beyond reach.

"Unless you give people some idea that a solution is possible, it is hard to motivate them to get them excited about implementing it. Folks like a challenge, but Mission Impossible is hard to sell," observes S. David Freeman, a man who has moved a number of utility companies away from a dependence on nuclear and fossil fuels to the use of renewable energy and energy conservation initiatives.

In writing *Eco-Pioneers* I deliberately chose to stick with the more mundane aspects of what it will take to provide practical solutions to environmental problems, rather than discuss the need for psychological or spiritual changes. I leave it to others, particularly the deep ecologists, to explore the profound change in values that humans will have to undergo in order to establish a sustainable culture on earth. This is not because I underestimate the importance of the psychological and spiritual transformation that is necessary before people rediscover a way of living that fits in with the natural order of life. Without such a change in values people will not demand goods and services that are produced sustainably and the work of the eco-pioneers described in this book is likely to be ignored.

I focused on the practical solutions to environmental problems because Americans are at heart a practical people and I believe they can be convinced to adopt eco-friendly technologies if they are convinced that these strategies will both work and save nature at the same time.

It is my sense that these stories may be useful in making the concept of sustainability more real to Americans, because they provide us with a growing body of evidence that eco-pioneers are ingenious enough to devise ways to meet our needs without destroying the natural world. The question remains unanswered, however, as to whether or not we, as a people, have the wisdom and willpower to create a political environment that favors the replication of these sustainable practices. Most likely, before the body politic is prepared to adopt sustainability as its central organizing principle, a critical mass of individuals convinced of its importance will be required. Here, the work of the eco-pioneers described in this book may play a critical role because they have the potential to inspire each of us to search out what our personal contribution might be to the creation of a more socially just and ecologically sustainable society.

Convincing others of the importance of sustainability will not come merely by preaching its theoretical merits. As David Crockett, the Chattanooga city councilman, reminds us, when a school superintendent wakes up in the morning he or she is not thinking about sustainability, but rather about how to improve student test scores. The likelihood is that haranguing the superintendent about sustainability will prove ineffective. More convincing to school officials would be the argument that using sustainability as a common theme throughout the curriculum could make academic material more exciting and relevant to students and potentially raise test scores.

Teaching the concept of sustainability by example is another powerful didactic tool. Eco-pioneers need not engage in long theoretical discussions about the definition of what constitutes sustainable activity; nor must they articulate the value system and spiritual beliefs that undergird their work. They need not preach the merits of sustainability because their actions speak for themselves: they have chosen to spend their lives solving environmental problems and teaching us all to live more lightly on the earth because they want to do something practical to preserve the glorious diversity of life.

But building a sustainable culture is a challenge that must engage all of us, not just a handful of eco-experts who make a living at it. This great human endeavor will draw on all our faculties and require all our skill and compassion. It will call for higher levels of cooperation than have ever been achieved before. It will necessitate a shift in consciousness from a human-centered to a biocentric worldview. And it will likely require the mobilization of human energy and resources that was previously reserved for the conduct of war. As earlier generations marshaled vast resources to contain fascism and communism, so ours will be called upon to bring to bear a large-scale effort to create a sustainable economy within the limits of nature.

Eco-pioneers, such as those described in these pages, are on the frontlines of this great effort and are among our best entrants in the race between the forces of environmental destruction and those of environmental restoration. As such, they should be honored, nurtured, and supported for the work they do for us all. After all, the preservation of the rich tapestry of life that provides for all our needs, while of international significance, is also an intensely personal affair.

During the six months my wife and I spent traveling together across the country, I watched our infant son, Benjamin, absorb wide-eyed the sights passing the windshield. As we drove through cornfields, mountain wilderness, deserts, suburban developments, bustling business districts, inner-city wastelands, and coastal vistas, I wondered how the landscape would change over the course of Benjamin's lifetime. It was heartening to think that a better future might be in store for him as a result of the work of the eco-pioneers we were visiting.

has had a busy schedule describing his new findings at various offices within the bureaucracy. He is a member both of the Committee on the Greening of the White House and the American Institute of Architects' Committee on the Environment; and a frequent visitor at the Environmental Protection Agency (EPA) and the Department of Energy (DOE).[1]

"He's given the whole [environmental] movement a more technical bent and a less touchy-feely direction," says James White, an EPA scientist.[2]

A couple of years ago, Fisk passed out samples of AshCrete to officials at the DOE where he pointed out that using it as a substitute for concrete had a number of environmental advantages. First, it could reduce carbon dioxide emissions, one of the key greenhouse gases responsible for global warming, because the manufacture of concrete generates an estimated 9 percent of CO_2 emissions globally. Second, it could reduce the waste stream of fly ash that pours out of coal-fired power plants.

AshCrete is also safe to use as a building material once the silica in the fly ash is bound up in the cement, Fisk asserts. Studies reveal that while fly ash contains slight traces of heavy metals, they are in quantities too minute to produce negative health effects according to EPA standards; and besides, the small amounts of heavy metals are stabilized within the concrete. Workers who manufacture the AshCrete from the fly ash, however, must protect themselves from inhaling its fine silica dust, which can cause respiratory disease.

Fisk's interest in using industrial and agricultural wastes as building materials extends beyond fly ash. For various building projects he scrounges aggregate from a nearby aluminum smelter that produces 800 tons of the stuff a day. The aggregate, which Fisk mixes with his AshCrete, is of an excellent quality, is totally safe, and is very strong, he asserts, "but we are the only people using it." Normally, much of the aggregate used in concrete comes from riverbeds and riverbanks and is extracted in a fashion that does damage to the river ecosystem, he explains.

Scrounging aggregate and recycling coal-fired power plant fly ash are not activities one might expect of someone whose name has the plutocratic ring of Pliny Fisk III. The descendant of a financial tycoon who made and then lost a fortune on Wall Street, Fisk's grandfather owned one of the largest banking houses in the country and has his name inscribed over the door of one of the fanciest eating clubs at Princeton. According to Fisk, his grandfather used to tie his yacht up to the yacht of J. P. Morgan during the holidays to share the joys of a vacation. Alas, while his grandfather's name passed down to him, none of the money came with it, he notes.

Instead of following in his grandfather's footsteps, Fisk pursues a vision of a radical change in the way we build and relate to nature. A graduate of the University of Pennsylvania, where he earned master's degrees in architecture and in ecological land planning, Fisk drove around the country in the 1970s in a Chevy pickup truck, visiting some of the early experiments with solar houses in New Mexico.

Now comfortably ensconced in his home on the outskirts of Austin, Fisk was recently spotted in his fabricating shop inserting an auger into a pipe with the help of two sons and some visiting school friends. As he stands surrounded by a useful profusion of tools, drill presses, lathes, and racks of clamps of all sizes, it is not hard to see that he belongs to the nuts-and-bolts school of environmentally responsible architecture. As he invents new ways to use local and recycled building materials, Fisk is engineering some of the technological breakthroughs that will make practicable a more ecologically sustainable style of architecture.

Fisk's interests in alternative construction materials and sustainable design are being put to a test in a project known as the Advanced Green Builder Demonstration, a 2000-square foot, $250,000 structure that he is constructing behind the five buildings that constitute his home, office, and laboratory. The demonstration building is designed as a structure for a typical family or business and is built out of a variety of byproducts of industry and agriculture and locally available materials. Photovoltaic roof panels, low-flush composting toilets, and a natural wastewater treatment system that purifies water using gravel and plants will allow the building to be off the utility grid for water, sewer, and electrical hookups. "It's a chance to jump ten years ahead," Fisk observes.

The foundation and post-and-beam frame of the building are made out of recycled steel, AshCrete, and aggregate from an aluminum smelter. All of the posts and beams that support the building contain hollows through which the plumbing, electrical, and communications lines can run. This makes it easy to change the location of rooms in response to users' needs over time, Fisk explains.

As Fisk began fabricating the hollow posts and beams out of rebar (a metal made largely out of crushed cars), he realized that he had invented a

Figure 1.1
Hollow posts and beams made out of rebar provide conduits for pipes and electrical
wiring. Courtesy of the Center for Maximum Potential Building Systems.

gigantic erector set that could be useful in many building applications. He promptly applied for a patent and called the system GreenForms. These hollow posts and beams, made of 90 percent recycled steel, contain built-in anchor points along their length on all four faces and on their ends, making them simple to bolt together. The anchor points can also be used to attach a scaffolding while the building is being built, a trellis on the outside for shade plants, stairs, or even furnishings such as shelving, desks, or canopy beds.

The hollow post-and-beam system, which Fisk calls the structure's endoskeleton, can be "wrapped" in a variety of materials. For example, when GreenForms are sheathed in a thin layer of wood they look like large wooden beams, but require only a small amount of wood in their construction. Or GreenForms can be wrapped in a cementitious material such as Ash Crete or colorful recycled plastics. Fisk is also experimenting with small amounts of precious woods as a kind of decorative inlay.

Posts and beams with multiple anchor points make it possible to install built-in furniture cheaply. A bench, shelving, a corner breakfast nook, a desk, or an "edu-tainment" center is easily installed between posts. GreenForms permit remodeling at minimal expense because walls built between posts and beams can be put up or torn down relatively easily. The house is also designed to grow or contract both vertically and horizontally. If the owner wants to add a couple of stories to the house, the hollow posts can be filled with concrete (or AshCrete) to provide additional strength; and insulation panels under the existing roof can be removed and reinstalled when the new roof is constructed.

Fisk registered another patent for a mobile kitchen system he calls Meals-on-Wheels. This innovative design for the kitchen permits the major appliances—such as the stove, sink, and serving cart—to be moved or docked for maximum efficiency. In Texas, outdoor barbecuing makes a lot of sense on hot days because it allows the homeowner to avoid heating up the house and placing an additional strain on the air-conditioning system. As a result, in Fisk's kitchen everything (including the stove and sink) can be wheeled out onto a prepared patio or breezeway, while the serving cart doubles as a mobile storage cabinet that contains racks that slide easily into the dishwasher.

Exasperated with wasteful architectural designs that call for two and a half bathrooms in a standard home, Fisk designed a single bathroom for the Green Builder Demonstration that is flexible enough to accommodate more than one person at a time, is easy to clean, and has rotating fixtures. In the center of the bathroom is a cylindrical column that the sink and shower revolve around. The whole room can be used as a shower (there is a drain in the floor) or the shower can be directed into a corner, permitting other family members to use the toilet or sink while finding some privacy behind heavy cloth screens—a configuration that may be ahead of its time in terms of being socially acceptable.

The exterior walls of the demonstration building are made out of a number of earthen materials with high insulation or mass value including adobe, rammed earth bricks that are made of sandy loam under pressure, stabilized earth that is made with a gluelike enzyme, and caliche (mixed with AshCrete to form CalCrete).[3] These earthen walls are then sealed with a mixture of wax and and linseed oil. A number of types of straw materials (all of which come from oat and wheat fields within ten miles of the site) are also used as walls and covered in a mix of caliche and fly ash plaster. Straw and mohair from local sheep are mixed to form wall panels; chopped straw is mixed with water and poured into wooden frames; and straw bales are bound with wire and staked together with bamboo. Fisk employs a number of industrial by-products in the construction of the house. Wasted wood fiber from a factory that extracts cedar oil from juniper trees is mixed with liquified, post-consumer plastic to form a wood substitute called AERT that can be used for fencing, decking, and window frames. Styrofoam from StarrFoam Enterprises in Fort Worth, Texas, is also used to form insulation panels.

Inside, Fisk combines a variety of native materials found in the five different vegetative and mineral soil zones that converge on Austin. For example, in one room, caliche will be used on the walls and a mesquite tile on the floor. Another room, which borrows from the temperate grasslands of Texas, features straw-based materials combined with oak and pecan woods. There is even a section of the building constructed of unstabilized adobe that can be plowed back into the soil after its useful lifetime. In fact, the Green Builder Demonstration has become a showcase for alternative and innovative building materials with twenty-six companies donating materials to the project. The sheathing on the roof is constructed out of straw panels and a recycled paper panel known as Homosote, while the roof itself will be covered with either metal or a membrane made from recycled tires.

A system of roof gutters and 13,000-gallon cisterns captures water on site for the use of the occupants. A former student of Fisk's conducted a study which shows that half the homes in Texas could collect adequate supplies of water on-site if they took appropriate measures. Because of the relatively dry location, Fisk is unsure if he will be able to collect enough water to supply four people, but intends to collect as much as he can to reduce dependence on an overtaxed river system.

As for energy, Fisk is installing a 1-kilowatt system that will supply enough power for an average home in a developing nation but nowhere near the 7.4 kW per household per day of energy that Americans consume. To make his low-budget energy system work, he has built stringent energy conservation measures into every aspect of the design. The 1 kW of power will be harvested by solar panels, while a backup generator will provide any excess power needs.

But here again, Fisk ultimately wants the backup generator for the building to be a hybrid electric car instead of the kind of generator you might buy at Sears. The U.S. Postal Service donated seven electric cars, which the center plans to convert into hybrid electric cars that run on a constant rpm (revolutions per minute) engine. With this extremely efficient engine, the hybrid electric car battery can be charged by plugging it into the house, and drawing from the energy it captures through its solar collectors. Or the relationship can be reversed: if there is not enough solar energy in the building to run the pumps, lights, and other electric appliances, the car engine can be used as a backup generator. In this case the house would be plugged into the car.

Deliberate landscaping around the house is also an essential element in making this design work. Fisk plans to lay out the flowerbeds, lawn, shade trees, fruit and nut trees, and shade vines in such a way that they can treat and absorb the wastewater and sewage from the house. "With this system, when it's August and everywhere else the lawns are parched, you have a beautiful lawn, your flowers are going like the dickens, and you are not using one smidgen of city water because it is your own wastewater going out into the garden," Fisk explains. He will also cover the outside walls of the house and a latticework over the windows with leafy vines to help shade the inside from the sun and protect it from the weather.

All of Fisk's architectural designs are grounded in the local environment. They are designed to accommodate the site's ability to absorb waste, har-

Figure 1.2
A system of roof gutters and 13,000-gallon cisterns captures water on site at the Green Builder Demonstration Project. Courtesy of the Center for Maximum Potential Building Systems.

vest energy and water and are built, where possible, out of native materials. One of a new breed of architects pioneering environmentally low-impact construction techniques, Fisk is an expert when it comes to building with straw and other earthen materials. He searches for high-clay soils that make good bricks, native trees that make good lumber, palm fronds that can be used to shade a roof, or bamboo that can replace rebar as reinforcement in cement foundations.

One of the central problems Fisk wrestles with is that our houses, factories, schools, and office buildings are made of materials and building systems that are unrelated to the environment in which they are built. As a result, building materials are often transported from distant locations, harvested in an ecologically destructive fashion, and put together in such a manner that they require much more energy to heat, cool, and light than is necessary. Furthermore, many buildings are built with little design flexibility, making it impossible to adapt them to changing needs. "We construct buildings and, on average, twenty-eight years later we slam them down with a wrecking ball," he observes. To remedy this, Fisk is looking for a way

of building with recycled materials, reducing energy consumption in buildings, and designing them so they are flexible enough to be adapted to new
uses or demolished at minimal environmental costs. "At the end of the
building's life you can unscrew the rebars and take them elsewhere, and
plow the rest of the structure into the soil," Fisk adds.[4]

Designing structures that place minimum demands on the environment
requires that Fisk identify what local building materials and methods are
available (or need to be developed) and learn techniques that allow people
to use them effectively. Far from viewing indigenous construction techniques as primitive, he regards some of them as environmentally sophisticated. Fisk's center has tested a variety of substances for their suitability as
building materials, including straw and clay, rammed earth, pozzuolana,
caliche, stabilized earth, sand and lime, sulfur, gypsum, adobe, laterite, and
alumina clay. If used on an appropriate scale, he argues, these earthen materials can be extracted without causing significant environmental damage,
and they can be purchased cheaply since they are widely available.

Fisk regularly finds resources where others would never dream to look.
For example, in Texas the clear night sky can be used as a heat sink if a
structure's roof area is designed to be equal to the floor area, he says. Cooling the house is accomplished by trickling water over a metal roof at night
and turning it off in the morning. This technique, coupled with the use of a
vine-covered trellis to shade the walls of the house, breeze-directing windows, and second-floor outdoor sleeping porches can substitute for air
conditioning, he continues.

Another material Fisk uses effectively in Texas is mesquite, a tree whose
roots penetrate 50 to 60 feet into the soil of arid sites unsuited to most
crops. Mesquite is viewed as a useless wood (other than as a source of char-
coal) because it does not grow straight enough to use for lumber. Because
mesquite uses up scarce water supplies in grazing areas, in the past some
Texan ranchers tried to eradicate it with Agent Orange, Fisk reports. But
after researching how mesquite is used as a construction material by people
who live in an environment similar to his own, Fisk found that inhabitants
of the Argentine and Uruguayan pampas cut mesquite into wooden tiles for
parquet floors. He successfully copied this practice and improved on it by
using mesquite sawdust as raw material for insulating block and blown
insulation.

Sulfur is also on Fisk's mind as a building block. "Did you know that sulfur is the fourteenth most available element on earth and we haven't been

taking advantage of its useful properties as a building material?" he asks. Blessed with a fertile imagination, Fisk envisions entire communities built out of sulfur, a thermoplastic material that can be shaped or repaired by heating it up or returned to the soil after its usefulness is over. Sulfur can be fireproofed by mixing it with its geologic neighbor gypsum, he adds.

While many of these building techniques are borrowed from others, Fisk has effectively assembled them into a regional "tool kit" that promotes a more ecologically sustainable type of construction than standard practices permit.

Fisk is an ardent advocate of mapping resources that can be used as building materials. One of those widely available in some 60 percent of Texas is known as caliche, a crusted calcium carbonate which forms on certain soils in dry regions. According to United Nations statistics it can be found on 13 to 14 percent of the earth's surface. This material appealed to Fisk because by mixing a small amount of cement with caliche he was able to reduce his use of Portland cement by two thirds.

Fisk used caliche extensively in a school for neglected and abused children he designed in Texas. Since the material was locally available it was possible to engage the children in making the bricks for their school. Not only did the children learn how to make building materials, they also learned why it was ecologically important to use local materials. When some visitors from California indicated that they wanted to buy some of these handsome caliche bricks, a twelve-year-old resident of the facility explained to them that the bricks were not for sale and that they should look for the ingredients for building materials in their own backyard instead of shipping them all the way from Texas.

But how does one determine what indigenous building materials are available locally? Often knowledge about where they are or how to use them has been lost as they have been replaced with modern construction methods and materials. As a result, learning how to use indigenous construction materials often requires painstaking research.

Fortunately, a network of groups around the world is piecing together a biogeographic map of indigenous construction materials.[5] The primary tool for creating this map involves dividing the world into fourteen distinct biomes or geographic areas whose climate, rainfall, soil, hydrology, vegetation, animal life, and a number of other factors are roughly similar. By looking at what construction techniques are used in biomes similar to his own, Fisk learned of technologies he could adopt. "The architecture that comes out of this kind of analysis takes into account the metabolism of the local environment," he observes.

Fisk gradually accumulated an extensive database about the environment in which he planned to build. He discovered, for example, that the region of Texas near Austin where he lives is part of the temperate grasslands biome where unused straw and clay is found in abundance. By looking at construction practices in other temperate grasslands, he learned an ancient technique that involves mixing a watery solution of clay into straw. He dribbled the clay batter onto the straw using a large ladle-like implement and stirred the straw with a pitchfork until it was coated with a thin layer of clay. Then he left the mixture overnight until the straw attained a noodle-like flexibility. The next day he tamped the straw-clay mixture into a wall mold and left it to cure.

Everything seemed to be going well until a few weeks later when, after the mold had been removed, he noticed that his experimental wall had sprouted and was growing like a bush. With a client coming to decide whether or not to build a straw-clay house, Fisk decided to prune the wall with a pair of shears. After examining the wall the client agreed to go along with the project, but Fisk was left with a dilemma: what would he say when his client's walls began to sprout?

Fortunately, a Dutch specialist in earthen building techniques saw Fisk's experimental wall and advised him that letting the wall sprout was an integral part of building with this particular straw method. The sprouts extract moisture from the wall and their root structure knits the wall together, he explained. When the sprouts shrivel up and die it signals that the wall is cured and ready to be plastered. The stems of the sprouts can then be bent over and used to form a rough lathe to anchor the plaster to the wall. "I felt like a real jerk for having pruned the wall," Fisk says, but at least he learned how to build a straw-clay wall that will last for hundreds of years.

"In the temperate grassland, where we live, straw and various other grasses have a turnover rate of two or three crops a year," Fisk says. By using abundant local materials like straw, he avoids using scarce wood products that must be transported from distant areas. By the time lumber arrives in Austin, the embodied energy costs in the wood are boosted by one third because of the energy expended in transporting it, Fisk calculates. As a result he uses wood sparingly.

In his investigation of earthen building materials Fisk experimented with various types of slump block machines such as the Mold Master and the Mud Cutter; rammed earth machines such as the one produced by Winget Works in England in the 1950s, the Hallomeca machine from France, and a U.S. hydraulic product fabricated by the M&M Metal Company; and cement block machines. While people in the construction industry, accustomed to pouring tons of concrete a day out of huge trucks might sneer at these techniques as archaic and slow, Fisk sees them as capable of reducing environmental damage and generating labor-intensive jobs in the local economy.

In addition to getting his hands into the soil making earthen building blocks, Fisk also is doing research for the government on how to determine which building materials are the most appropriate to use in different regions. To this end, the Center for Maximum Potential Building Systems signed a $250,000 cooperative agreement with the EPA to devise an information system that will provide agency officials with data on which they can base policy about how to guide the construction industry into environmentally sustainable practices.

Fisk believes that builders should try to meet their needs for energy, water, materials, and waste absorption capacity locally before placing these demands on more distant areas. For example, architects should attempt to treat graywater and sewage in a manner in keeping with the plants that grow in a region, which are in turn determined by the climate and soils. If graywater and sewage treatment cannot be accomplished on-site, then planners may be forced to treat the waste on a slightly larger scale that encompasses a cluster of houses; or, in some instances, a neighborhood waste treatment facility might be logical.

A similar scenario holds true for building materials. If there are not enough wood fiber or appropriate soils on-site to build a structure, then architects must look farther afield to see what materials are available in the region. However, without considering the larger picture, mistakes can easily be made. Take the practice of building with adobe bricks. While adobe bricks sound environmentally benign, in fact building with them is not always without environmental costs. The soil that produces adobe block is sandy loam found in excellent agricultural lands. As a result, building with adobe in some parts of New Mexico, for example, removes prime

Figure 1.3
Pliny Fisk III, co-director of the Center for Maximum Potential Building Systems in Austin, Texas. Photograph by Steve Lerner.

agricultural lands in a desert area where sandy loam soils are precious. Deciding what is the most ecologically sound material to build with in a particular area can involve complex calculations, Fisk points out.

While he is increasingly involved in research projects such as the one for the EPA, Fisk keeps his hand in as an architect who designs buildings that incorporate environmentally sustainable features. Building in an environmentally friendly fashion requires more than finding locally available building materials; it also involves evolving a design that fits in with nature rather than fighting it, Fisk asserts.

In his masterplan of a facility for the Tejas Council of Camp Fire, an organization that provides environmental education for Camp Fire girls in Waco, Texas, Fisk designed a new kind of campus. "Kids from Dallas and other cities don't know where anything comes from, where it is going, or how they fit into the scheme of things," he observes. So his design integrates the operation of the facility into the surrounding ecosystems to teach the children how to relate more harmoniously to the environment.

For example, in designing the dining hall, Fisk started by asking how many children will eat how many meals. He then calculated how much of the food could be grown locally to minimize the necessity for transporting it from distant sources. Plans were made for the processing facilities that will be needed to wash and store the food as it comes in from the fields or surrounding farms as well as a system for composting the kitchen wastes.

His blueprints show a flow of resources through the facility that is considerably more complex than the standard architectural master plan. The architectural drawings are filled with arrows showing how resources move through the facility to the point where it is sometimes hard to distinguish where buildings begin and end. This underscores the point that the design of a building does not stop at the exterior walls, Fisk observes. Trees planted to shade a house are an integral part of the design, as are the leach fields, the water-catchment opportunities outside the house, and the energy-harvesting possibilities on the land.

On the 400-acre Camp Fire facility, the roofs of buildings and other impervious surfaces take on a new significance in Fisk's design because they permit the on-site collection of water. Fisk will use the impermeable surface of the parking lot to collect water that can be used for flushing toilets or for other nonpotable services. Once water is collected on these surfaces it will be piped to the central plaza of the campus where it will be pumped up into a water tower disguised as a clock tower. A bubbling fountain will not only be an attractive focal point in the plaza, it will also provide an opportunity to purify the water using an ozonating system supplied with electricity by photovoltaics.

Fisk intends to use various icons and color codes to help visitors understand how resources cycle through the campus. For example, the tiles covering the water line from the parking lot will be color coded blue so that students will be able to understand how the campus water system works. The blue tiles will also be removable so that the water line can be easily serviced.

Furthermore, the young people who come to the camp will be invited to join in a kind of green manufacturing process where they help provide needed services and reconfigure the materials that move through the facility. This goes one step beyond recycling. Instead of simply separating waste and sending it off for reprocessing, residents will be asked to work with the materials that come through the facility to create something useful. In a central market these products and services will be exchanged for "Sustain-a-bills," which will replace dollars as the local currency.

In Laredo, Texas, Fisk designed, engineered, and oversaw the construction of the Blueprint Farm/Rio Grande International Study Center, which borrows from Israeli techniques for dry-land farming. Located between an arid desert and temperate grassland on the Rio Grande, the farm complex includes five buildings which house offices, classrooms, workshops, and storage areas. Each of these structures features a distinctive set of cooling towers, the design of which is borrowed from Persia. The cooling towers, covered with metal rooftops, work in pairs that permit both a downdraft and an updraft. Working as a passive solar air conditioner, the updraft tower (painted black) expels hot air, while the downdraft tower (painted white) sucks cooler air into the building.

Low-energy materials used in the construction of these buildings include straw bales, pozzuolana, caliche, iron ore, mesquite, stucco made from fly ash, and recycled oil well drilling stems. Tensile steel cable structures provide continuous shade to reduce open space temperatures for certain crops and the insectary for breeding beneficial insects. The project won the 1991 Environmental Protection Distinguished Appropriate Technology Award from the National Center on Appropriate Technology.

Wherever Fisk travels he teaches people how to identify local resources that can be obtained cheaply and transformed with relatively low-tech equipment into useful building materials. In 1985 he did this for the Miskito Indians who live on the Caribbean coast of northeast Nicaragua and southwest Honduras. As a first step he did a biome search of tropical savannas to gather ideas from other cultures on how to build with indigenous materials. During an eighteen-week visit to the Miskito tribal lands he conducted an inventory of locally available resources, among which he found high-alumina clay kaolinite, which can be used for fired clay and cement production. He also identified deposits of limestone and seashell as sources of masonry and cement; rice husks used in cement manufacture; coconut palm and bamboo, which could be used as a fibrous reinforcing material; pinewood for lumber; and abundant rainfall for on-site water catchment.

Fisk brought with him a minimal amount of laboratory equipment that permitted him to test the composition of these raw materials to see if they would stand up as building materials. He also built a prototype sawdust panel press, a mold to make fibercrete corrugated roofing, and a slipform to pour septic tank cisterns. Unfortunately, the political situation became so chaotic that no village was ever built using these techniques, but they were incorporated into the manufacture of some houses, he says. When Fisk's life was threatened he left promptly.

Closer to home, Fisk was called to Crystal City, a largely Mexican-American community in southern Texas where the utility company had shut off the gas after the community failed to pay its bills. Health problems associated with lack of hot water for cooking and cleaning began to arise. Asked what he could do to help, Fisk started building solar water heaters out of locally available materials. This involved collecting thousands of fluorescent tubes and emptying them with a vacuum cleaner. He then tied the fluorescent tubes around the circumferences of old solar water heaters using pieces of garden hose to cushion them. He also scrounged printing plates to use as reflectors. When installed on the roof the sun would hit the empty florescent tubes and be directed right into the water heater rather than bouncing off. Fisk was able to produce these solar water heaters for $18 apiece and the community affectionately nicknamed them "las bombas." He also designed mesquite-burning stoves for the town, some 800 of which were eventually installed. Finally he helped put together a solar collector factory that sold solar collectors to five surrounding towns.

Fisk's work with local and recycled building materials won him a contract helping to rewrite the architectural and engineering guidelines for the state of Texas. This gave him an opportunity to insert language into the guidelines that calls for publicly funded projects to be built using both energy conservation measures and a variety of materials containing recycled content.

"What we discovered was that it was extremely important for people to understand where materials come from and where they end up so that consumers can be responsible for their effect on the environment," Fisk says. To this end, he helped create a system that rates building materials according to their environmental costs. The rating system calculates the environmental costs of manufacturing or harvesting a building material,

Figure 1.4
Straw bale buildings, such as this one designed by Fisk, are so well insulated that they reduce heating and cooling costs substantially. Photograph by Ken Haggard.

whether a material is safe to use, how long it will last, whether it is useful, whether someone really needs it, and whether there is a place to get rid of it once its usefulness is over.

Over time the rating system evolved into Austin's Green Builder Program, which certifies green homes on a scale of one to four stars with more stars indicating that more green features are built into the home. "Initially people thought this method was somewhat far-fetched," Fisk reports. But the Green Builder Program was honored by the United Nations at the Earth Summit as one of the twelve exemplary local government initiatives around the world.

Some local builders find that building in an environmentally sensible fashion and getting their houses certified through the Green Builder Program is a great marketing tool and, to date, some 250 houses have been built using the rating system. The city of Austin now applies the rating system to city buildings and has developed a rating system for commercial structures that will soon go into effect. What surprises Fisk is that some of the more radical implications of the rating system, such as building with earthen materials, have caught on. "I can send you to five builders whose main livelihood is straw bale buildings; I didn't think we would see that for the next ten years."

To provide a resource for builders and designers, Fisk's wife, Gail Vittori, co-director of the Center for Maximum Potential Building Systems and former head of Austin's Solid Waste Advisory Commission, assembled a 500-entry database and library complete with samples that architects and engineers can refer to when looking for recycled materials and energy conservation techniques. The center is currently offering a number of seminars in Texas aimed at stimulating the use of environmentally sustainable practices in the construction and retrofitting of state buildings. The seminars familiarize designers and contractors with a range of environmental practices involving the use of passive solar and natural daylighting design, landscaping design that serves to cool the house and treat graywater, building techniques that improve indoor air quality, innovative solid waste processing systems, and alternative and recycled building materials.

"I used to think that all you needed was a good idea and everyone would just start to do it," Fisk says, but he has since learned that shifting building practices to a new paradigm requires a series of incremental steps to bring them into common practice.

Bringing Native Plants Back to the American Lawn

It is not hard to spot Lorrie Otto's front yard. Compared with the manicured lawns and clipped hedges of her neighbors, Otto's two acres stick out like a tramp at a garden party. Covered with an overgrown profusion of wildflowers, prairie grasses, and berry-laden shrubs, Otto's yard contrasts sharply with the well-tended gardens of Bayside, Wisconsin, a wealthy suburb north of Milwaukee.

In the past, some neighbors complained about the "weeds" growing in Otto's yard. But from the perspective of a bird, butterfly, snake, frog, turtle, dragonfly, or fox, her property is an oasis of safety where small animals can find food, shelter, and cover from predators that could easily target them on the close-cropped lawns of her neighbors.

Now seventy-six years old, Otto is a long-boned, blue-eyed woman who has a pilot's license and used to be a specialty welder on Liberty ships during World War II. Often found in her garden wearing a floppy straw hat, she has spent the past two decades planting and propagating a wide variety of species native to Wisconsin and the prairies, some of which are rare or in danger of extinction. Her yard is awash in prairie bluestem and dropseed grasses mixed in with a jumble of pale-purple coneflowers, black-eyed Susans, blue spiderwort, and nodding onion.

In his book *Landscaping with Wildflowers: An Environmental Approach to Gardening*, Jim Wilson features Otto's yard as "one of the country's best-known private prairie gardens."[1] A long-time environmental activist who helped organize the banning of DDT in her state, Otto's interest in saving prairie plants was spurred in 1970 when she first saw photographs of endangered and threatened wildflowers at a prairie conference in Madison. A shock of recognition ran through her as she realized that "these were the flowers of my childhood." Flowers that she had last seen along the borders of her father's dairy farm were described as "relic prairie plants." Distressed to learn that only one in five of these prairies plants were being saved from extinction, she decided to turn her own small slice of the earth into a sanctuary for these fast-disappearing old friends.

Since then, Otto, a founding member of the Citizen's Natural Resource Association of Wisconsin, has traveled with a shovel in her car, stopping on back roads to ask farmers' permission to dig up such wildflowers as compass-plant, cup-plant, and prairie-dock so she can transplant them into her garden and propagate them. This led to a hodgepodge collection, but gradually Otto became more selective about what she planted. She weeded out the non-native in favor of the native plants and learned about companion plants that thrive when grouped in small communities. Her passion for native species eventually led her to cut down the sixty-eight Norway spruce trees on her land, leaving the native white pine.[2] In their place she planted a meadow of wildflowers, including red milkweed, Jerusalem artichoke, and eight varieties of native sunflowers.

Otto is part of a growing movement around the country of people who are gardening with native plants. In back yards across the nation, home-owners are intentionally letting their lawns revert to the wild and reintro-ducing native species. An increasing number of nurseries, seed companies, and experts are supplying this new generation of gardeners with the re-sources and knowledge to reintroduce native plants, some of which are scarce owing to habitat loss from development and recreation, overgrazing, competition from non-native species, and erosion. Of the 20,000 plants na-tive to the United States, some 4200 plants are threatened with extinction and approximately 750 of these extinctions may occur in the next decade.[3]

Recognizing that natural gardens are currently no more than a sprinkling of islands in a sea of some 24.3 million acres of lawn, Otto envisions a day when isolated back yard bioshelters begin to link up and create a ribbon of native plants that ties suburban plots together by creating corridors for wildlife migration.[4]

Otto is not alone in entertaining this vision. Sara Stein, in *Noah's Garden: Restoring the Ecology of Our Own Back Yards*, writes:

America's clean, spare landscaping and gardening tradition has devastated rural ecology. . . . Suburban development has wrought habitat destruction on a grand scale. As these tracts expand they increasingly squeeze the remaining natural ecosystems, fragment them, sever corridors by which plants and animals might refill the voids we have created. To reverse this process—to reconnect as many plant and animal species as we can to rebuild intelligent suburban ecosystems—requires a new kind of gardening, and, I emphasize, a new kind of gardener.[5]

This idea of restoring habitats for wildlife in suburbia is more than just wishful thinking. Already, over 12,000 gardeners have joined the National Wildlife Federation's Backyard Wildlife Habitat Program and had their yards certified as meeting strict criteria for supporting wildlife.[6] Kenny Ausubel, founder of Seeds for Change, an organic seed company, sees American gardeners as a potential "volunteer army in the service of biodiversity." Gardening is the largest leisure time activity in the United States and many gardeners are becoming more environmentally aware and are questioning the indiscriminate use of chemical fertilizers, pesticides, and herbicides. While this vision of a regenerated suburban wilderness is appealing, natural gardeners are still a minority and face an uphill struggle against the chemical-laden, overwatered American lawn.

Even Otto was a lawn lover as a young woman and she kept her yard just as tidy as her neighbors. When she and her husband first moved to Bayside forty years ago, they purchased a house that looked like a Swiss chalet surrounded by Norway spruce. To Otto the orderly grounds looked romantic and she loved to trim the lawn with a hand-mower, reveling in the fragrance of the cut grass and the snicking of the revolving blades.

Otto was not alone in her fondness for a putting-green lawn. Americans have a love affair with their lawns that dates back over a century. They find something comforting about a lawn and see it as an oasis of domesticated nature, a place where kids can play safely, and a setting for the outdoor barbecue. Looking out across their lawns, many Americans feel a sense of satisfaction about the piece of earth they tend. They are proud to have tamed a piece of nature and shaped it to their liking. It also doesn't hurt that a well-kept lawn adds to the value of their homes.

In fact, Americans love lawns so much that they have cleared vast swaths of the country. Most of us think of a lawn as small thing. Yet, while individual lawns are relatively small, if stitched together American lawns would occupy some 50,000 square miles—an area slightly larger than Pennsylvania. This figure is even larger if one includes areas covered with turf grass around schools, colleges, factories, cemeteries, churches, parks, golf courses, playing fields, and along highways.[7]

"Most Americans have become so alienated from the land that they don't know how to treat it," Otto says. When it comes to the vegetation in their own yards, they mow it, whack it, and poison it until nothing but grass grows. She doesn't much like the American lawn. "When I look at a lawn it fills me with great sadness because I know it adds to the precipitous loss of butterflies and bird life."

Lawns and non-native plants also place a strain on the environment by requiring frequent watering and large inputs of chemical fertilizers, herbicides, and pesticides, Otto continues. It is estimated that homeowners use 5 to 10 pounds of pesticides per acre of lawn each year, or some 150 to 300 million pounds of pesticides annually.[8]

According to the National Coalition Against the Misuse of Pesticides (NCAMP), "of the 36 most commonly used lawn pesticides, 13 can cause cancer, 14 can cause birth defects, 11 can interfere with reproduction and 21 can damage the nervous system." A National Cancer Institute (NCI) study linked exposure to 2,4-D, the most common lawn care herbicide, to increased rates of lymphatic cancer in the Midwest. And the Environmental Protection Agency (EPA) "classified benomyl, a common fungicide applied to lawns, as a possible human carcinogen, and propoxur, an insecticide designed for home and lawn use, as a probable carcinogen."[9] Lawn pesticide applications can also prove toxic to birds, fish, and wildlife. Furthermore, studies show that gasoline-driven lawn mowers "produce as much pollution in half an hour as a car would driven 172 miles."[10] While manufacturers of chemical lawn-care products dispute many of these studies, it seems clear that America's love affair with the perfect lawn is exposing both humans and wildlife to increased levels of toxic chemicals.

In contrast, native species are intrinsically low maintenance because they have adapted themselves to available rainfall and existing soils and they have also evolved defenses against local predators and diseases. Landscaping with prairie plants can also prevent stormwater runoff and erosion because their long roots penetrate deep into the soil.

Lawns are so popular that they have become big business and are represented in Congress by their own lobby groups and industry associations. The number of people involved in lawn and turf grass care is mind-boggling: an estimated 56 million Americans take some part in caring for their own lawns, according to the National Gardening Association. For many of them, maintaining a lawn is the only intimate contact they have with the earth. Turf grass is a $25 billion industry in the United States employing more than 500,000 people, according to Eliot and Beverly Roberts at the Lawn Institute in Pleasant Hill, Tennessee.[11] Sales of do-it-yourself lawn care products have skyrocketed to $8.4 billion, up 50 percent in the last five years.[12]

In some communities the lawn enthusiasts are so influential that they can enforce conformity and require their neighbors to keep the grass cut at a certain height. Otto has had several encounters with the "weed police."

It all began during the 1950s when a marsh area near her home was cleared for a development. It was then that Otto decided to allow one corner of her yard to revert to the wild so that her children would have a place to play, pitch their tents, and hang their hammocks. When she stopped mowing in this area a number of native perennials grew back from seeds that had lain dormant in the soil. Soon, her thicket of overgrown native grasses and plants became a popular hangout for the neighborhood children. It wasn't long before she became quite attached to her little wilderness.

Otto might have remained a quiet suburban resident, with a patch of wilderness in her back yard, had the town weed commissioner not sanctioned the cutting of the overgrown portion of her lawn. Returning from a canoe trip to Lake Superior with her husband and children, Otto was down in the basement doing the laundry and had just bundled wet clothes into a basket to take up the cellar stairs to hang on the line when she heard her eight-year-old son yelling "Mommy, Mommy, they have cut it all down."

When she ran upstairs she found that her son was right. A neighbor who "didn't like looking at all those dandelions" had called the village and they had sent a man out to mow her wild plants. Otto was so incensed at the destruction of her property that she called the village offices and asked that the mayor, the village lawyer, and the weed commissioner come out to her place immediately.

Before they arrived, Otto consulted a list of so-called nuisance weeds posted on a telephone pole near her house. She wanted to satisfy herself that none of the plants that had been cut down were on the list. When the village officials arrived she took them on a tour of the dismembered plants and told them their names and how they had been used by Native Americans. She was well versed in this knowledge because she taught a wild plants class to the local Brownie Girl Scout troop. At the end of her lecture one of the men said to her, "Gee, Mrs. Otto, it sounds like we cut down a whole museum." Otto agreed and insisted that the town pay for damages. Thus ended the opening skirmish.

Otto soon found herself involved in a second campaign against the weed police. With the money she received from the town for damages to her wild plants, Otto bought a painting from a well-known local artist who lived in Milwaukee. After they became good friends, the artist decided to follow Otto's example and allow *her* yard to grow wild. Her city garden was so inviting that the editors of the *Milwaukee Journal* decided to do a photo feature of it on their front page.

Just before the feature was published, Otto's artist friend received a warning that if she didn't mow her lawn within 48 hours the city would cut it for her. Otto called city hall and discovered that the rationale given for the enforced mowing was that the plants were putting too much pollen into the air. She argued that the trees were putting more pollen in the air than the plants and asked if the trees would also be cut down. She was referred to the sanitation department where she was told that the man who collected her garbage had reported that her yard was an eyesore. Finally, Otto told the mayor that the newspaper was about to come out with photos of this beautiful yard and that they would be interested in a story about why the city was insisting on cutting it down. It was then that the mayor called off the weed police.

More recently Otto's wildflowers suffered an attack from a neighboring property that was undergoing a chemical lawn treatment. A gust of wind blew the herbicides over onto her yard killing some of her wildflowers and prairie plants. Again she insisted that she be remunerated for the damages so that she could buy replacement plants.

Given this environment, what is remarkable about Otto within the natural gardening movement is the success she has had introducing this gentle art into a suburb like Bayside where the traditional style of American gardening is firmly entrenched. Located on the shores of Lake Michigan, Bayside is host to well-kept brick and stone houses with asphalt driveways, two-car garages, and weed-free lawns bordered by neat rows of ornamental flowers and potted plants.

While Bayside does not look like the sort of suburb where the back-to-nature movement would take root, Otto's enthusiasm for gardening with

Figure 2.1
Lorrie Otto in her garden of native wildflowers in Bayside, Wisconsin. Photograph by Steve Lerner.

native plants has proved contagious and a number of gardeners in her community have decided to let their own yards grow wild. Otto convinced the school her son attended to plant native species on the grounds so that children would grow up familiar with local vegetation. Through classes she gave at a local nature center and a series of cable television shows, Otto spread the word about the joys and ecological benefits of landscaping with native plants. (She received the Margaret Douglas medal for conservation education from the Garden Club of America in 1991.) After attending some of Otto's lectures, a number of local residents started a group called the Wild Ones that maintains a network for some 800 natural landscapers and publishes a newsletter. Now, twenty years after having let her own lawn go to seed, Otto leads tours of numerous natural gardens in her community.[13]

In fact, there are signs that gardening with native plants is coming to be accepted, even by local officials. "We no longer make people mow their lawns because we have yards here and there with long grasses and native plants," the Department of Public Works superintendent is quoted as having said. Otto sees this as an important shift. "My God, I have lived long enough to hear a maintenance man say 'native plants.'"

Elsewhere around the nation a more natural approach to lawn and turf care is gradually emerging. In Maryland, for example, the State Highway Administration instituted a "Grow Don't Mow" program aimed at reducing pollution and maintenance costs. "Since Maryland's program began four years ago more than 3,100 acres have been returned to a natural forest state , saving more than $400,000 in annual maintenance costs," *Washington Post* correspondent Caroline E. Mayer writes.[14] Private companies are also cashing in on rising demand for a chemical-free lawn or at least one blanketed with a reduced load of pesticides. Companies such as NaturalLawn of America and Barefoot Grass and Lawn Service in Maryland are cropping up to meet the new demand.[15]

For do-it-yourselfers a number of books are available to help lawn buffs kick the toxic chemical habit, including *The Chemical-Free Lawn* by Warren Schultz (Rodale), *Redesigning the American Lawn* by F. H. Bormann, D. Balmori, and G. Geballe (Yale University Press), and *Energy-Efficient and Environmental Landscaping* by A. S. Moffat, M. Schiler, and the staff of Green Living (Appropriate Solutions Press).

Otto's involvement with native plants is rooted deep in her childhood. Her father was a local dairy farmer who hand-terraced his farm to prevent erosion. From an early age she remembers having her hands in the soil of the vegetable garden that her German grandmother kept and her feet in the furrows her father ploughed with a horse. Her father paid her a dollar a year to pull up all the pernicious weeds on his farm so she became proficient at identifying weeds. She recalls setting off into the fields for the day with a lunch basket packed by her mother, a small shovel, and her Scottish terrier for company. Her early interest in plants blossomed after she married a physician who was also an amateur botanist. She and her husband made it a habit to spend part of every Sunday identifying plants in the woods with their two children.

Now regarded as one of the pioneers of the natural gardening movement in her area, Otto wants to help preserve some of the unique features of the ecosystem that evolved in Wisconsin. Much of the country is coming to look the same with standard plots of grass, shade trees, and hybrid flowers,

Otto observes. "I think when people come to Wisconsin they ought to be able to look around and tell by the trees and flowers that this is Wisconsin or at least the Midwest and that this is what grows here."

Otto's wildflowers and native shrubs have brought back a host of animals to her property, including a red fox. "I don't prune my shrubs like a sculpture, but rather let them grow into a thicket so that the birds can find a safe place to nest. Catbirds and indigo buntings have nested here and the red-eyed vireo comes to visit."

Of her own plantings she says, "You feel so much better when you are in an area where the plants, birds, and insects match up. . . I get such a wonderful feeling when I see birds feeding here and I know that this is such a safe place for them to feed and breed."

Greenhouse Treatment of Municipal Sewage

A ruddy-faced man smeared with pond muck and sweating profusely stamps into a greenhouse in Providence, Rhode Island, carrying a bucket brimming with algae, snails, microorganisms, and other critters. Without ceremony he dumps the bucket into one of the forty-eight translucent tanks where carefully balanced ecosystems are busy digesting sewage. "Now let's see what happens," he says with a laugh.

The pond scavenger is John Todd, a fifty-six-year-old Canadian-born marine biologist who is pioneering a new science that uses what he calls "living machines" to treat sewage. A water doctor who restores unhealthy or polluted waters to ecological balance, Todd takes as his patients municipal sewage and septage and industrial wastewater, as well as lakes and rivers that suffer from eutrophication, acid rain deposits, toxic chemicals, and heavy metals.

The son of a 3M company executive, Todd began his training for this work as a high-school truant. It was only after he read a series of books by Louis Bromfield about restoring an impoverished, rundown farm in Ohio that Todd says he found his life's work—land restoration. "After reading *Malabar Farm, From My Experience,* and *Out of the Earth,* I knew I had to do this environmental repair work," he recalls. He subsequently made up for lost school time by earning two agricultural degrees at McGill University and a doctorate in animal behavior and fisheries at the University of Michigan before settling down to teach at San Diego State.

But Todd was not satisfied with just teaching. The urge to do hands-on ecological work led him to join in founding a series of nonprofit organizations to conduct experimental work. In 1969 he moved his family to Cape Cod where he took a job at the Woods Hole Oceanographic Institution.

That year, he and his wife, Nancy Jack Todd, and aquatic biologist William McLarney founded the New Alchemy Institute to investigate new applications of solar energy and to "design food producing systems based on renewable sources of energy."[1]

By 1973 Todd was building transparent silos in which he raised large numbers of fish; in the same tanks he grew lettuce and tomatoes to clean up the waste the fish generated. It was his first foray into mini-ecosystem design and construction. Later, at the urging of Margaret Mead with whom he had worked in Indonesia, he joined in founding Ocean Arks International, a nonprofit also on Cape Cod, in Falmouth. The first project of Ocean Arks was to design and build a prototype solar and wind-powered, energy-efficient fishing boat suitable for use in developing nations.[2]

Initially, Todd had envisioned a large, sail-powered, floating greenhouse—a kind of a "biological Hope Ship"—that could deliver "seeds, plants, trees, and fish to impoverished areas with the hope of reviving the biological support base and thereby improving the means of the human population to support itself." When this project proved impossible to finance, Todd and his colleagues scaled the project down and built an "ocean pickup," a smaller, fast-moving catamaran that could be used in developing nations for fishing.[3]

After returning from South America in the early 1980s Todd found that his family had to buy bottled water because the tap water was unhealthy. It was then that he decided that "if I could purify water for fish I could bloody well learn how to purify water for people."[4] To design a system of bioremediation that simulates the way nature cleans polluted water, John and Nancy Todd founded the Center for the Restoration of Waters at Ocean Arks in 1989. Since then they have been building living machines that use plants and animals to treat municipal sewage and industrial wastewater.

In many respects Todd's living machine wastewater treatment method is more elegant and efficient than conventional treatment that depends on the use of chemicals and costly mechanical devices to coagulate, skim off, or settle out impurities in the water that subsequently must be disposed of through burial or incineration. In contrast, Todd's more organic method uses less energy, minimizes the production of sludge, and eliminates the need for toxic chemicals in the treatment process. Todd faults conventional sewage treatment for displacing instead of solving the sewage problem by generating large amounts of sludge laced with heavy metals that must be dumped in landfills or incinerated.

Instead of trying to detoxify sewage with massive machines and large doses of toxic chemicals, Todd designed a treatment system that mimics nature. "By harnessing biological processes, living machines imitate the way nature purifies water, but more quickly and thoroughly," writes Nancy Jack Todd in the publication she edits, *Annals of Earth*. The basic design of the Todd's sewage treatment system is a food chain in which small organisms consume the nutrients in sewage and are in turn consumed by larger, more complex organisms.

Like many other environmental pioneers, the key to Todd's success lies in his ability to turn an environmental liability into an asset. From the outset he recognized that one of the components of sewage that makes it environmentally destructive is its nutrient richness. The reason for this is that unlike a cow's stomach, the human stomach is inefficient at digesting nutrients and produces a nutrient-rich waste. If these wastes are discharged into a freshwater stream or pond, the undigested nutrients cause algal blooms. While this may sound harmless, when the algae fall to the bottom of the lake they are digested by bacteria that use up large amounts of oxygen dissolved in the water. The bacterial action uses up so much of the oxygen that fish and other organisms become oxygen starved and suffocate.

Todd's living-machine greenhouse brings this eutrophication process into an environment where it can do no harm. The sewage water, rich in nitrogen and phosphorus, flows through a gravity-feed system past floating, hydroponically grown plants. As the sewage circulates and recirculates among the roots of these plants and trees, and amid fish and snails, the nutrients, heavy metals, and toxic chemicals in the wastewater are taken up or broken down by a wide variety of organisms. In this fashion, algal blooms that would kill fish in streams and lakes are encouraged to bloom within the confines of the greenhouse tanks where they can safely use up the nutrients in the sewage. Air is pumped into the water to further speed up decomposition and keep the nutrients in circulation until they are absorbed.

In contrast to a conventional sewage treatment plant, which generates large volumes of toxic sludge, the chief byproducts of Todd's living machines are plants, trees, snails, and fish. These byproducts can then be sold as compost, ornamental plants, or baitfish. A study, funded by the Environmental Protection Agency, demonstrated that what sludge does remain (about a quarter of the amount generated by a conventional system) can be safely placed on outdoor marshes to further decompose or be sold to turf farmers as an soil amendment.

Since 1989 Ocean Arks has tested living machines as a sewage treatment system on a strip of land down by the docks in Providence, Rhode Island, an area adjacent to the conventional sewage treatment plant, the dog pound, and a scrap metal yard. There, in a 3600-square foot greenhouse, Todd demonstrated that he can detoxify 16,000 gallons a day of sewage (the sewage from about 150 households) without the use of toxic chemicals.

Inside the greenhouse, forty-eight cylindrical tanks (or cells), each with its own ecosystem, are configured into four separate "rivers" of twelve tanks each. Raw sewage, which has already had the large clumps of grit screened out of it, is pumped into the greenhouse where it flows into the first of these 1300-gallon tanks, each of which contains a unique recipe of bacteria, algae, snails, and a variety of plants and fish.

Where the raw sewage enters the greenhouse, the first ecosystem to receive it is quite primitive. Todd describes a visit to the life forms in the first tanks of the living machine as a trip back in geologic time to an era when simple forms of life such as blue algae predominated. In these first treatment tanks there is a preponderance of water hyacinths and other hardy plants and organisms that can survive harsh conditions. The wastewater then passes into the first of two man-made marshes composed of sand, gravel, and bulrushes where more of the solids are removed.

As the sewage moves from tank to tank its toxicity is reduced and an environment is created where more complex organisms can survive. As one walks toward the front of the greenhouse, along the duckboards that run between the tanks, cattails, watercress, mollusks, fish, and 40 varieties of plants, including iris and tomato, join the food chain that digests the nutrients in the sewage.

In addition to taking up nutrients, the plants and trees—such as Eucalyptus and willow—also absorb phosphorus, cadmium, copper, and lead. When the trees outgrow the greenhouse they are transplanted outdoors.[5] This provides only a temporary solution to the disposition of the heavy metals because the tree sequesters the cadmium and lead for its lifetime, returning them to the soil when it dies.

The fish, mostly shiners, a baitfish, are small enough to migrate from tank to tank through pipes. As a result, if a slug of highly toxic sewage enters the system, the shiners are able to evade it by swimming downstream until the upstream plants have had a chance to do their purifying work. From this safe haven, the fish can then recolonize the upstream tanks. If the snails climb out of the wastewater onto the rim of the tanks, it is another sign that a slug of highly toxic material is passing through, Todd explains.

Each mini-ecosystem in the greenhouse is also designed to be self-balancing so species keep one another in check and prevent any one group from dominating. The food chain begins as bacteria consume suspended organic matter and convert toxic ammonia into nitrites and then nitrates. This waste product of bacterial action makes a feast for algae and duck-weed. Snails and zooplankton then gobble up the algae. Fish eat the zoo-plankton. Duckweed becomes a cover crop that shades out the sun so that algae will not grow in the later stages of the treatment process. Finally, the wastewater moves through a man-made marsh (contained in galvanized tubs) that further filters the water before it exits the greenhouse.

Over the last five years the Providence facility has demonstrated that it can successfully remove nutrients, toxic chemicals, and heavy metals from city sewage. This is no easy feat considering the witches' brew of chemicals and heavy metals that arrives for treatment. The costume jewelry capital of the world, Providence generates sewage that carries large quantities of heavy metals associated with jewelry manufacturing. The wastewater that Ocean Arks purifies, however, is not released into coastal waters. Todd never applied for a discharge permit because the Providence facility was built for research and not treatment, he explains. As a result, the treated wastewater is shunted down a drain and re-treated by the conventional sewage treatment plant next door.

"We've just taken nature and sped it up, using a lot of high light and sup-plemental aeration and some real interesting ecological engineering to min-imize the amount of space required to purify," Todd explains.[6] He foresees a day when sewage will be routinely treated with "ecological digesters."

After five years of research, the Providence living machine proved it could purify municipal sewage to the point where it met EPA's advanced waste-water (tertiary) treatment standards. Specifically, it met drinking water standards for heavy metals in the discharged water; swimming water stan-dards for pathogens; and advanced wastewater treatment standards for nitrogen, biological oxygen demand (BOD), and total suspended solids (TSS).

Following the Providence experiment, Todd improved the living machine design and soon had a chance to put this modified system to the test in Frederick, Maryland. There, with EPA funding, he built a new living machine enclosed in a greenhouse next door to the Ballinger Creek Sewage Treatment Plant, a high-tech conventional treatment facility. The new facility is undergoing EPA validation studies that include an independent review.

Opened in January 1994, the Ocean Arks facility in Frederick is more efficient than the Rhode Island greenhouse in a number of respects. Not only will it treat 40,000 gallons per day of sewage, roughly three times as much waste as the Providence facility, it does so using only half as much energy while occupying one-third the space per gallon of sewage treated.

This increase in efficiency is made possible by a number of innovations, the first of which is that sewage at the Frederick facility initially enters an

Figure 3.1
Living Machines digesting municipal sewage in an Ocean Arks International greenhouse near Frederick, Maryland. Photograph by Steve Lerner.

anaerobic treatment tank set into the ground outside the greenhouse. This anaerobic phase acts as a buffer so that if a particularly concentrated slug of sewage or septage enters the system, the bacterial activity breaks down much of the waste before it enters the aerobic tanks in the greenhouse. As the sun beats down on the airtight, black plastic that covers the anaerobic tank, the bacterial activity inside increases dramatically.

A second innovation is that wastewater in the Frederick greenhouse circulates more often and more rapidly than it did in the Providence facility. Greater circulation gives organisms more opportunities to absorb or adsorb pollutants in the wastewater. To take maximum advantage of this increased circulation, Todd introduced what he calls "ecological fluidized beds" which consist of large quantities of pumice stone, a light volcanic material that is riddled with cracks and crevices. The myriad interstitial spaces in the pumice beds increase the surface area to which algae and microorganisms can cling, creating a biologically alive surface for the sewage to pass through, further increasing the efficiency of the system. Finally, the sewage passes into an indoor marsh that serves as what Todd calls "a catcher's mitt" to take up any pollutants that evaded the earlier treatment process.

Ocean Arks employees are always on the lookout for organisms that take nutrients and pollutants out of the waste stream. Part of their job involves collecting aquatic plants and organisms with which to seed the greenhouse. Some of their star performers include the bald cypress, which is efficient at sequestering heavy metals for several hundred years; papyrus, which not only takes toxins out of the water but can also be made into paper; duckweed, which is a good cover crop and high in protein; as well as pennywort, parrot's-feather, arrowhead, hawthorn, black currant, and elephant ear taro.

Ocean Arks aquatic experts are also experimenting with ways to make some of the byproducts of the waste treatment system marketable. In a big tank with duckweed on the surface, which clarifies the treated sewage water before it empties into a final marsh area, Ocean Arks employees are breeding goldfish for sale. And just at the end of the treatment process, where the quality of the finished water appears to be completely clean, is a pond with a fountain where carp swim amid flowering lilies. These fish are now sold to Lily Ponds Inc., a well-known water-garden company. While sales from byproducts of the system are not on a scale where they can

Figure 3.2
Ocean Arks International staff worker planting a tree in one of the living machines inside the greenhouse at the Frederick, Maryland facility. Photograph by Steve Lerner.

support the operation of the living machine, Todd foresees a day when they can.

Creating living machines to treat sewage makes sense for four basic reasons, Todd asserts. First, the greenhouse system costs less to build than a conventional treatment plant and about the same to maintain. Second, it uses less energy because it depends on photosynthesis and a gravity-feed system instead of large and complex mechanical devices. Third, it does not burden the environment with the chlorine used in conventional treatment. And fourth, it creates much less sludge than does conventional treatment.

But not all the problems with treating wastes using living machines have been worked out. A number of challenges remain before living machines become a closed system that recycles all the wastes contained in sewage. The two thorniest problems that remain to be solved are the removal of high levels of phosphorus in the waste stream and the generation of small volumes of sludge.

At the Providence facility the living machine was unable to take up all of the phosphorus in the sewage. Phosphates are efficient cleansers that are used in many products, but once they go down the drain they are hard to treat. Plants store and use only small amounts of phosphorus. Of course more phosphorus could be removed from the water by harvesting the plants and growing new ones, but this is an expensive, labor-intensive practice. Thus, while plants and other organisms can sequester a good portion of the phosphorus load, they cannot pick up enough to allow the treated water to be released into a freshwater stream.

The living machine at Frederick reduces phosphorus levels about 50 percent to about 5 milligrams per liter (mpl), while the target level is 3 mpl. Once it is fully mature, however, the Frederick facility is expected to reduce phosphorus to acceptable levels. A number of experiments are going on to achieve these reductions using materials that adsorb phosphorus such as slate and limestone. Experiments using filamentous algae, which operate as sponges for phosphorus, are also being conducted in South Burlington, Vermont, at a facility that is overseen by Todd but run by a new company called Living Technologies, which specializes in building living machines that treat high-strength organic waste streams from industry.

"We still haven't totally licked the sludge problem," Todd admits, although the living machine only generates about 25 percent as much as the conventional facility. At the Frederick facility, the bulk of this sludge

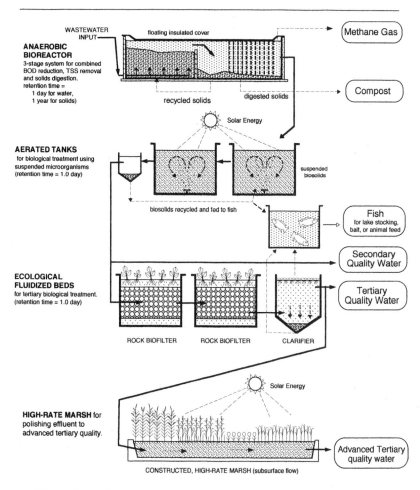

**PROCESS FLOW DIAGRAM FOR THE
BALLANGER CREEK LIVING MACHINE**

A demonstration project of an innovative ecological wastewater treatment technology.
The system is designed, and managed by Ocean Arks International. The project is funded U.S. EPA.

Figure 3.3
*Schematic of the Living Machine at Frederick, Maryland by Living Technologies,
Burlington, Vermont. Courtesy John Todd.*

collects in the new, outdoor anaerobic reactor. "My solution, which we are demonstrating at a facility in Waco, Texas, is to pump the sludge on a batch basis into an outdoor, plastic-lined, reed bed," Todd says. When the sludge dries between pumpings it cracks and mineralizes, dramatically reducing its volume, he explains.

In other areas the living machine is amazingly efficient at detoxifying the waste stream, despite receiving some highly concentrated wastes that include septage, a dense form of sewage that is pumped out of septic tanks. In fact, living machines are superior in their performance to all conventional wastewater treatment facilities in the United States that treat water to "secondary" treatment standards, Todd notes. When the water exits the living machine, it is reservoir-quality water that you can swim in but not drink, he adds.

Having achieved this standard, Todd is now attempting to meet the even stricter "advanced primary" treatment standard. While he has not achieved this in all areas, the water treatment performance of the Living Machine at Frederick between May 1995 and May 1996 was excellent with the exception of phosphorus, Todd reports. The system proved remarkably stable in the treatment of the principal performance measures: biological oxygen demand (BOD), chemical oxygen demand (COD), and total suspended solids (TSS), he continues. On average, by the time the water exited the living machine, its BOD was 4.8 mg/l (target 10 mg/l); COD was 27 mg/l (target 50 mg/l); TSS was 1.4mg/l; total kjeldahl (TKN) was 3.8 mg/l (target 10 mg/l). However, the living machine came close to but failed to meet the targets for a number of other performance measures: total nitrogen (TN) in treated water was 10.6 (target 10 mg/l); ammonia (NH_3) was 1.7 mg/l (target 1 mg/l); and fecal coliform was 265 col/dl (target 200 col/dl).[7]

Summing up the implications of these results, Todd writes: "The Maryland Living Machine performed well during its last year of operation. Despite stronger than expected influent waste, the system performed at close to design expectations, with the exception of phosphorus. . . . The Living Machine for sewage treatment at Frederick demonstrated the high potential of ecological engineering in the future of waste water treatment."[8]

A study of the system done by the EPA, however, concludes that higher plants do not add to the performance of the Living Machine.[9] Todd disagrees with this finding. His own observations, made after the EPA study, showed a reduction in ammonia and fecal coliform levels in the train of water treated with higher plants compared with the one devoid of plants. Unfortunately, the performance of the Living Machine during the EPA study was not typical of its overall performance—because of abnormally high concentration of pollutants in the influent that exceeded the designed capacity of the system and operational issues, not because of a design flaw, Todd writes. Further EPA studies of the performance of living machines in operation in South Burlington, Vermont, will take place in 1997, Todd says.

Living machines are now in operation in thirteen states around the United States and in seven countries, but Todd is by no means the only one using nature to treat sewage. Many small towns, residential subdivisions, and facilities use a system where sewage is channeled into reservoirs, churned, and aerated to separate out solids before being filtered through sand and disinfected. The wastewater, still rich in nutrients, is then sprayed on lawns, golf courses, crops, soccer fields, or even turned into snow for ski slopes. It is estimated that there are some 900 water-reuse, land-application systems operating in the United States.[10]

In the 1960s Bill Wolverton, a NASA scientist, designed a "closed ecological life support system" for treating waste aboard spacecraft. To solve a more immediate problem, he constructed a marsh that used water hyacinth to clean up the sewage generated by 4000 people at NASA's National Space Technology Laboratory in Mississippi where he was working. Then he went into business for himself and built "microbial rock filter marsh systems" for a mobile home park in Pearlington, Mississippi, which treats 10,000 gallons of wastewater a day. His system involves digging a shallow trench, lining it with plastic, and filling it with stones and wetland plants such as canna lilies, water iris, and elephant ear. As the sewage flows through the stones and the roots of the plants, microorganisms consume the nutrients in the sewage. A similar system has been installed in Denham Springs, Louisiana, that treats 3 million gallons of sewage a day.[11]

But unlike the marsh system, which processes sewage that has already undergone primary treatment, Todd's system treats more concentrated raw sewage. Furthermore, by bringing living machines into a greenhouse, Todd has expanded the climatic range in which the natural treatment of sewage can work. Since the greenhouse can keep an ecosystem alive during the winter, the living-machine wastewater treatment system can operate year-round in cold climates.

Todd's research at Providence and Frederick was preceded by a series of smaller trials (some more successful than others) that took place over a number of years. The first test of this fledgling technology occurred in 1987 when Ocean Arks installed a raw sewage treatment system at the Sugarbush ski resort in Warren, Vermont, a system that worked but was eventually scrapped in favor of a conventional treatment system.

In 1988, Ecological Engineering Associates (a group Todd founded, but which subsequently broke off from Ocean Arks to concentrate on marketing commercial-scale living machines designed to treat municipal sewage) built an ecologically engineered sewage treatment system in Harwich (pop. 9000) on Cape Cod to handle one-third of its highly concentrated septage waste. Previously the septage was buried in the ground where it polluted the groundwater. Tests of the living machine at Harwich revealed that it removed 100 percent of thirteen of fourteen of the EPA's top-priority pollutants; and it removed 99 percent of the fourteenth pollutant, toluene.

Although the treatment system was validated by Massachusetts inspectors, a permanent facility was never built because the town ultimately opted to send its septage to a conventional facility built in a neighboring community. But with the technology already validated, Ecological Engineering Associates (EEA) shifted its focus and built a living machine to treat wastes from boats that dock at Marion, Massachusetts, on Buzzards Bay. This facility has since been expanded to treat all the town's septage. Subsequently, EEA built commercial ecologically engineered municipal sewage treatment systems in Mexico, Nova Scotia, and elsewhere in Massachusetts.

Todd also designed (and Living Technologies Inc. built) a living-machine wastewater treatment unit mounted on an eighteen-wheeler tractor-trailer truck that can travel to different sites to test the system's effectiveness in treating various types of waste. This mobile living machine, which includes a fluidized bed component, is now in San Francisco where it is treating 50,000 gallons per day of secondary sewage that it upgrades to advanced (tertiary) treatment standards. The unit is designed with a small laboratory at the front- and back-ends of the system so that operators can monitor the level of pollutants in the water they take in and release.

Despite the advantages of living machines, municipalities are not rushing to install them. Todd attributes this lack of enthusiasm to bureaucratic inertia. The Clean Water Act spawned a water treatment industry that is

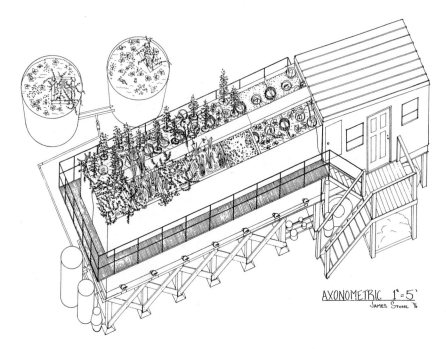

Figure 3.4
Drawing of the Living Machine in San Francisco. Courtesy John Todd.

addicted to chemical treatment where the rate of innovation is very low, he observes.

While municipalities have been slow to adopt environmentally engineered sewage treatment systems, a number of industries are experimenting with them. Not all of these experiments, however, have been successful. For example, in 1991, EEA, Gardner Supply Company, and Ben and Jerry's Inc., the ice cream manufacturer, embarked on a joint venture to test the ability of living machines to treat dairy wastes. The experiment was designed to treat 10 percent of Ben & Jerry's waste stream at its Waterbury, Vermont, site.

Gary Audi, environmental coordinator of Ben & Jerry's Waterbury site, who operates its wastewater pretreatment plant, observes that while one could claim that the system worked (dirty water came in one end of the system and cleaner water came out the other end), the system did not make sense for Ben & Jerry's at the time because it was not capable of replacing the waste treatment system it had in operation.

Figure 3.5
John Todd, co-founder of Ocean Arks International, holding flasks of influent and effluent water amid his Living Machines at Harwich, Massachusetts. Photograph by Dann Blackwood.

EEA faced a series of problems at Ben & Jerry's, Audi continues. First, the waste stream kept changing as the company came out with new products. "Treating our high-strength dairy waste proved to be more of a challenge than they expected," Audi says. "Our wastewater was definitely the ruggedest stuff they had ever come up against." While the septage they had been treating had a BOD of maybe 4000 mpl, Ben & Jerry's waste stream had a BOD of 10,000 mpl.

Furthermore, in order to protect their own lagoons, an elaborate pretreatment system was already in place, which removed 90 percent of the butterfat and half of the BOD from the waste stream. This significantly reduced the load the greenhouse had to treat. As a result, almost all the waste components that were difficult to treat were removed before they entered the greenhouse. "Any residual grease that did pass through the pretreatment process caused operational upsets in the greenhouse," Audi recalls.

"Our cold winters also made optimal conditions difficult to achieve in the greenhouse," Audi continues. It took an undetermined but significant amount of energy to keep the twenty-by-eighty-foot greenhouse lighted and heated during the winter; providing energy for a full-scale cluster of greenhouses covering two acres (which would have been necessary to treat all of the wastes) would have been prohibitively expensive, he says. Perhaps in a warmer climate the system works better, he concludes.

Clearly, there has been a steep learning curve in the development and application of living machines to various waste streams. Since the less-than-successful Ben & Jerry's experience, the living machine has been further refined and Living Technologies Inc. is now treating industrial waste at an M&M Mars candy plant in Waco, Texas, where Snickers, Skittles, and Starburst candies are manufactured.

Norm Burgess, site engineer at the plant, is impressed with the ability of living machines to treat Mars candy wastes. The company's conventional, on-site, activated sludge treatment plant was producing 130,000 gallons of wastewater a day they had to pay the city to treat, and 50,000 gallons a week of sludgy water that had to be hauled away in tanker trucks. "The sludge was costing us big bucks to get rid of," he states.

Living Technologies, the Vermont firm that Todd remains affiliated with, is solving the Waco plant's waste problem through a two-phase approach. First, the water leaves the facility's conventional, primary treatment plant and empties into an outdoor wetland. The wetland is made up of horizontal trenches lined with thick plastic, filled with pea gravel, and studded with "marshy plants that like to get their feet wet," Burgess explains. As the wastewater passes through the gravel, contaminants are removed by the roots of the plants and the purified water is collected in a pond. Once he has the requisite permits in hand, Burgess plans to use this recaptured water in his cooling (evaporation) tower, which consumes 100,000 gallons of water a day as it cools the candy plant under the hot Texas sun.

Instead of hauling the sludge, which is 98 percent water, off in tankers at great expense, Living Technologies is calling for it to be piped onto gravel beds where it dries out and the remaining solids are composted. Reeds and cattails are grown in the gravel beds to keep the sludge broken up and permit percolation. The sludge ultimately composts into a topsoil that will be applied as a soil amendment on the facility's forty acres of lawns and plant beds.

Burgess is so impressed with the pilot project, which is currently treating 5 percent of the plant's wastes, that he recently opted to expand it into a full-scale treatment facility that will handle all of the plant's wastewater and sludge. He calculates that the living machine wetland and sludge beds will pay for themselves within 1.6 years, providing a 30 percent yield on the company's investment. By using the living machine, Mars will no longer have to (1) pay the city to treat its wastewater, (2) pay to have the sludge hauled away, or (3) buy water for its cooling towers.

"Lots of people are impressed by the environmental aspect of this system, but what gets me really excited is that it makes business sense. I am not an environmentalist, I am a businessman, and this is a good business opportunity. The fact that it is environmentally correct is just icing on the cake," Burgess concludes.

Today, most sewage treatment facilities are located in industrial areas far from residential communities where the dirty work of treating waste can be done out of sight. In fact, communities tend to group all of the least attractive facilities—prisons, landfills, dog pounds, sewage treatment plants—in some wasteland on the wrong side of the tracks. To reverse this pattern, Todd envisions either designing houses to treat their own wastes or infiltrating neighborhood greenhouse treatment of sewage into residential areas.

Todd says. Benthic life forms recolonized the lake bottom, biodiversity increased threefold, and beaches appeared in places where they have not been seen in 20 years, Todd reports.

In all this work what is remarkable about Todd is his humility in the face of nature, says Scott Sargert, who has worked at the Ocean Arks Providence facility since 1989. Todd looks at himself more as a steward than as an inventor or engineer, Sargert observes. He speaks of building structures that will harbor and nurture ecosystems that purify water as opposed to controlling them. He does not pretend to know all of what is in the tanks or how they do what they do: he invites organisms into the tanks and some of them stay and some die. "He is one of the few scientists I know who expects to fail a certain percentage of the time and is not downcast by failure but rather sees it as part of a learning process," Sargert notes.

Todd teaches that natural systems cannot remain stagnant and isolated from nature, Sargert continues. Man-made ecosystems need to be renewed by, and stay in touch with, the larger outdoor ecosystems by constantly introducing new organisms into the tanks. The reason for this is that ecosystems in nature are continually evolving to deal with changing waste streams, so living machines must keep pace with them, Todd says.

For their work Nancy and John Todd have won numerous awards, including the Chico Mendes Environmental Merit Award from the EPA, but it is not for the awards that John Todd works. He seems to be satisfied mucking about in a swamp looking for species proficient at detoxifying water that he can propagate in his living machines.

The New Frugality Movement Promotes Living Better by
Consuming Less

"The single most important contribution any of us can make to protecting
the environment is a return to frugality," says Vicki Robin, president of the
New Road Map Foundation in Seattle.[1] "Money is a lien against the resources of the planet. Every dollar in your wallet is an IOU against Mother
Nature. When we buy a green pen we don't think about what Mother Nature had to cough up to produce it. We don't think about the energy, natural
resources, and toxic emissions involved in making it. We just think: I have
a dollar, I like that green pen, and I will buy it," says Robin, who is coauthor with Joe Dominguez of the best-selling book *Your Money or Your
Life: Transforming Your Relationship with Money and Achieving Financial Independence.*[2]

By living more simply, purchasing fewer goods, buying used merchandise
whenever possible, and taking care of what we own, fewer trees will be
felled, fewer minerals mined, and less energy burned to produce new products. Overall the through-put of resources will slow down and stress will
be reduced on nature. Simply put, spending less money means consuming
fewer of the planet's resources.

But how can Americans be convinced to buy less at the height of their
record-setting shopping spree? How can consumer mania be curbed when
young men sport T-shirts emblazoned with the message: "No Man with a
Good Car Needs to Be Justified" or "The One Who Dies with the Most
Toys Wins."

To date, several strategies aimed at cutting American consumption have
failed. For example, the "shame game" of attempting to make Americans
feel guilty about their high-consumption lifestyle has fallen flat. Environmentalists who argue that we should all consume less are widely perceived

as fun-suckers and killjoys: no one likes being told that what gives them pleasure and status is destructive of nature.

Furthermore, in the media market the environmentalist message is simply outgunned. In a pamphlet of statistics they compiled entitled *All-Consuming Passion: Waking Up from the American Dream*, Dominguez and Robin point out that American teenagers are now exposed to some 360,000 advertisements by the time they graduate from high school and that this bombardment of messages urging them to buy, buy, buy is highly effective.[3] Today 93 percent of teenage girls report store-hopping as their favorite activity.[4]

What can we do about runaway consumption? Some argue that if we only had decent political leadership we could cut American consumption radically. But even a presidential appeal to the public to reduce waste failed. When President Carter appeared on television wearing a cardigan sweater and urged citizens to conserve energy by turning down the thermostat and wearing warm clothes at home, the media derided his message as overly austere. Such efforts to reduce stress on the environment by depriving Americans of their comforts and luxuries have not played well on Main Street. But alternative approaches to convincing us to reduce our high rates of consumption exist. For example, while we bridle at being told to use less, we are more willing to experiment with ways of saving money and stretching a buck.

In fact, hunting for a bargain and getting your money's worth is a cherished, all-American pastime. Attesting to the popularity of the stretch-a-buck message is the proliferation and success of newsletters such as Amy Dacyczyn's *Tightwad Gazette*. In 1990 there were only a half-dozen newsletters focusing on how to save money, but now there are over forty, Robin notes.

Also signaling the growing strength of the frugality movement in the United States ("the shrift to thrift") is the number of Americans interested in simplifying their lives so they can establish a better balance in the amount of time they devote to their job and the rest of their lives. Many people are now reassessing their way of life and asking themselves if they would not enjoy a simpler, slower-paced, less expensive lifestyle.

Evidence of this can be found from a number of sources:

• The Trends Research Institute of Rhinebeck, New York, named voluntary simplicity one of the top ten trends of 1994 and predicted that by the year 2000, 15 percent of the nation's 77 million baby boomers will be part of a "simplicity market" for low-priced and durable products.

• A 1994 *Health* magazine survey states that one in three people questioned said they would trade a 20 percent pay cut for fewer work hours.

• A report entitled *Yearning for Balance: Views of Americans on Consumption, Materialism and the Environment,* published in July 1995 by the Merck Family Fund reveals that 82 percent of Americans agree with the statement "We buy and consume far more than we need," and that 67 percent agree that Americans cause many of the world's environmental problems because we consume more and produce more and waste more than anyone else in the world. The study also found that Americans perceive a connection between the amount we consume and concern about environmental degradation. Twenty-eight percent of those polled also said that they had downshifted and voluntarily made changes in their lives (other than retirement) during the last five years that decreased their earnings.[5]

Growing numbers of people are beginning to realize that while they may have a career they don't have a life, Robins observes. But is this sense of being overworked just yuppie whining? In her book *The Overworked American,* Harvard economist Juliet Schor suggests that it is not.[6] Schor calculates that two-income families now work 1000 hours more a year than couples did 25 years ago. In response to the growing interest in finding a good balance between having a career and having a life, a whole line of books is now available, including. *Downshifting: Reinventing Success on a Slower Track* (New York: Harper Collins, 1991), by Amy Saltzman; *Muddling Toward Frugality* (San Francisco: Sierra Club Books, 1978), by Warren Johnson; and *Simple Living: One Couple's Search for a Better Life* (New York: Viking, 1992), by Frank Levering and Wanda Urbanska; and *Voluntary Simplicity* (New York: William Morrow, 1993), by Duane Elgin. A *Simple Living Network* has also opened on the Internet that reviews products and provides tips for simple living.

Amid this new literature, *Your Money or Your Life,* by Dominguez and Robin (which has sold 500,000 copies) takes a refreshing approach to coaching Americans on how to figure out what constitutes enough. "Most Americans have bought into the myth that more is better. But if more is better, then you never have enough," notes Robin.

Rather than adopting the failed strategy of blaming the individual for excessive consumption, the authors deal with the consumption issue indirectly by talking about a subject Americans are obsessed with—money. They write about how we can improve our lives through frugality rather than preaching to us that we must consume less to save resources for future generations.

Unlike most of the other popular books on money, Dominguez and Robin are not touting a new way to make a killing in real estate or the stock market. "Most books in this area are about how to make more money, but ours is about having more money by plugging up the hole in your wallet," Dominguez says. Instead of pushing a new get-rich-quick scheme, the authors describe the joys of saving money by spending less while enjoying life more. A byproduct of this simpler lifestyle is that it will also help protect nature, they add.

Most of us do not have a mature relationship with money nor do we realize what we sacrifice while earning it, the authors argue. What we fail to grasp is that we trade time and life energy for money, and this failure to understand what we sacrifice for money leads to an inexact sense of its value. Without an accurate idea of what money is worth, we often spend it irrationally and receive less than full satisfaction from our purchases, they continue.

It is the irrational or ill-considered purchase that the authors want us to avoid. Far from being anti-materialistic, they encourage us to pay more attention to the goods we need by comparison-shopping for price, quality, and durability, as well as taking care of what we own. If people kept track of what they spend (and whether or not they received satisfaction commensurate with the work they do to pay for their purchases), then their rate of consumption will automatically decrease, the authors observe.

Both Robin and Dominguez have spent the better part of their adult lives practicing what they preach. Having stepped off the more-is-better treadmill, they demonstrated that it is possible to live cheaply, save money, and at the same time enjoy life more. "Life gets better when one discovers enough," they insist. Both have lived comfortably for the past twenty-five years on $6000 to $7000 a year. Their stories are important, not because we should all strive to live on a similar amount, but rather because they point to ways we can all reduce our consumption relatively painlessly.

Figure 4.1
Vicki Robin and Joe Dominguez, authors of Your Money or Your Life, *at the offices of The New Road Map Foundation (TNRMF) in Seattle, Washington. Courtesy TNRMF.*

Joe Dominguez grew up poor in Harlem. His family was supported by welfare after his Cuban-born father, an accountant, fell ill with tuberculosis and was confined to a sanatorium. A child of immigrants, Dominguez was brought up amid expectations that his life would be better than that of his parents.

Today, Dominguez at age fifty-eight, is financially independent. He has enough money socked away in long-term treasury bonds that he can live off the income. But his is not the typical Horatio Alger story of escape from the ghetto. Dominguez did not get rich, he just got smart. Instead of making a whole lot of money and living high on the hog, he saved assiduously.

Dominguez came by his frugality out of necessity. Since his family did not have a lot of money, he learned to live cheaply. If he needed a desk, he built it. He sought out hobbies and forms of entertainment that were inexpensive. He bought his food at the supermarket and cooked it at home. While at school he also held down a number of part-time jobs, starting out as a newspaper delivery boy. It was while delivering newspapers that he noticed that the more prosperous clients on his route did not seem to be any happier than the less affluent.

Like many children of immigrants, Dominguez escaped the ghetto by climbing the educational ladder. He was accepted at one of New York City's elite public schools, the Bronx High School of Science, and subsequently studied science and engineering at the City College of New York. At age eighteen he moved out of his family apartment and found his own rent-controlled apartment in Harlem.

After spending a summer traveling across the United States with a friend, Dominguez dropped out of college and found a job as a messenger at a small brokerage house on Wall Street. There he worked his way up in the firm until he became a technical analyst who wrote and edited *Technical Notes,* a newsletter for institutional investors. As he moved up the ranks on Wall Street, his salary increased, but instead of buying expensive clothes and moving to a ritzier neighborhood, Dominguez maintained his low-rent lifestyle and stashed his savings. "I couldn't bring any of my coworkers home with me because they were terrified of Harlem, but, hey, I was comfortable," he recalls.

While working long hours on Wall Street, Dominguez concluded there must be something more to life than a 9-to-5 job until age sixty-five. But how could a poor boy get out of the rat race? After reviewing his options, he decided that by maintaining a modest lifestyle and committing himself to a dedicated savings plan he could afford to retire at age thirty. Following this plan to the letter, by the last day of his thirtieth year, Dominguez had saved $85,000 which he invested in long-term treasury bonds. The treasuries provide him with $5000 to $6000 a year in interest, a sum equal to his annual expenses. Financially independent and with no need to hold down a job, he bought a Chevy van, outfitted it for camping, and headed west.

Not far from where Dominguez grew up in Harlem, Vicki Robin was following her own path toward a frugal lifestyle and financial independence.

Robin grew up in Hempstead and Manhasset, Long Island, the daughter of two professionals. Her father was a radiologist and her mother, a psychotherapist. Her parents were well-off and had all the 1950s accoutrements of affluence, including a nice house, a ride-along lawn mower, and a convertible with tail fins. Then, when Robin was thirteen, her father committed suicide. Along with the tragic personal loss came an indelible lesson that affluence and happiness are not necessarily linked.

When Robin tells the story of her frugality, she begins with the lessons that she learned from her mother whose parents endured hard times during the Great Depression. Robin learned to shop at discount stores, only later discovering the joys of thrift stores and garage sales where a bag of clothes could be bought for a dollar. While at a rustic camp in Maine she also learned about the pleasures of a simple life close to nature.

It was not until she went to Brown University and took her junior year abroad, however, that these lessons in frugality paid off. Robin calculated that she could spend a year at the University of Madrid and six months traveling in Europe for the price of her tuition at Brown if she was thrifty. While traveling in Spain, she learned she could knock on the door of a convent in the evening and spend a night on a straw mattress for 25 cents. She also discovered that she had more interesting adventures and more fun traveling cheaply in third class than when she spent a lot of money.

After graduating from Brown, Robin moved to New York City where she worked as an extra on soap operas until she decided to sell her car and furniture, buy a van, and start traveling. As she drove across the United States, Robin recalls feeling as if she were playing hooky and was vaguely guilty about taking time out from pursuing a career. This sense of guilt abated when she met retired people camping in national parks who told her she was smart to take time off from work to travel while she was still young. They told her that while they were having a great time in their retirement, they regretted waiting until they were old and losing their health before giving themselves permission to step out of their workday jobs to explore and enjoy life.

As she traveled, Robin was spending down her savings when she met Joe Dominguez in a trailer park in Mazatlán, Mexico. Robin was intrigued by this thirty-year-old man who didn't have to go to work during the day because he was already financially independent. Dominguez explained how he had achieved financial independence and Robin decided to invest her savings and a small inheritance in long-term U.S. treasuries, in effect becoming the first graduate of the Joe Dominguez School of Frugal Living and Dedicated Savings. Since then, for the last twenty-five years, Robin has lived on $7000 a year.

What Dominguez and Robin discovered was that by living frugally they could drop out of the job market at a young age and devote the rest of their lives to work that they wanted to do; they could follow their hearts (not their jobs) and be of service to others. "When I worked on Wall Street I saw that most of the people there were not making a living, they were making a dying. They would come home from work a little deader than when they started out in the morning. I was determined I would not make the same mistake," recalls Dominguez.

What prevents many people from developing a mature relationship with money is that once they examine their spending patterns, they feel guilty about their expenditures but find it hard to change. Again and again they make efforts to rein in reckless spending and put themselves on a budget. But, repeatedly, they succumb to some temptation in the marketplace and then feel guilty about their purchase.

The first step in developing a mature relationship with money is to let go of your sense of guilt about your spending patterns, Dominguez and Robin advise. After all, it is your money; you earned it, and you have the right to spend it any way you please. But you owe it to yourself to see that you get satisfaction out of your purchase that is commensurate with the amount of life energy you traded for it. To make this calculation you must first figure out how much you make an hour so you know how much life energy you exchange for some trinket that catches your eye.

That's easy, you say: just divide the amount of money you make a year by the number of hours you work. But what about the hours spent commuting to your job? What about the money spent on costuming yourself for work, paying taxes, and on childcare? What about the expensive lunches you eat at restaurants with your coworkers? Robin and Dominguez recommend adding in all the unofficial hours you spend on your job and subtracting the costs associated with your job to arrive at a more realistic hourly wage.

These calculations (part of a nine-step program to financial independence) can be very revealing for those of us who have never had a mature relationship with money. Suddenly you have a way to judge purchases in a more realistic light. For example, is a dinner out really worth five hours on the job? Is a new car worth nine months on the job? Is a new house worth seven years on the job?

The next step is to keep track of how you spend every penny of your money. Most people find that monitoring their expenses causes them to spend about 20 percent less than before. These savings can then be put into long-term treasury bonds that offer risk-free interest. The interest from these bonds reduces the pressure to make more money and frees you up to pursue your interests rather than the demands of a job.

Having hooked the readers with the simplistic yet profound revelation that they can improve their lives by saving money, Dominguez and Robin add a subversive twist by challenging their readers to ask themselves if their expenditures are in accord with their values.

This calculation is a quick path to a midlife crisis, Robin jokes. If you ask yourself if your spending is in alignment with your value system, suddenly you are faced with some tough questions. First, you have to figure out if you have any values and if you do, whether or not you are using your time and money in a fashion consistent with those values. Some people find that taking a lower-paying but less demanding job frees them up to live a life more congruent with their values. Learning to live a simpler life and spend less money relieves the pressure to make a lot of money, and with lower monthly bills one can be more flexible in pursuing a fulfilling life.

To minimize the stress your lifestyle places on the environment, there are some simple rules that generally hold true, Dominguez and Robin say. For example, stop trying to impress people with your purchases of fancy clothes, cars, and houses; don't go shopping when you don't really need something; live within your means; take care of what you own instead of throwing it out and buying something new; use up, wear out, or make do with what you have; buy used goods whenever possible; anticipate your needs; and research the products you are thinking of buying to find those with the best value and durability. If you begin to follow these simple axioms and add some of your own, you can downshift into a simpler, thriftier, less cluttered lifestyle that is more rewarding, Dominguez and Robin assert.

Taking responsibility for how you use both money and resources is time-consuming work that requires monitoring your behavior. But it could be that the protection of the health of the ecosystems we all depend upon lies in the way we handle the mundane details of our lives.

Robin uses the example of a toaster that breaks. It may not cost a lot to buy a new one, but once you toss out the old one, the energy and resources that went into its manufacture are lost. A life cycle analysis of a product makes clear that keeping it in service as long as possible is one of the most environmentally friendly things you can do, Robin observes.

Unfortunately, mundane calculations about how to handle your purchases in the most ecologically sustainable manner possible are sometimes complex. For example, if you own an old car and can afford to buy a new one that gets better gas mileage, does it make sense to trade it in? Those who do this in the name of the environment are often pennywise and pound-foolish when you calculate the amount of resources and energy embodied in the old car, Robin argues. Is the gas mileage you save worth more ecologically than preventing a new car from being built by refusing to upgrade your equipment? Will the new car emit significantly less pollution? Similarly, buying green products is better than buying ones that are manufactured in a manner that is insensitive to the environment. But best of all would be to cut down on consumption instead of buying more products, however environmentally virtuous they may be. These are subtleties that require careful examination.

For those who are willing to explore the ethical implications of their lifestyle, one of the most troubling issues is that we Americans consume more than our fair share of the planet's resources. Statistics that demonstrate this fact abound. For example, the amount of energy used by one American is equivalent to that used by three Japanese or 531 Ethiopians.[7] The per capita emission of carbon dioxide in 1991 was 19.53 metric tons in America, 8.79 tons in Japan, 2.2 tons in China, and 1.03 tons in Africa. Furthermore, when one looks at the number of families with two cars in the United States it is sobering to realize that only 8 percent of the people in the world own a car and that most of them are in North America or Europe.[8] "On average, residents of industrialized countries use nineteen times as much aluminum, eighteen times as many chemicals, fourteen times as much paper, and thirteen times as much iron and steel as do their Third World counterparts," writes John Young in an article entitled "The New Materialism: A Matter of Policy."[9] Even intergenerational comparisons within this country reveal that current consumption by Americans has mushroomed. People now own twice as many cars and drive two and a half times as far every year as people did in the 1950s. Furthermore, the median size of a house has doubled in the last forty-five years.[10] In fact, if you look at the statistics, it becomes clear that during previous generations many Americans lived a frugal lifestyle.

Frugality is not something new in American history, Dominguez points out. It would be possible to assemble a hall of fame of frugal Americans that would include such luminaries as Benjamin Franklin, Henry David Thoreau, Ralph Waldo Emerson, and Robert Frost, all of whom championed the virtues of thrift. And these luminaries were hardly an isolated group: Americans of past generations were considerably more frugal than we are today. In fact, it is really only in the last generation or two that consumer spending has increased exponentially and advertisers have convinced us to buy things that we may be induced to want but really don't need.

Americans are brought up to believe that progress means that each generation should live better than the last. That made sense when it involved switching from outhouses to indoor plumbing, but it makes less sense when it involves upgrading the bathroom with a Jacuzzi and gold-plated fixtures, Robin observes. We have confused an increased standard of living with a higher quality of life, Dominguez adds. "I found that I could lower my standard of living and that my quality of life went up. Suddenly I had more time for my friends and to do things I wanted to do when I wasn't busting a gut making a living." The real challenge is to find out what is enough for you and what constitutes a good life.

To their surprise, many people find that when they monitor their expenditures, their spending drops without a lot of pain or sense of sacrifice. Rather than feeling deprived people feel empowered. When people know what is enough and demand satisfaction for their expenditures, they feel smart, more in control of their lives, and find their spending patterns are more in harmony with their values. In short, they have a greater sense of wholeness and consistency about their lives. As the Chinese proverb suggests: "He who knows he has enough is rich."

Spending less also makes it possible for people to reduce the clutter and excessive baggage in their lives. "You should have everything you need to be happy but no excess," Robin says. In her own life she has come to the conclusion that she doesn't own things, things own her. Once you buy a car, you have to pay for its insurance, registration, maintenance, and fuel, and you have to clean it. This takes a lot of time and money. "Things have to earn a place in my life," she says.

That doesn't mean Robin doesn't buy anything. But she does it with her eyes open. When she bought her four-wheel-drive Toyota Tercel, she comparison-shopped at dealerships within a 100-mile radius, found the car with the options she wanted, bought the demonstrator model, and paid with cash.

Not throwing money at problems such as the need to fix the plumbing, build a bed, or maintain a car can also provide an opportunity to learn new skills that can make you more self-reliant. For example, instead of paying a mechanic, Robin learned how to maintain her own car and became proficient enough to build a motorcycle from a kit. She also learned to grow her own food and forage for food, as well as raise, butcher, and can meat. "That doesn't mean I preserve meat and forage for food all the time, but at least I know I can do it," she says. Knowing how little you can live on has a liberating effect, Robin continues. It gives you confidence that you can survive in relative comfort without having to make a lot of money.

In addition, living a simpler life and being more frugal has the potential to reinvigorate a person's relationships with family, friends, and involvement in the community. One frightening statistic Robin and Dominguez came across indicates that on average Americans spend six hours a week shopping and only forty minutes a week playing with their children. Less time spent shopping, eating out, going on expensive vacations, and paying for entertainment can mean more time within one's community making friends, educating children, and improving the community.

Were frugality to catch on as an attractive lifestyle, it could lead to a renaissance in communities across the country, permitting people to shape their lives so they had time to devote to community building. Currently, in many communities working people are only at home at night and during the weekends. With more time to spend with family and friends, community events might flourish and bring about an improvement in the quality and safety of community life. For example, instead of purchasing entertainment by going out to the movies or renting a video, more people might entertain at home. Robin describes how she enjoys using her skills as an actress to put on local productions of skits about sustainability, such as *The Lifestyles of the Frugal and Obscure.*

If all of this sounds grim to you, then you have missed an important part of the message. Dominguez and Robin underscore the point that no judgment is involved in figuring out what is enough for you. If you like expensive vacations in Hawaii or get a lot of satisfaction out of eating at fancy restaurants, then go for it. There is no approved, one-size-fits-all frugal lifestyle. Each of us has to work out what is enough and what is satisfying for us. The fact that Dominguez and Robin live on $6000 to $7000 a year should not be taken as a goal, Robin says. "After all, we have our Ph.D. in frugality. Each of us has to work out what is enough."

Environmental Solutions to Inner-City Problems

SCHEME urges the sign hanging on the wall over Scott Bernstein's desk at the bustling, multiracial, inner-city Center for Neighborhood Technology (CNT). And it is with the organizing savvy of an old-style Chicago pol that the forty-five-year-old Bernstein mobilizes local residents to take care of their neighborhoods' own needs. A consummate coalition builder with an uncanny ability to analyze local problems and craft innovative solutions, Bernstein thinks of himself more as a community organizer than an environmentalist.

"I live in Chicago, not a bioregion," asserts Bernstein, who founded the CNT in 1977. "For some screwy reason I see myself as a Chicagoan and I believe there is a Chicago economy and a Chicago environment. What distinguishes us from the rest of the environmental movement is our concern about making cities affordable, livable, and equitable places," he says.

Despite the distinction he draws between himself and other environmentalists, Bernstein's work—finding pragmatic solutions to neighborhood problems—fits neatly with the aims of those calling for a just and environmentally sustainable society. For example, Bernstein launched a drive for energy conservation to keep affordable housing available, a program to save inner-city manufacturing jobs by helping factory owners comply with environmental regulations, and a campaign to expand the use of mass transit so that low-income residents could commute to suburban jobs. He is also working with small printing and dry-cleaning businesses to help them convert to environmentally friendly technologies. Recently Bernstein helped build a coalition to promote "location-efficient mortgages," which prevent urban sprawl and traffic congestion by allowing people to qualify for higher mortgages if they save on transportation expenses by purchasing a house near where they work.

One of CNT's first initiatives was to look for ways to increase the stock of affordable housing in Chicago. To this end, Bernstein and his colleagues analyzed the cause for the disappearance of low-income housing from Chicago's inner city. What they found was that energy costs had risen 214 percent over a decade, forcing many landlords to raise rents beyond what local residents could afford.[1] Predictably, vacancies skyrocketed and buildings were abandoned, giving once vital neighborhoods that bricked-up look that characterizes modern urban wastelands.

To find a solution, Bernstein looked at the problem from the apartment owner's perspective. Landlords, he recognized, had few options. They could renovate and try to attract wealthier tenants, but this was a risky investment in a rundown neighborhood that was unlikely to appeal to the affluent. And even if this strategy proved successful, it would dislocate low-income families. Another approach was to brick up windows and wait for more prosperous times, but this diminished the stock of affordable housing. Or, at the extreme end of the spectrum, landlords could pay an arsonist to torch their buildings, collect the insurance money, and move to Florida. Bernstein found that most landlords did not want to resort to unscrupulous tactics; they wanted to be reasonable but were finding it difficult to keep their buildings profitable.

Rather than vilify them as slumlords, Bernstein decided to help owners of multiple-family dwellings reduce energy costs and keep their rents within reach of low-income families. To this end, CNT established a "one-stop," inner-city, energy consulting service to assist landlords by arranging for energy audits, bank loans, and renovations.

To date, the center's energy efficiency program has consulted on the retrofit of over 1000 buildings, from low-income housing to the Sears Tower. CNT helped create the Chicago Energy Savers Fund—a coalition of eleven community groups, the city of Chicago, and People's Gas—which has loaned $17 million to retrofit 10,000 units of housing since 1985 and brought in an annual savings of $1.5 million in energy costs.[2]

Bernstein also approached the Amoco Corporation about a grant to help community organizations make energy-saving investments. He pointed out that a grant for an energy-efficiency retrofit would be akin to an endowment by providing an annual payback in energy savings. Amoco gave $4.7 million to help retrofit 170 buildings belonging to 150 community organizations.[3] Energy efficiency investments were made in schools, hospitals, community colleges, churches, and nonprofit groups. At a community center called Casa Central in the Hispanic community, the CNT program helped install better storm windows, insulated the roof, put in new boiler controls, and saved 51 percent of Casa's operating budget. The energy retrofit paid for itself quickly.

The CNT offices were also built to be a model of energy efficiency. "We believe in demonstrating what we advocate," Bernstein asserts. When Bernstein first found the building on West North Avenue that was to become the CNT office, it was an abandoned weaving factory where tablecloths, place mats, and curtains were made. To transform the old factory into a superinsulated, energy-efficient, and nontoxic office space, a number of measures were taken. In the retrofit, locally fabricated energy-efficient windows were installed and the building was heavily insulated. To keep indoor toxics from becoming a problem, no foam insulation was used in the office, Styrofoam was banned, the furniture and carpets were purchased secondhand (the carpet pad is felt), and only nontoxic wood finishes were used. As a result, the air is relatively clean and the 6000 square feet of floor space costs only $590 a year to heat.

The CNT is also helping to provide what Bernstein calls "sustainable manufacturing jobs" for the residents of inner-city Chicago. One of the leading providers of walk-to-work jobs for African American and Latino residents are Chicago's 200 electroplating and metal finishing job shops. Most of these are family-owned businesses that operate on a narrow profit margin and employ an average of twenty workers each.[4]

Because electroplating involves a number of highly toxic chemicals and heavy metals, it has come under intense pressure from environmental regulation. In recent years, costly pollution controls mandated by the Environmental Protection Agency threatened to close many electroplating shops. The owners of these factories were required to buy expensive pollution control equipment which they could not afford. Many of these small electroplating firms in Los Angeles and other cities were driven out of business.

When Bernstein began to explore the electroplating industry in Chicago, he found that a number of shops had gone underground. "This is an industry where people had taken down their signs and unlisted their telephone numbers. They figure everybody is out to get them," Bernstein says. The metal finishers' trade association, feeling itself under attack, was one of the most litigious around, he continues. It also had a penchant for issuing ridiculous statements, Bernstein recalls, insisting, for example, that Coca Cola is more toxic than the compounds used in electroplating.

Environmentalists often stereotype people who run small electroplating businesses as "midnight dumpers," Bernstein says, but CNT took a markedly different approach. "Sure, some of these guys are midnight dumpers, but most electroplaters aren't. To treat them that way is wrong and frankly the environmental movement has not yet come off that attitude. Grassroots environmental groups are not as well trained to facilitate the solution to the problem as they are to point fingers and ascribe guilt. Part of this comes from the inability to distinguish the difference between the way you treat GM and a small, struggling, minority-owned business firm."

Bernstein found his first job was to get the electroplaters to trust him, a task that took the better part of two years. That sense of trust was enhanced when CNT staffers began to testify before various government agencies on behalf of the electroplaters. Then CNT offered electroplaters environmental and energy audits to help them spot new ways of meeting environmental regulations.

To keep electroplating jobs available in Chicago, CNT set up a program that offers electroplaters technical assistance to comply with regulations and find capital to retrofit their businesses. "We helped with waste reduction and material substitution, we got the plight of the electroplating industry on the public agenda, and a lot of resources were devoted to modernizing metal finishing." CNT employees have conducted 100 site visits and its chemical and electrical engineers have consulated at forty metal working shops. These consultations are aimed at improving pollution prevention techniques as well as helping with waste reduction and material substitution. In the future, CNT plans to help job shops secure loans to upgrade their equipment.

Bernstein would like eventually to provide the equivalent of an "HMO for industry" where job shop owners pay an annual fee that provides access to low-income financing, technical assistance, information about new technologies, and job training. Like any other HMO, CNT would require factory owners to schedule an annual checkup where they would learn, "whether they like it or not, how well they are doing with respect to best available pollution control technology and best practices, as well as where they stand in relation to regulations. At the same time they would be given advice on how to meet the regulatory standards."

By providing assistance to small-scale electroplating operations, Bernstein is demonstrating that instead of just pointing the finger of blame, environmentalists and community organizers can help firms obtain the resources they need to do a responsible job of running their operations while keeping inner-city jobs available. In the process, he is helping to clean up a highly polluting and highly decentralized industry.

Bernstein is taking a similar approach to working with the dry-cleaning industry. When he heard from a friend that someone in London had figured out how to dry-clean clothes without the use of toxic solvents, CNT dispatched its chemical engineer to investigate. "We like to joke that we sent him to London with a bag of dirty laundry to get his clothes cleaned," Bernstein says, but in fact the effort was considerably more involved than that. What CNT staffers did was to take a number of different types of materials and mark them up with various kinds of dirt, grease, and stains, sending one batch to be cleaned using the innovative process in England and the other to be dry-cleaned.

Figure 5.1
Scott Bernstein, president of the Center for Neighborhood Technology (CNT) in Chicago. Photograph by Thom Clark.

The new British technology, renamed "wet cleaning," was in fact an old technique known as "spot cleaning," which had been abandoned when the use of solvents became widespread. Often dry cleaning can be avoided because many water-based stains can be removed by placing a garment in a dryer that dries out the stain, permitting it to be brushed off, Bernstein explains. Grease-based stains can be removed by spot cleaning with soap and water followed by steaming and then pressing. "This is a highly labor-intensive process," Bernstein admits, "but so what?" It is no more expensive than the current, highly toxic process, and wet cleaning has the advantages of being safer for workers, less harmful to the environment, and capable of generating more jobs. CNT's engineer reported that there was a direct tradeoff of labor, soap, and water on the wet-cleaning side for chemicals, equipment, and capital on the dry-cleaning side.

The EPA is funding a consortium to test and certify this process that includes CNT, Greenpeace, the Massachusetts Toxic Use Reduction Project, and the Telis Institute. CNT worked with a local dry cleaner who opened the first wet-cleaning business, which is now operating in Chicago under the name Greener Cleaner. Ten similar ventures have opened in the New York City area and the idea is beginning to catch on. "It really works," Bernstein claims.

To understand Bernstein's pragmatic approach to solving interlinked, inner-city environmental and economic problems, it helps to know something of his background. He started out as a researcher at Northwestern University's Center for Urban Affairs, an incubator for social change, where he looked at the causes of illness in an inner-city neighborhood.

At the Center for Urban Affairs, Bernstein led a group of twenty students who went into local hospitals and dug into a year's worth (22,000) of medical records. "We tracked the causes of illness, hired some people, gave them titles like health protector, and turned them into grassroots epidemiologists." The job involved reconstructing the causes of illness and trauma in an effort to see what kind of community action was necessary to prevent them. How did an arm get broken? Was a pedestrian hit by a car at an intersection with no stop sign? Was the patient intoxicated? Was there a fight?

The study revealed that the top causes of hospitalization were assault and battery, traffic accidents, alcohol and drug abuse, respiratory ailments (many from occupational exposures), fires, falls, venereal disease—and dog bites.

Bernstein recalls that the first campaign his students took on was aimed at reducing dog bites. Residents of the inner city acquire dogs to protect themselves, but when times get tough they abandon them to the streets. The abandoned dogs then congregate in packs and frequent alleys where the garbage is not being picked up. When people pass by these packs of hungry dogs they sometimes get bitten. To solve the problem, Bernstein's health protectors invented a sort of lanyard that enabled kids to catch the dogs, turn them into the police, and receive a $5 bounty. Before long the incidence of dog bites went down and Bernstein had achieved one of his first small wins.

In another neighborhood, Bernstein recalls, there was a very high incidence of traffic accidents where an expressway bisected the community. The reason was that traffic lights were not working, stop signs were down, curbs had crumbled, and the area was full of drunks. "We started something called 'the workable streets project' and got the National Traffic Highway Safety Administration to pay for repairing lights, signs, and changing the direction of streets."

Another campaign involved an assault on disease caused by poor nutrition. In one community of 45,000 there was only one supermarket and an ex-Jewish deli turned soul food restaurant, Bernstein recalls. "Corned beef on rye with sweet potato pie for a buck is a great deal but you can't live on it," he observes. "So we set up food co-ops. Someone bought a refrigerated truck that served as a mobile grocery store for people who were shut-ins. It was at that point that we started to think about building a greenhouse on the roof of a local school."

Bernstein wanted to build the greenhouse on top of the school so that local residents could have access to cheap fresh vegetables and children could learn about growing food. It sounded like a good idea, but no foundation would provide the seed money. "Then a corporation with some heart came along and gave us some help: McDonalds." Bernstein admits that when they went into the greenhouse business they didn't know much about it. "We oriented it north-south instead of east-west and it was uninsulated, but in a funny way it worked. There was enough light, it ran off the waste heat from the uninsulated school, and we produced 10 pounds per square foot of tomatoes, green peppers, and cucumbers. Local politicians started coming around to sample the fresh vegetables and admit that maybe solar solutions could work in Chicago." Eventually, the green-house project made the headlines in *The New York Times* and the *Christian Science Monitor.*

Since then the CNT has moved on from greenhouses to bigger projects that link economic and environmental issues. For example, Bernstein was instrumental in establishing the nation's first rail-based "reverse commute" program, making it possible for unemployed residents of inner-city Chicago to work in the suburbs. In looking at why unemployed Chicagoans found it difficult to find jobs in the more affluent suburbs, he discovered that the commuter trains that left the city in the morning did not stop in the poor districts. To solve this problem, he persuaded authorities at Metra, Chicago's suburban transit system, to schedule a new train on the Milwaukee West Line that would stop in the heavily African American and Hispanic districts of Humboldt Park and Westown, pick up workers, and deliver them to the suburbs in time for the morning shift. Enough people now use the train that it pays for itself, Bernstein reports. As to how many people use it to commute to work in the suburbs, he says he has no idea. "But at least people now have access to mass transit that makes it possible for them to work at jobs in the suburbs."

Bernstein has also been working on getting communities more actively involved in transportation planning. With a coalition calling itself the Chicago Land Transportation and Air Quality Commission, CNT is lobbying to shift transportation investment away from building new roads that encourage development at the fringes of the city and further urban sprawl. "Instead of an ever-outward march through farmland, our transportation dollars should be serving existing communities where two thirds of the region's people now live," he told the *Chicago Tribune.*

Another way that CNT is addressing the problem posed by urban sprawl and the continued disinvestment in the inner city is by promoting the use of "location efficient mortgages," which provide additional credit for people seeking to buy a home close to where they work.[5] A study done by the Natural Resources Defense Council (NRDC) in the San Francisco Bay Area found that there was a $396 a month savings per household for people who lived near where they work. Normally, lenders look at prospective home-buyers' income and their fixed expenses (such as principle, interest, taxes, and insurance costs) to decide how much money to lend them to buy a house. But if they add into that calculation the fact that a family will spend $4750 a year less on transportation if they buy a home near where they work, then lenders can increase the size of the loan.

Often a couple cannot get a mortgage large enough to buy a home near where they work, but can find financing to buy one out in the suburbs. The reason for this is that most lenders are not calculating the fact that buying the house far from work the couple will probably buy two cars and their transportation expenses will soar. Even from the lender's perspective, a larger loan for the purchase of a home near where the homebuyers work makes a lot of sense, Bernstein says. It also cuts the amount of energy the family uses on transportation, lowers the amount of pollution they generate, and reduces congestion on Chicago's overcrowded highway and road system.

Location efficient mortgages make sense for other reasons, Bernstein continues. "There are studies that show that one of the few ways that poor people get out of poverty is by owning real estate," he notes. The location efficient mortgage is useful because it is a way to lend would-be homebuyers more money without any subsidy; it helps get people out of poverty, increases home ownership, reduces the demand for transportation, and improves air quality.

Location efficient mortgages have been adopted as one of numerous proposals being made by the Clinton administration's national home ownership strategy: Partners for the American Dream. A coalition of groups, including CNT, NRDC, and the Surface Transportation Policy Project, are working on additional research to prove the financial rationale for lenders to back location efficient mortgages. If the research pans out, a number of banks have promised to give location efficient mortgages a trial run, Bernstein reports.

To say that there is not enough money to solve urban environmental problems is to miss the big picture, Bernstein insists. The fact is that billions of dollars are spent on energy generation, water and sewage treatment, garbage disposal, transportation, and environmental cleanup. The money for these initiatives already exists, it just needs to be redirected to the local level, he argues.

For example, Chicago, like many other cities, collects all of its sewage and wastewater and pipes it to one big centralized treatment plant. When it rains hard the combined sewage treatment plant is overburdened and raw sewage is released. An alternative approach would be to decentralize some of the sewage treatment within the neighborhoods and build settling ponds to redirect rainwater back into the ground. This would solve an environmental problem at the local level and generate local jobs. Similarly, the burden placed on water treatment plants could be reduced if money was spent locally installing water-saving devices.

There are many variations on this theme. Instead of building more power plants, Bernstein suggests that utilities use funds to make buildings in the inner city more energy efficient. In this fashion the demand for energy would be reduced, less pollution would be produced, residents would pay less on their utility bills, local contractors could be hired to do the work, and the investment money would be "pushed down to the community level where it is needed most." Huge investments now spent on new incinerators and landfills could also be rechanneled to the inner-city neighborhoods where local residents could be hired to set up recycling and material reuse stations, thus lowering the volume of solid waste and the need to treat it.

The environmental movement must begin to place a higher priority on finding solutions to urban environmental problems such as these, Bernstein argues. "It is the Chicagos of the world that are placing the largest strain on the environment because it is in the Chicagos of the world that most of us live," he observes. "What exists all over the world are cities in all their glory, with their toxic waste, and their garbage. You have to accept the right of these cities to exist. You have to provide incentives for the right kind of behavior in these cities. You can't just stand there and assert that the world should be different. You have to provide the resources for change."

Making a city like Chicago more environmentally sustainable is a tall order. But Bernstein's work as community activist demonstrates that campaigning for affordable housing and jobs is compatible with efforts to create ecologically sustainable cities.

A Utility Company Switches from Nuclear Power to Energy
Conservation, Renewable Energy, and Electric Vehicles

Looking like piglets suckling a sow, a half dozen electric cars are plugged
into a solar-powered electric vehicle charging station in a parking lot in Sacramento, California. As the morning sun beats down on the overhead array
of photovoltaic cells, the batteries of the cars are charging. Soon, officials
from the Sacramento Municipal Utility District (SMUD) will arrive at work
and use these quiet, nonpolluting vehicles on their daily rounds.

This fleet of electric vehicles, now totaling 110 when combined with
those from McClellan Air Force Base, is the first sign that SMUD is not an
ordinary purveyor of energy. In fact, under the leadership of general managers S. David Freeman and now Jan Schori, SMUD is fast becoming the
most ecologically sustainable utility in the country. It has earned this distinction through a combined strategy of investing in renewable sources of
energy, operating an aggressive energy conservation program, and converting its fleet of vehicles from gas guzzlers to electric cars. In the process
SMUD is demonstrating to other utilities that "going green" is both practical and profitable.

The man who pioneered this effort, S. David Freeman, a dramatic fellow
with dark curling eyebrows, a bushy white mustache, and a mischievous
look, has made a career out of being a maverick on the cutting edge of utility
reform. He describes himself as a "utility repair man" who helps power
companies make the transition from a dependence on fossil fuel and nuclear energy sources to a softer energy path comprised of conservation
measures and renewable energy.

Freeman—who keeps a collection of hardhats, baseball caps, and cowboy hats in his office—grew up in Chattanooga, Tennessee. It was there
that he discovered firsthand, as a boy, the connection between pollution

and energy generation. On Sundays, when it was his turn to stoke the coal-burning furnace, his white shirt would turn gray with soot before noon, he recalls. Eventually, his family was forced to move from their home near the railroad tracks to a more rural area because Freeman suffered from asthma that was exacerbated by soot emitted from passing steam engines.

By his own account, however, Freeman was not always an environmentalist, but rather underwent an environmental awakening that shaped his career. "I can tell you exactly when I became environmentally aware," says Freeman, who in 1968 was working in the White House science office when President Lyndon Johnson asked him to coordinate energy policy for the government. At the time he was studying the siting of power plants and trying to figure out how to convince utility companies to tell the government in advance where they planned to locate future generating facilities. It was during that period that Freeman recalls meeting two women from New Hampshire who objected to the proposed construction of a nuclear power plant near where they lived. In the course of their research, these two women became convinced that the plant was not needed—that in fact there was a surplus of power in the region. Their presentation stimulated Freeman's thinking about the rate at which Americans were using energy: "It became obvious to me that we were using energy the way the proverbial drunken sailor spends money. And it just hit me that part of the answer to the pollution problem as well as to energy planning was to move towards a more efficient use of electricity," he recalls.

Prior to his job at the White House, Freeman had worked at the Federal Power Commission in the 1960s where he promoted the greater use of public power. "At the time we were gung-ho for nuclear power. We just felt that electricity was an unmitigated good. And the more we used of it the better," he says. The United States was more or less in a race with the Russians to see who could use the most energy, Freeman recalls.

By the early 1970s, Freeman became one of the first (along with Amory Lovins) to seriously address the energy efficiency question. He talked the Ford Foundation into investing $4 million in the first thorough research on energy efficiency and published the results in a book he wrote entitled *Time to Choose*. The book hit the bookstores just as the oil embargo peaked, giving it a broad circulation.

One of those who read the book was President Jimmy Carter who adopted it as his energy bible. Carter invited Freeman to the White House

to talk about energy efficiency and the President recalled more details from the book than the author, Freeman says. Then Carter appointed him to run the nation's largest public power authority, the Tennessee Valley Authority (TVA). During his tenure at TVA, Freeman terminated funding for eight nuclear reactors that were under construction because he judged them to be too expensive. "The expense of building these nuclear reactors would have put the TVA in the S&L category," he comments. He also spent a total of $1 billion on installing scrubbers in coal generating plants, and reached 1 million homes with his energy efficiency program.

Since his TVA days Freeman has helped a number of utilities make the transition from nuclear power and dependence on fossil fuel sources to a softer energy path. Interviewed while he was general manager at the Sacramento Municipal Utility District, prior to becoming the director of the New York Power Authority, Freeman said: "Our idea is that we will be a solar utility in the future."

The Sacramento Municipal Utility District began moving in this direction in 1984 when Sacramento residents voted to close Rancho Seco, the nuclear power plant that previously generated 914 MW (megawatts) of electricity annually and provided the lion's share of Sacramento's power supply. Rancho Seco had been plagued with technical problems over its fourteen-year history and had operated only 40 percent of the time. The issue of whether or not to close the plant was placed on a ballot and SMUD customers voted to shut down their nuclear reactor, becoming the first community in the nation to vote to close an operating nuclear power plant.

With Rancho Seco mothballed, SMUD officials had to conjure up some new sources of renewable energy quickly. To kick-start the process they built two of the world's largest, utility-owned, photovoltaic (PV) power plants—twenty acres of solar panels—that surround the now defunct nuke. During summer months these two solar power plants (PV1 and PV2) generate 45 MW of electricity—enough to power 660 homes in the summer when demand peaks. That is not much when compared either with the 914 MW generating capacity of a Rancho Seco nuclear reactor or with Luz International's giant solar power plant located in the Mojave Desert which uses parabolic trough collectors to generate enough electricity to power 170,000 homes. But the sight of SMUD's twenty acres of active solar panels encircling an inactive nuclear power plant speaks volumes about the

Figure 6.1
Solar panels surround the Sacramento Municipal Utility District's (SMUD) now defunct Rancho Seco nuclear power plant. Photograph by Steve Lerner.

new direction that Freeman took SMUD to meet the energy needs of its customers.

In 1994 SMUD opened its Hedge Solar Station, the lowest-cost, utility-scale, sun-tracking, photovoltaic system ever built. The 550-kW solar station's 4800 photovoltaic modules (there are 38 solar cells in a module) provide an alternative energy source to boost output during hot summer days when energy demand is at its peak, driven by the use of air conditioners. The Hedge Solar Station was built close to a large concentration of consumers to cut down on the power loss that occurs as electricity is transmitted over long distances.

More revolutionary yet is Freeman's plan to further decentralize energy generation by mounting solar collectors on his customers' rooftops. Although this initiative is still in an early stage, SMUD technicians have already installed 2- to 4-kW photovoltaic panel systems on the rooftops of

250 customers who volunteered to be "residential PV pioneers." As a first phase of this program, SMUD officials plan to construct a total of 350 of these mini-power plants. Another program for "commercial PV pioneers" has equipped the rooftops of five churches, and one VFW hall with 18- to 38-kW systems. The utility owns, installs, and maintains the solar panels on the homes of these Sacramento residents for $4 a month, so that the homeowners themselves will not have to become solar equipment experts. Elsewhere, on the roof of SMUD's 59th Street warehouse, a "solar concentrating system" using ribbons of solar cells instead of photovoltaic panels, will generate another 40 kW of energy using an innovative system that decreases the use of expensive silicon by a factor of ten by substituting an acrylic material.

SMUD is also harvesting the wind by investing $50 million with Kenetech Windpower (previously U.S. Windpower) to purchase the first commercial-scale, utility-owned wind farm in the United States. Located on 4100 acres in the Montezuma Hills Wind Resource Area in Solano County near Rio Vista, California, the first phase involved the construction of seventeen giant wind turbines with rotor diameters of 108 feet and blades weighing 2100 pounds. These towering wind turbines, which operate in variable wind conditions ranging from 9 to 65 miles per hour and use sophisticated computer software, produce four times more energy than conventional wind turbines and will generate 5 MW. After a testing period, if all goes well, the utility will purchase another 150 turbines that will generate 45 MW of power by 1997—enough energy to provide power to 12,000 homes. SMUD is financing the wind farm with tax-exempt municipal bonds that will bring the cost of wind-generated energy down to 5 cents per kilowatt-hour, making it SMUD's most cost-competitive renewable resource.

Cogeneration of energy will also supply Sacramento residents with more than 500 MW of electricity by the year 2000, generating the yearly energy demands of 316,467 homes and replacing much of the electricity lost with the closing of Rancho Seco, SMUD officials estimate. At the same time the generating of this energy will provide steam to run local industries. Cogeneration is 65 percent more energy efficient than a typical power plant and will produce power at 3.7 cents per kilowatt-hour, well below the 5 cents per kilowatt-hour for a conventional power plant.

The efficiency and savings come from the fact that the byproduct from generating electricity—steam—is used by a manufacturing plant located next to the power plant. The industrial "host" buys the steam from SMUD at a lower cost than they could make it using their own inefficient boilers and uses it to bring down manufacturing costs and reduce air emissions, as well as to heat or cool the factory.

The new 97 MW cogeneration power plant, Cogen One, will provide enough power to meet the energy needs of 54,464 households served by SMUD. It will also serve the Sacramento Regional County Wastewater Treatment Plant; and Glacier Valley Ice Company, owned by Carson Energy. The overall design is ingenious. The cogeneration power plant will generate energy using both purchased natural gas and methane that is a byproduct of the wastewater treatment plant. As sewage decomposes it generates methane that is usually burned off. Instead of wasting the methane and creating air pollution in the process, it will be used to generate electricity and steam that will run the refrigeration equipment at the ice plant, which is expected to produce 400 tons a day of bagged ice. This arrangement reduces the cost of power generation and also results in a net air quality benefit to the community. SMUD plans to build three additional cogeneration plants that use previously wasted energy to do double duty.

SMUD is also experimenting with fuel cells, a technology conceived in the 1880s, but not demonstrated until the 1960s when it was used to provide electricity aboard the Apollo space missions. Fuel cells operate by mixing natural gas with steam to release hydrogen that reacts with oxygen in the air to produce electricity and pure water. Fuel cells can run on a variety of fuels, including natural gas, methanol, ethanol, propane, hydrogen, or biogas. Since they work by an electrochemical reaction rather than combustion, fuel cells are noiseless, odorless, and nonpolluting. SMUD has installed fuel cells at its headquarters and at the Kaiser Medical Center South in Sacramento where they are producing 200 kW of energy.

While some renewable sources of energy are more expensive than electricity generated in highly polluting coal-fired plants, which provide 40 percent of the electricity sold in the United States, it is important to recognize the superior value of an energy source that does not pollute the air or pose the danger of radioactive contamination, Freeman says. SMUD officials polled customers about whether or not they were willing to pay a modest premium for renewable energy, and the majority said yes.

Currently SMUD produces about 54 percent of its energy from renewables and conservation (41 percent from hydroelectric power). "I think we are going down the green road," says Freeman, "but it is not just green. . . . What we are doing makes a good deal of common sense and it is cost effective: we are reducing the electric bills of our customers."

For all his enthusiasm about renewable energy, Freeman depends even more heavily on energy conservation to meet the electrical needs of customers in Sacramento. "If we didn't have our conservation program we would have to build an extra power plant," he says. SMUD officials projected that they would spend 7.2 percent of their operating revenues on energy efficiency in 1995 and lend their customers $45 million for energy efficiency improvements. A total of 558 million kilowatt-hours of energy were saved by this program by the end of 1995. SMUD officials intend to meet all their load growth over the next twenty years, 800 MW of power, with what Freeman calls a "conservation power plant" that will provide enough electricity for 640,000 customers. This places SMUD firmly in the forefront of the movement to improve energy efficiency in the utility sector.

The use of conservation programs to meet load growth is part of a growing trend. Utilities now spend over $2 billion annually on "cash rebates for efficient appliances or compact fluorescent lamps, low interest loans for home weatherization improvements or industrial retrofits, and even rebates for the purchase of solar water heaters."[1] Projections suggest that by the end of the decade the annual investment by utilities in this type of Demand Side Management will rise to $10 billion. Experts estimate that utilities will have invested a total of $50 billion in energy efficiency during the decade of the 1990s.[2]

One of the advantages of investing in energy conservation measures instead of paying to build new power plants is that conservation programs neither pollute nor break down on a hot day the way power plants do. Furthermore, meeting the need for energy through conservation programs is actually more cost effective than providing customers with renewable energy. Freeman calculates that SMUD can help its customers save 20 to 25 percent of the energy they use.

How can a utility make money by convincing its customers to buy less energy? The answer to this apparent paradox is that some regulators, who set the rates utilities can charge for electricity, have begun to assign higher rates for energy-efficiency investments than for investments in new power

plants. This provides a powerful incentive for energy conservation. Furthermore, twenty-six states now require utilities to calculate the indirect environmental costs of building a new power plant (such as the cost of acid rain damage caused by coal-fired plants) when assessing the best way to meet their customers' power needs.[3] This internalizing of environmental costs also favors energy conservation strategies.

In fact, lowering demand for electricity through a variety of conservation measures allows the utility to save money by building fewer power plants and charge its customers less for electricity. In this fashion it actually costs less to save electricity than to supply it. "Both the customer and the utility win with energy conservation," Freeman observes. "We buy the energy that our customers waste." For example, rebates are given to customers who trade in old refrigerators for new models that use about half as much energy. Since the refrigerator replacement plan began in October 1990, SMUD has given rebates on the purchase of 47,015 refrigerators. A total of 37,033 refrigerators replaced in this manner have been recycled as scrap metal after the ozone-depleting chemical Freon was drained from them and sold back to DuPont, the company that made it. Freeman calculates that by spending $100 to $175 per refrigerator on rebates, SMUD pays its customers about 4 cents per kilowatt-hour of energy compared with paying 5 to 8 cents per kilowatt-hour for electricity from various other sources. The program was changed in September 1995 to cover only chlorofluorocarbon-free refrigerators. Since May 1992, SMUD has also provided an $850 rebate with 8.5 percent financing over ten years to convince 2900 customers to replace their inefficient electric hot water heaters with solar hot water heating equipment. "What we are basically doing is buying energy from the cheapest source: our customers," Freeman observes.

Trees are also part of SMUD's energy conservation strategy. Through the Sacramento Tree Foundation, SMUD has given away 200,000 trees since the program began in October 1990, so that Sacramento residents can shade their houses from the sun and save on their air conditioning bills. "I think that I shall never see an air conditioner as lovely as a tree," Freeman recites, taking liberties with Joyce Kilmer's poem. But he is serious about his tree-planting program: two or three shade trees properly located on a property will grow to be a natural air conditioner in ten years, he notes. A local tree foundation provides the volunteers for the tree-planting program and SMUD provides the money. Drought-resistant trees are donated to customers who are taught how to maintain them.

Unwilling to confine himself to building solar collectors, buying wind energy, planting trees, and replacing refrigerators, Freeman wants to take on the big oil companies in a head-to-head competition over who will fuel the nation's vehicular fleet. "I want to put Exxon and the other big oil companies out of business," he says modestly. A visionary, Freeman forsees a time when electric vehicles will take over the roads from those powered by fossil fuels and he wants SMUD to be positioned to take advantage of this shift. The demand for clean air will drive this change, he claims. Sacramento's air is now the tenth most polluted in the nation, but Freeman aims to change that: "We want to make Sacramento the clean air capital of the nation."

As a first step SMUD opened twenty-nine electric vehicle recharging stations with a total of 184 outlets—the first in the West—in an effort to make Sacramento a city that welcomes the use of electric cars. "We are going to build charging stations all over town and get the infrastructure in place so Sacramento can be one of the first cities to use electric cars on a massive scale," he says. SMUD has also established special off-peak rates for electric vehicle owners. To date 3000 residents of Sacramento have signed up to be part of the SMUD Electrical Vehicle Pioneer Program.

To demonstrate its own commitment to electric transportation, SMUD and McClellan Air Force Base, in conjunction with the Sacramento Electric Vehicle Transportation Coalition, have purchased 110 purpose-built electric cars. Mechanics in the garage basement of SMUD's headquarters also tore the guts out of 30 Chevy pickup trucks and reequipped them with electric engines and batteries using a conversion kit developed by SMUD engineers.

Freeman is trying to interest other large employers in the Sacramento area in using electric vehicles. He loaned the Sacramento airport an electric van to use in shuttling people from the airport to town, calculating that once they use it and find that it costs 2 cents per mile to fuel they will begin to order their own. SMUD is also participating in a joint research and development program with McClellan Air Force Base to develop the use of electric vehicles on defense installations in California and elsewhere around the country.

Not afraid to dream, Freeman would like to see electric car manufacturers set up shop in the Sacramento area. To attract prospective electric vehicle manufacturers he has offered SMUD's giant Rancho Seco nuclear power plant facility as a potential site for such a venture. He concedes, however, that the best hope for an electric car being marketed in the future is for the Detroit automakers to come out with a commercially viable electric vehicle. "We don't want to go into the car manufacturing business," Freeman says. But if Americans do not come out with attractive electric cars and a roofing material that collects solar energy, then the Japanese or Germans will beat us to it, he says. If that happens, Freeman sees the United States becoming a second-class economy. "Take photovoltaics: that's a trillion dollar market . . . If we develop it first we will have a lot of high-tech, well-paying jobs; if we don't we could be the next Brazil," he continues.

A new federal industrial policy will be required to help American industry develop a commercially viable electric car and solar roofing materials, Freeman asserts. There needs to be government money spent on research and development of these products. Once the basic research has been done, American manufacturers will have a competitive product that can be sold both domestically and internationally.

In March 1994, Freeman continued his peripatetic journey as an electric repairman when he was appointed president and chief executive officer of New York State Power Authority. Freeman amazed officials at Consolidated Edison Company when he suggested that New York State is "awash in power" and that plans to build huge, environmentally destructive dams in Quebec were unnecessary.

Freeman relishes the chance to provide practical solutions to environmental problems rather than just bemoan the destruction of the environment. It is essential to show people that we can solve our environmental problems, he says. "Unless you give people some idea that a solution is possible, it is hard to motivate them to get excited about implementing it. Folks like a challenge, but Mission Impossible is hard to sell."

Breeding Naturally Colored Organic Cotton Eliminates the
Need for Toxic Dyes and Pesticides

It looks like a Halloween prank: fields of cotton spray-painted brown and
green. But this is no autumnal hoax. Sally Fox, a pioneer cotton breeder,
has painstakingly bred varieties of cotton that are naturally colored. In the
process she is revolutionizing the cotton industry by permitting manufacturers to skip the highly toxic cotton-dyeing process. Raising her cotton organically also means that no pesticides or chemical fertilizers are used. The
end products are proprietary varieties of naturally colored cotton that recently earned her business, Natural Cotton Colours, Inc., $5 million a year
in sales.

But big breakthroughs with potentially important ecological conse-
quences, such as Fox's breeding of a naturally colored cotton, do not come
about easily or overnight. Fox's fascination with textiles began at the age of
twelve when she started to hand-spin yarns for knitting and weaving. "I'd
spin just about anything I could lay my hands on," says Fox, a 38-year-old
San Francisco native whose parents were in the real estate business. She
even hand-spun the long white hairs from her family dog and the cotton
balls found in pill bottles, as well as silk, linen, and mohair.

The sixth of seven children, Fox studied entomology at California State
Polytechnic University at San Luis Obispo. She subsequently witnessed the
massive misuse of pesticides in Gambia as a Peace Corps volunteer before
going on to earn a masters degree in integrated pest management from the
University of California at Riverside.

Fox's first glimpse of naturally colored brown cotton came when she was
attending the Southern California Handweavers Conference in Santa Maria. It was then that she first asked herself why the naturally colored cotton
that came from Guatemala couldn't be grown in the United States.[1]

By 1982 Fox was working for a plant breeder looking for pest-resistant strains of tomatoes and cotton. The breeder gave her a brown paper bag containing some cotton seeds and brown lint he had been given by a U.S. Department of Agriculture (USDA) breeder who thought the brown cotton might have pest resistant properties. The seeds were from a variety of brown cotton that had been grown for centuries in the United States, Mexico, and Central and South America. Intrigued by the natural color of the cotton, Fox asked why it wasn't grown commercially. She was told that white cotton was more easily dyed than naturally colored cotton and that the fibers of the colored cottons were too short to spin on commercial machines. "The cotton breeders all said it was impossible to improve this fiber," she recalls. But this was not enough to deter Fox who had grown up hand-spinning wool that was black, brown, and gray. Why couldn't the same be done with this cotton?, she asked herself, after having seen the beautiful natural colors of cotton from Central and South America in a textile museum. It was then that she decided to do some experimenting on her own.

Fox took the bag of seeds to her parents' house in Menlo Park and planted them in six plastic pots in the back yard. The problem with the fiber they produced was that it was only about three-quarters of an inch in length, while it needed to be at least an inch long to spin. So Fox began to grow out the seeds and select the ones with the longest fibers and the darkest and most vibrant colors.

By 1983 she had thirty brown cotton plants growing on the back porch of her apartment in Bakersfield. A year later she planted a half-acre in a friend's farmyard and by 1986 she negotiated a deal with a farmer who allowed her to plant 2 acres of her colored cotton on his land in return for cleaning up all of the trash on his land, a job that took three weeks of work.[2]

On the 2 acres, Fox did all the weeding, cotton picking, and plant selection by hand. After discovering a turn-of-the-century gin mill owned by the USDA, which she thought would work on her cotton, she convinced a retired machinist to build a replica for her. Having grown, picked, ginned, and spun the cotton, she founded a mail-order business through which she marketed her naturally colored cotton to hand-spinners and weavers. She named the business Vreseis Limited after her great-grandmother who spun cotton in Crete. In her first year of business she made less than $1000. As she became more deeply engrossed in cotton breeding she learned that her maternal grandfather was a Greek cotton merchant who had initially come to this country with the idea of trading cotton for mohair but ended up establishing a Greek grocery instead. "Obviously I have cotton in my genes," she says.

As the fibers of her colored cotton grew longer, Fox tested her cotton's spinning characteristics at Texas Tech University. By 1988, through selection and crossbreeding, she produced machine-spinnable cotton with natural color. This was a significant breakthrough because it meant that a colored cotton that did not need to be dyed could be used by manufacturers. Fox recognized that her organically grown, naturally colored cotton would have a powerful appeal to environmentally conscious shoppers. Her cotton could be marketed as ecologically sensitive clothing to the 15 percent of consumers *American Demographics* describes as "visionary greens" who regularly buy environmental products.

That same year three Japanese textile mills bid on her 12 acre crop of colored cotton which they wanted to use for bath towels. The winning bid from Yanbo Mills resulted in the purchase of 2000 pounds at $5 a pound, considerably above the prevailing price of 70 cents per pound for conventionally grown cotton. This prompted Fox to quit her $30,000-a-year job with Sandoz Crop Protection and found her own company, Natural Cotton Colours, Inc. in Wasco, California.

In 1989 Fox grew 100,000 pounds of colored cotton and sold it to a Japanese mill for $279,000. She filed for a trademark for FoxFibre and subsequently earned plant variety protection (PVP) certificates (a sort of farmer's patent) for three varieties of naturally colored cotton; a fourth is pending. To breed a variety of open-pollinated cotton takes years of selecting plants with the desired characteristics. It is a much more difficult process than producing a hybrid that either cannot reproduce or does not breed true, Fox explains. USDA inspectors require that plant breeders go through a three-year certificate process to demonstrate that their proposed variety is different from any other and that less than one plant in 10,000 is off-type. This means that a PVP certified variety must grow out to be uniform in height, color, and the way it puts out its leaves, Fox continues. The next year, 1990, the Japanese mill wanted 800,000 pounds of FoxFibre worth $4 million.

With bigger and bigger orders rolling in, the prospects for Natural Cotton Colours looked good. But not everyone was enthusiastic about colored cotton. It was at this point that Fox ran into the Acala Cotton Board, the San Joaquin Valley cotton industry association, which refused to give her a permit to grow 2000 acres of colored cotton. Their action was based on a 1925 California law prohibiting farmers from growing more than 200 acres of colored cotton. The law was established to protect California's highly valued, pure-white cotton from being contaminated with strains of colored cotton through cross-pollination.

The limit on the amount of colored cotton she could grow in California set Fox off on a mad scramble to find growers in Texas and Arizona. She initially found enough farmers to grow 400 acres of her cotton, which sold for $600,000. Subsequently, she convinced forty farmers to put naturally colored cotton into production on 1500 acres and by the end of 1992 she had 3000 acres under contract in these two states.

Fox's confrontation with Acala made the news, and designers at Levi Strauss & Company, the largest clothing manufacturer in the world, purchased some of her bales for research. In 1991 and 1992 Levi Strauss signed contracts to buy her Coyote brown colored cotton for a new line of Levi's Naturals denim jeans and jackets. Esprit also placed an order. Before long she was selling to over twenty companies and products made of FoxFibre were carried in mail-order catalogues such as L.L. Bean and Seventh Generation. By 1992 Fox's business sales reached $1.6 million.

That year Fox moved her business to western Arizona because the number of acres she was allowed to grow in California was cut once again from 40 to only 10 acres. Being forced out of central California was hard on her because her location permitted her to hop in her truck and visit customers at Levi Strauss, Patagonia, and Esprit. Furthermore she was close to her family and friends. Fox had also invested a good part of her profits in a farm in central California and equipped it with an expensive water-conserving irrigation system.

Facing a hostile community of California cotton growers who were worried that her breeding program threatened the purity of their crop and by extension their economic security, Fox says she spent so much energy defending what she was doing that it reduced the amount of time she could spend on her breeding program. Equally irritating were the personal encounters with angry cotton farmers at the grocery store and elsewhere.

Fox moved to a 1000-acre farm that is certified for organic crops in western Arizona 25 miles from the town of Wickenburg. "My first two years

Figure 7.1
Sally Fox, founder of Natural Cotton Colours, selects the seeds that produce the longest fibers and bear the brightest colors at her farm in Wickenburg, Arizona. Photograph by Carey S. Wolinsky.

here I felt miserable," she confides, but since then she has come to enjoy Arizona where she is not shunned as an agricultural pariah. "When I go into town here people treat me like a local hero," she says. Farmers are interested and delighted that she is breeding a new variety of cotton. They stop by and look at her operation and make suggestions about techniques that might work for her. A neighboring farmer recently decided to experiment with organic methods on a 200-acre plot.

An unexpected benefit of moving her breeding program to Arizona is that the climate is good for her selection process, she says. Since the summers are very hot in Arizona, Fox must breed plants that are heat tolerant, a quality that will help her develop cotton that can weather almost anything. It was at this point that Fox also began to crossbreed her colored cotton plants with organically grown, long-fiber white cotton. To do this she took the pollen from flowers of colored cotton and pollinated the flower of a regular cotton plant. When the crossbred plants matured she selected the seeds that had produced the longest fibers and bore the strongest colors. She then planted these seeds and once again crossbred these plants with long-fiber white cotton.

The business has also survived the transplant well. In 1993 Fieldcrest-Cannon started using Fox's naturally colored brown cotton, mixing it with undyed white cotton. That year farmers under contract with her grew 5000 acres of her cotton, a crop that brought in $5 million and a pretax profit of roughly $1 million. By that time FoxFibre was being used for upholstery fabric, table linen, hosiery, sheets, towels, and clothing.

By 1995, Fox had enough cotton in the warehouse that she could afford to devote the entire growing season to producing new varieties of seed instead of concentrating on cotton production. She grew 1000 acres of experimental strains of colored cotton she thought would increase the color range as well as the length, quality, and yield of the cotton fibers. A good part of the growing season was devoted to inspecting 250,000 cotton plants and selecting the ones with the best characteristics. The results have been magnificent, she says, and next year she expects to have enough seed to plant 10,000 acres of organic colored cotton if there is a demand for it.

But 1996 has not been a good year for organic cotton. "There has been a tremendous shakeout in organic cotton this year," she reports. The reasons have to do with the international market and the price fluctuations as a

Figure 7.2
*Fox with her harvest of naturally colored organic cotton. Photograph by Carey S.
Wolinsky.*

result of which conventionally grown cotton prices reached a 100-year
high. For convoluted reasons, some 70 percent of the farmland previously
devoted to organic cotton, particularly in California, was planted in conventional crops and farmed with agricultural chemicals. To put this land
back in organic cotton once again will require that farmers repeat the three-
year, organic grower's certification process, Fox says. The upshot in the
shakeout of the organic cotton industry for Fox is that, now, manufacturers
who want to mix her colored organic cotton with organic white cotton
will discover the latter hard to find. "It is a very rocky road right now," she
says.

"I was heartbroken because we had worked so hard to convince farmers
to grow organic cotton," Fox says. When she started out she couldn't find
cotton growers who would talk to her if she used the word "organic." But
gradually she convinced some farmers to experiment with a 40-acre organic cotton crop. To facilitate the transition, she hired a full-time entomologist to travel to the farms growing her cotton organically to provide them
with expert advice on how to deal with pests and other problems. After a

couple of years the entomologist had enough organic cotton growers who could pay him for his services that Fox no longer needed to pay his salary.

Despite the disruption in the supply of organic white cotton, Fox says that she continues to make progress in her sector of the industry. Last year, once again, she successfully widened the pallet of naturally colored cotton. And the fibers of her cotton are becoming longer, stronger, and more spinnable, she claims. "I am excited that the cotton did so well this year and that it was all grown organically," she enthused. Last year she produced a chocolate-brown with a red cast that she thinks will be popular. And she is equally enthusiastic about the greens. "The greens are like a dream: they are glossy and fabulous to spin. One is a yellowy green that designers love to work with and the other is a deeper green."

Naturally grown cotton has a range of colors similar to that of wood, spanning the spectrum from green to yellow, to red, to ebony, she explains. FoxFibre is currently available in a limited number of colors—shades of brown, green, and tan. These colors are used full-strength or blended with unbleached white cotton to create a lighter variety of hues. Orange and yellow are under development as is a blue-green and brick-red.

To convince manufacturers to buy her cotton, Fox has had to not only grow it but also design yarns and fabrics out of it in order to demonstrate to clothing makers its commercial potential. She designs her own fabrics and has come up with plaids, flannel, ticking, gingham, and denim. Recently, Fox designed a denim that mixes in the same yarn three of her certified organic, naturally grown colors: Coyote, a light brown; Buffalo, a dark brown; and Palo Verde, a green that is named after the Arizona state tree because its color resembles the tree's bark. "It has these subtle flecks of green, red-brown, and chocolate-brown," she notes. She also designed the yarn that Fieldcrest-Canon made into sheets and came up with a handsome brown canvas as well.

After the crop is in, Fox takes some of the cotton and experiments with it, hand-spinning and weaving it as she looks for new products she thinks will be of interest. When she finds something she thinks has promise she sends it to a research textile mill she works with in Texas. The fabric is then shown to large clothing manufacturers, or is sold through a mail-order business to hand-spinners and weavers to defray research expenses. In this fashion, Fox "pulls the cotton through" from the field to the hand-spinners, to the manufacturers, to the retail stores, showing business people along the way the potential of her naturally colored cotton.

To further make the connection between her breeding program and the potential commercial use of the varieties of cotton she comes up with, Fox holds an annual field day at her farm in Arizona. At the height of the 1995 growing season 130 people showed up, including organic cotton growers from the high plains of Texas and flatlands of Arizona, clothing manufacturers, retailers, furniture manufacturers, bankers, and even a papermaker who uses the vividly colored lint from her cotton to make beautifully textured paper. Those who attend the field day are put up at an abandoned mine that has been restored as a hotel 6 miles from Fox's Farm where they set up displays of the goods made with her cotton. They are then bused to the farm where they board horse-drawn wagons for a tour of the 50-acre nursery where Fox is crossbreeding the naturally colored cotton varieties of the future.

FoxFibre is an important ecological breakthrough in the cotton-growing and textile fields because naturally colored cotton has the potential to reduce the use of agricultural chemicals in the cotton industry. Most of us don't think of cotton as being environmentally hazardous. After all, what could be more environmentally correct than wearing a fiber that grows out of the ground and that has been around for centuries? It sounds both organic and wholesome. But a closer look reveals that these days there is nothing very organic about the way most cotton is grown or manufactured.

The conventional method of growing cotton is highly dependent on agricultural chemicals. Cottonseed is treated with a fungicide before planting. Then the cotton fields are sterilized with a general fumigant such as aldicarb, a well-known groundwater pollutant, or ozone-destroying methyl bromide.

Next come the pesticides. During the growing season the cotton is sprayed up to ten times with pesticides for aphids, mites, and other pests. Anywhere from 16 to 20 million pounds of pesticides are used on U.S. cotton crops every year.[3] An estimated 25 to 50 percent of the pesticides used in the United States every year are sprayed on the cotton crop.[4] A portion of this pesticide load joins the other agricultural chemicals that are known to pollute the groundwater in thirty-eight states.

Nor do the environmental costs of cotton stop there. Plant growth regulators are applied, as are water-soluble fertilizers, which run off the land after it rains causing algal blooms that choke rivers and lakes. In the fall a defoliant, such as paraquat, is used to strip the leaves, which otherwise would leave a stain on white cotton. This potion of pesticides, herbicides, fungicides, and defoliants is so strong that it takes anywhere from three to five years of working and building up the soil before earthworms will return to it and an organic cotton crop can be grown.[5]

Dyeing cotton also pollutes the environment with a witches' brew of chemicals. The dyeing process uses more toxins than any other phase of the textile manufacturing business.[6] Phosphates, phenols, nonbiodegradable surfacants, volatile organic chemicals, and heavy metals are some of the chemicals used in commercial dyeing. Conventional methods used to dye textiles on a commercial scale also depend on smog-producing volatile solvents.[7] Because cotton is difficult to dye, half the petroleum-based dye is wasted and goes down the drain. Furthermore, the conventional, synthetic dyeing process uses up to 85 percent of the energy used in commercial textile processing.

Even the so-called natural dyes made from berries, roots, insects, and recycled coffee grounds tend to fade or run when washed. To fix the colors and keep them from fading, chrome, arsenic, and copper are used.[8]

Fox's naturally grown organic cotton reduces the use of toxic chemicals by demonstrating the feasibility of growing organic cotton and by eliminating the need for toxic dyes. In addition to reducing the amount of pesticides and dyes used in producing cotton textiles, another advantage of FoxFibre is that instead of fading like other dyes, it actually become darker and more vivid with each washing until it gradually reverts to the color it was when it grew in the field. Nor does FoxFibre require the use of defoliants to get rid of the leaves that stain white cotton bolls, because her colored cottons don't stain. Farmers who grow her cotton simply wait for a frost to kill off the leaves.

Fox's colored cotton also has qualities that make it somewhat pest resistant and easier to grow organically or with a reduced use of pesticides. Most of FoxFibre is organically grown cotton and bears a Colorganic label. In 1993, 80 percent of her 3 million pounds of FoxFibre cotton was grown organically. To grow organic cotton, farmers plant cotton when the

Figure 7.3
Growing naturally colored organic cotton eliminates the need for pesticides, chemical fertilizers, and chemical dyes. Photo by Carey S. Wolinsky.

weather is suitable, use lots of compost, cultivate out the weeds, and use biological control techniques to reduce damage from pests. Since this is more labor intensive and risky than conventional cotton-growing techniques that depend on toxic chemicals, her organic cotton costs more to produce—normally about $2.50 a pound compared with 72 cents a pound for undyed cotton. However, those who choose to buy white cotton and then dye it brown face two costs instead of one: first they must buy the white cotton and then pay an additional $2.40 a pound to dye it, Fox says. This is what makes her naturally brown cotton cost competitive.

Naturally colored cotton also has significantly reduced environmental hazards associated with it. For having pioneered this environmental breakthrough, Fox won the United Nations Environment Programme Award; the 1992 Edison Award for Environmental Achievement by the American Manufacturing Association; *Green Housekeeping* magazine's Green Award; and the 1993 IFOAM Organic Cotton Recognition Award.

Despite all the publicity, Fox remains remarkably grounded in her work. At her farm she can regularly be seen riding out on horseback accompanied

Figure 7.4
Fox inspects her cotton crop on horseback, accompanied by her dogs Eli and Whitney. Photograph by Carey S. Wolinsky.

by her dogs Eli and Whitney to survey her colored cotton in the field. The view must be satisfying. Once naturally colored cotton was considered a "folk crop" that could not be spun on commercial machines. But now Fox has transformed this ancient breed of cotton into a raw material that works for the modern fashion industry. In the process she has given a boost to sustainable agricultural practices and provided an example of how to reduce cotton's toxic legacy by eliminating the need to dye it.

Mining the Discard Supply

"Waste is not waste until it is wasted," says Daniel Knapp, president of Urban Ore, Inc., one of the largest and most diverse reuse and recycling outfits in the country.

A former professor of sociology with a piano mover's build, Knapp swapped the lectern for the landfill some seventeen years ago when he started wading through windrows of trash to rescue valuable discards from the waste stream. Capitalizing on his expertise at scavenging, Knapp founded Urban Ore, a unique marketplace for a wide variety of discarded materials, which now earns annual revenues of $1.4 million.

A visit to Urban Ore's sprawling 2.2-acre Discard Management Center in the industrial section on the west side of Berkeley, California, reveals the mind-boggling variety and value of what Americans throw away. Here, contractors, artists, and curio hunters browse amid a well-organized inventory that includes toilets and telephones, motors and metal roofs, computers and chicken wire, plastic pipes and pinball machines. At Urban Ore, used computers, art, kitchen cabinets, doors, windows, lumber, and bricks find new owners and uses; broken window glass is sent to a quarry where it is ground into sand; and scrap metal is separated into twenty-five categories before being sold to metal buyers.

Other than a resident cat named Stinky, salvaged from the nearby transfer station, what cannot be sold or traded at Urban Ore is dismantled for recycling. As a result, less than 1 percent of the 4000 tons of materials handled annually ends up in a landfill.

"Our long-term vision at Urban Ore is that we want to do our part to end the age of waste. Our mission is preventing landfilling and that's why we are located close to the transfer station. Anyone who's going there can drop by

Figure 8.1
Reusable plumbing fixtures for sale at the Urban Ore facility in Berkeley, California. Photograph by Steve Lerner.

Urban Ore and get rid of a lot of stuff they would otherwise have to pay to dump," Knapp says. "Haulers and contractors easily understand the difference between paying $60 a ton to get rid of something vs. being paid $200 a ton for it if it is brought to us clean and ready for reuse or recycling."

In a sense, Urban Ore is a way station for reusable goods that are temporarily homeless, Knapp explains. Contractors discover that if they unscrew or lever kitchen cabinets off the wall, instead of tearing them off and damaging them, they will get credit in cash or trade. Independent contractors and carpenters often find that they can avoid paying a tipping fee at the landfill, make some money by selling to Urban Ore materials recovered from their last job, and purchase secondhand materials they need for their next job at reduced prices.

"Lumber sells as soon as we get it at 75 or 80 percent of new lumber prices," Knapp continues. Urban Ore recently made a single sale of $13,500 worth of lumber taken from a closed military base. "Used brick is sometimes worth more than new brick because of the appearance," he adds. Miscellaneous items are bought by antique and secondhand dealers who mark up goods two or three times for resale.

The emphasis on diverting reusable discards from the landfill distinguishes Urban Ore from other recycling companies. Unlike many secondhand sales operations, which sell small amounts of high-quality merchandise, there is a deliberate policy at Urban Ore to deal in a large quantity and range of materials. To make this possible, goods are priced for fast turnover.

"We believe that reuse should be the first priority for disposal because it retains the most value," says Knapp, taking a break in an employee lunchroom decorated with a velvet painting of the rock group Kiss, a metal music stand, and a worn bamboo shade. "I truly believe that you can reuse or recycle nearly everything," he says. "What is truly unrecyclable should be banned."

Knapp is a bit surprised to find himself in this line of work. "Rummaging through garbage was not my first choice of what I wanted to do with my life." But he felt uniquely qualified. "I'm pretty big so I can handle working at a landfill. And I come from a humble background so I didn't have a lot to lose." The son of an aluminum factory foreman, Knapp was the first in his immediate family to earn a doctorate, but he gave up his tenure-track position as an assistant professor in the seventies when he saw the connection between a faulty collection system for discards (commingling) and an escalating demand for landfill space. Rather than continue to work within the classroom, he decided to become what sociologists call a "participant observer," with the disposal system as his area of inquiry.

It was while he was teaching at the University of Illinois at Springfield that Knapp first took his Twentieth-Century Homesteading students for a drive in his 1948 GMC flatbed truck to cruise the alleys scavenging the best of what people threw away. "I was groping around trying to find out the value of these discards because I was fascinated that there were large numbers of valuable things being thrown out. We kept asking ourselves why people would throw this stuff away," he remembers.

After six years in academia, Knapp resigned and moved back to Eugene, Oregon, where his teaching career had started. There he joined a food cooperative and, since he was the only member of the co-op with a flatbed truck, he volunteered to make the dump runs. Before long he found himself bringing at least as many materials back from the landfill as he took there. "When I got to the dump and stood on the bed of the truck I had an excellent view of what people were throwing out. The more I went there the more I felt revolted by the whole spectacle. I wanted there not to be a dump. I thought it was gigantic, organized stupidity," he recalls.

Knapp was puzzled by the fact that while the dump was almost full and would soon be forced to close, signs were posted prohibiting salvage, a rule the management enforced erratically. One day, while he was ignoring the warning signs and loading discarded lumber onto his truck, one of the dump operators attempted to run him over with a bulldozer. "I thought it amazing that these people were willing to kill to enforce this rule that everything dumped there had to be wasted. What's going on here? Why is it such an important principle that we fill up this dump?"

In 1977 Knapp's interests included recycling, organic gardening, small-scale farming, windpower, solar energy, and community organizing. "I was building solar greenhouses and teaching classes about how to build them, but my money ran out and I had to find a job," he recalls. Fortunately, a job opened up in Lane County, Oregon, as co-director of their Office of Appropriate Technology (OAT). His first project was to evaluate a new garbage-grinding plant designed to produce shredded garbage, which when burned with wood chips would produce steam used to generate electricity.

It didn't take Knapp long to report to the county commissioners that the plant was only the latest version of a series of garbage separation plants that broke down frequently, produced large quantities of air pollution, and generated materials that were of such poor quality they were impossible to sell. Knapp's research also uncovered a study of explosions in similar plants, which suggested it was virtually inevitable that a shredder would eventually produce a spark that would ignite a container of flammable liquid in the garbage or, in dry conditions, the dust that routinely filled the shredder building. He warned the commissioners that the plant was "essentially a concrete-and-metal–clad bomb waiting to go off." Knapp was prescient. In December 1980, an explosion ripped the sides and parts of the roof off the shredder building.

At the time, however, his warnings were ignored and the grinding machine churned on despite a host of problems. The shredder was often broken, the internal conveyers were gummed up with garbage turned into papier-mâché that clung to the moving parts, expensive motors burned out, fuel hardened in the bins, abrasive components in the waste stream wore out the bends in the high-speed transport tubes, and the air transport system sprang a leak and blew tons of finely shredded paper and plastic all over the site. What scrap metal was produced could only be sold as lowest-grade mixed metal because it was contaminated with foreign substances, Knapp recalls.

"Some wit at the OAT office hung a crushed can dangling from a long black sock on the bulletin board," Knapp continues. "The shredder had mashed one end of the sock into the can so tightly it wouldn't come out. The mixture of sock and can symbolized our discontent with the high-tech, whiz-bang approach to recycling."

Turning to other issues, Knapp and his staff designed a research project to find out if it made sense to clean and separate by hand donated scrap metals into about two dozen scrap grades to maximize income to the county. The alternative was to continue the county staff practice of dumping the metals into a single bin. At the time, the single-bin method resulted in a payload worth about $20 a ton. Since the new sorting process required adding a single worker, the operational question was whether enough value could be realized to pay the worker's salary and benefits. To make the new system pay for itself, the OAT staff tried to capture more supply by informing transfer station customers about the new recycling option. To make it

Figure 8.2
Daniel Knapp, president of Urban Ore, Inc. Photograph by Steve Lerner.

user-friendly, an OAT person received, dismantled, upgraded, and sold the material in full public view. While these efforts to divert materials from the waste stream suggested that a more labor-intensive recycling system would pay for itself, county officials were unenthusiastic about this approach and Knapp and some of his coworkers decided to market their expertise outside the county structure and form a new business called Oregon Appropriate Technology.

The first project they took on was a composition study and feasibility analysis for a multimaterials recovery facility at a small rural county transfer station near Dexter, Oregon. "This was a big break because we could study everything, not just metals. Since the discard supply was small, we were able to observe and categorize virtually everything people were dumping," Knapp recalls. He and co-researcher Tom Brandt devised a set of eight categories to record what was in the supply.[1]

Much later, in Berkeley, Knapp expanded the eight categories to twelve, which he called master categories to distinguish them from hundreds of subcategories commonly used by the scrap industry. "To be a master category the material has to be something that can be reused or recycled; its physical makeup can't overlap significantly with another master category; and the set of categories together has to be small enough to remember and big enough to define everything in the supply, with nothing left out and nothing left over," Knapp explains.

In the Rattlesnake Road Transfer Station study, one of the eight categories used was reusable goods. "It was the first time such a category was ever used in a discard composition study to my knowledge. Even today there are only a few who do this," Knapp says. While gathering observational data, the researchers worked closely with a project architect to redesign the dump site to accommodate a handling system for containers for metal and glass, scrap metals, and reusables. It was during the analysis phase that Knapp discovered the economic value of reusables. "What we had was this long list of individual objects that Tom and I had recorded as they were dumped into the bins. In all our sampling periods, we named all the separate things and conservatively estimated their sales value item by item. I was astounded when the numbers showed it was likely that by marketing the reusable goods a big part of the operating costs for a multimaterials recycling facility could be paid for by reuse sales."

After the report was turned in, however, it became apparent that the county had no intention of actually building the proposed facility. Disgusted and discouraged, Knapp hitchhiked to Berkeley with $40 in his pocket. "I went to Berkeley first because a close friend lived there and I could stay at his place, and because I had moonlighted there as a consultant to a Berkeley recycling company," Knapp recalls. Within days after he arrived, he found a new job at the dump as a metal upgrader and scavenger. "The starting pay was only $4 an hour," he recalls, "but to me the prospect of actually getting to touch the materials and get them to market without interference from waste bureaucrats was exhilarating."

At the dump he found parts of an old bicycle he pieced together so he could commute to work; and a bag to hold his tools, most of which he found in the trash. Soon his friend's house became an organizing hub for a small but effective group tracking the city of Berkeley's plans to build an incinerator. By 1981 the group had gathered enough information so that Mary Lou Van Deventer and Knapp could put together a booklet entitled *The Berkeley Burn Plant Papers,* which was distributed to the press, politicians, and activists.[2]

By year's end the anti-incinerator activists realized they could not win on a city council vote, so they took the issue to the voters in the 1982 general election. They wrote a citizen's initiative for a five-year moratorium on building the incinerator and gathered enough signatures to get the initiative on the ballot. They then launched a Give Recycling a Chance campaign to educate the public about the issue. When the votes were tallied the incinerator was soundly defeated. This was the first time a mass-burn incinerator was stopped by citizen action after the politicians had approved it, according to Knapp. "People all over the country wanted to know how we had pulled it off."

During the years spent building a business that diverts discards that formerly went to landfills and incinerators, Knapp bumped heads with a number of solid waste professionals. "The waste management establishment is more or less committed to waste as both a present and a future disposal technique," Knapp says. "They find it hard to transcend the limitations of waste as a metaphor. They think of waste as something we have had with us for the past 4000 years and will have with us for another 4000 years. They don't have a hopeful view or long-term vision of a zero-waste society."

But there is more than just a difference in outlook between entrepreneurs involved in reuse and recycling and those in the traditional solid waste sector. A vicious struggle for control of the waste stream is now intensifying between these two groups.

While most of us do not think of waste as valuable stuff, in fact, large businesses are competing for waste streams because collecting, hauling, and burning waste is a lucrative business. Until recently waste haulers who serviced landfills and incinerators had a lock on the disposal business. But over the last decade recycling and reuse businesses have begun to capture a significant market share. As a result, the large waste processing companies that invested heavily in landfills and incinerators are fighting to protect what remains of their supply. "They lost market share to us in the 1980s and so far the trend has accelerated in the 1990s," Knapp says.

The true percentage of discards diverted from the landfill is not reflected in federal statistics, according to Knapp. "I know all such studies are estimates, but the EPA numbers have big errors built into the study design," he adds. The Environmental Protection Agency uses a contractor who annually calculates the volume of goods consumed in the United States, estimates how long they last, and assumes they become wastes as soon as they are discarded, regardless of whether they are recycled or not. The amount of actual waste taken to landfills and incinerators is not measured directly, nor is the amount of recycling done by major industries. In fact, much of recycling is excluded from government calculations either by definition or by inadequate categorization. As a result, the recycling rate is grossly understated in the government's main indicators, Knapp asserts.

"You have to infer the decline of wasting and the rise of materials recovery from other clues," Knapp says. Stock analysts began warning stock buyers back in 1991 that the waste industry was headed for trouble because of declining market share and cited recycling as a major cause. Subsequently, the biggest waste company's stock declined sharply in 1992 and has not recovered since, Knapp says. The second largest also has suffered declines. Anecdotal evidence Knapp collected also shows declining waste tonnage in various localities and regions. In Urban Ore's home county, for example, waste declined 23 percent from 1990 to 1993, from 2.1 million tons to 1.6 million tons.

The attempt by waste disposal interests to monopolize the discard stream through something called "flow control" is another sign that they are feeling the heat. To protect their supply, waste disposal companies together with municipalities that have invested in waste disposal systems are asking the courts and Congress to sanction their contention that wastes from a given community can be directed to a particular landfill or incinerator. A number of recycling companies have successfully challenged this practice.

"Flow control isn't designed to keep waste out of landfills and incinerators," Knapp observes. "Its purpose is to keep wastes coming in. In this industry, flow control means government protection for wasteful disposal technologies; it forces haulers to use certain facilities, and pay whatever the going rate might be, or face fines. It is insurance for the people who finance these waste facilities. This effort to impose flow control tells us that reusers, recyclers, and composters are very effective competitors."

The flip side to this phenomenon is the exploding growth of the recycling industry. A recent study in North Carolina alone tallied 474 reuse, recycling, and composting businesses statewide. The average size of these companies was $1 million in sales annually, with an average of fifteen employees making an average of $9 an hour, and 80 percent of these businesses were locally owned and operated. The rate of employment growth in materials recovery was six times the average for the state's economy.[3] "If North Carolina is typical, and there is no reason to think that it isn't, then there may be as many as 20,000 reuse, recycling, and composting businesses in the United States working away on shrinking the discard stream," Knapp estimates.

Much of the persistence of the waste problem stems not from a lack of appropriate technology for dealing with discards, but rather from linguistic confusion, Knapp contends. Since the way we think influences the way we act, it is important to get our terminology and concepts right, he continues. For example, things that are discarded are not automatically wastes; they only become wastes after indiscriminate mixing by what Knapp calls the "garbage manufacturing infrastructure."

"Its quite Orwellian when you think about it," he muses. "In the name of sanitation, waste collectors clean and restore order to our living areas, but what they do to the discards after they get them is the opposite of clean and orderly. By the time they are done, everything is contaminated with everything else and it is all degraded and useless. Whatever value there was has been eliminated."

The word "disposal" is another important term that the garbage industry has appropriated. "If you look up the word *disposal* in a Webster's dictionary, it is defined as the orderly disposition of goods. That sounds more like recycling than what happens at the trash dump," Knapp observes. Yet in the professional literature, "disposal" is usually substituted for the much more descriptive terms "landfilling" and "incineration."

"You see, waste disposal is a service industry. It has no product other than a filled-up landfill, which is a liability, not an asset. But it is a very lucrative business, or it used to be. It's variously estimated to be a $25 to $30 billion dollar industry in the United States. The money is sometimes paid by the customer as tipping fees, and is sometimes part of tax levies, but the service of disposal is always paid for, often under a contract that guarantees a hefty profit of 15 percent or more," Knapp says.

Waste professionals have not been blind to the fact that they are losing market share to the reuse, recycling, and composting companies and they are scrambling to get into the recycling business themselves. Unfortunately, when waste professionals go into the recycling business they frequently make the critical mistake of commingling discards before sorting them out again for recycling. This creates a business niche for more efficient business competitors who know the value of source separation and can work with supply customers to achieve it.

"What is most amazing to me is how little real learning takes place in the field," Knapp says. A number of plants with expensive equipment designed to turn garbage into fuel—such as the Lane County plant in Oregon and another operation in El Cajon, California—failed in the 1970s and 1980s. Despite these and other failures, a $132 million garbage sorting plant was built in 1994, only a few miles from the defunct El Cajon plant, by San Diego County. This mega Materials Recovery Facility (MRF) failed within six months of opening, leaving the taxpayers stuck paying off the bonds. Amazed that this pattern is permitted to repeat itself time after time, Knapp asks: "Why doesn't the credit for these crazy contraptions dry up? Why should anyone bail them out when they go bankrupt? Why should they get protection in the form of a guaranteed supply of waste?"

While Knapp appreciates the need for machinery, he suggests that the equipment needed for recycling need not be particularly complicated as long as there is good upstream separation of materials so the right feed stocks arrive at the right place in a clean condition. Good site design, an effective and flexible rate structure, and attention to labor needs is more critical to the success of an operation than a massive investment in heavy equipment, he adds.

Ideally, where space is available, a resource recovery depot should be located next to a transfer station to facilitate the diversion of as much of the discard stream as possible. However, in Berkeley, no single large location was set aside for this purpose. Nevertheless, a wide variety of materials recovery facilities have organized to provide these services. City crews offer a commercial curbside service and a yard debris pickup program; the nonprofit Community Conservation Center has two drop-off and one buy-back facilities; the nonprofit Ecology Center offers weekly curbside residential recycling on the same day as garbage collection; and there are also a number of reuse yards and a big soil blender called American Soil Products. The sum of these parts amounts to what has come to be known as a "serial materials resource recovery facility." Elsewhere, more centralized facilities are sometimes described as integrated resource recovery facilities, recycling parks, or recycling estates. What they have in common is three main modules: reuse, recycling, and composting.

A study carried out by Urban Ore and funded by the Center for Neighborhood Technology (CNT) found that in 1992, the six organizations active in the Berkeley cluster employed ninety-four people full-time while diverting approximately 83,000 tons per year of materials from landfills; generated a cash flow of $8.5 million annually from service fees and product sales; and produced $335,700 in taxes on revenues.[4] By comparison, the city of Berkeley's refuse division employs ninety people full-time, hauled 90,000 tons per year to the landfill, and had an annual budget of $11 million.[5] In other words, the serial MRF was almost as big as the city's waste operation.

This type of disposal system works because local residents, contractors, and commercial operators learn that if they separate their discards they can get rid of unwanted materials more cheaply (or actually get cash for some goods) through one of the recovery operations. The alternative is to mix everything up in a single load, and pay a tipping fee of $57.50 per ton at the municipal transfer station (dump).

"In a situation where you have high disposal fees for handling materials as waste, you can build these source separation facilities as a competitive way of doing business. With a combination of product sales and tipping fees you can divert resources from the landfill while at the same time lowering the overall cost of disposal," Knapp explains.

Knapp is prepared to move beyond reuse operations and take on the more demanding job of proving that an integrated resource recovery facility can radically reduce the amount of discards that must be landfilled. Already he has made a number of efforts to get a demonstration project started. Urban Ore's design group developed plans for facilities in Sonoma County, California, and in West Virginia that would maximize reuse and recycling. But both of these plans were "slowed and partially thwarted by entrenched waste interests," Knapp says. Similarly, the Urban Ore design team's efforts to reconfigure the Berkeley Transfer Station to minimize wastes have often been frustrated.

Part of the problem is that the system is "hard-wired" to give design contracts to traditional waste planners and processors instead of innovative reuse and recycling outfits. Requests for Proposals (RFPs) are often written in such a way that innovative approaches are automatically rejected, Knapp says.

But government procurement is not always tilted toward wasting. In Canberra, Australia, for example, the government is planning a mammoth, 200-acre "recycling estate" designed to eliminate the commingling of discards and reroute them into on-site upgrading and remanufacturing enterprises. There will even be space for manufacturers who use the refined feed stock and make things on-site. "It is a big country over there and they have enough space to spread out and do things right," Knapp observes. "Australia is poised to assume world leadership in sensible, affordable materials recovery," he adds.

Another governmental intervention that has sometimes been effective is direct market intervention. A number of states—including New York, Illinois, and Kansas—are offering low-interest loans and grants for entrepreneurs willing to manufacture goods with recycled materials. Eighteen states now require that newsprint contain some recycled paper and Oregon now requires that glass containers be made with 35 percent post-consumer recycled glass.

To avoid a confusing patchwork of conflicting state regulations, there is growing interest in federal legislation that would require the use of recycled materials in packaging and other products. The Clinton administration is amenable to this approach and has worked to expand the use of recycled materials by federal agencies. President Clinton also indicated that one of his goals is to "create and expand markets for recycled products by providing revenue-neutral tax incentives that favor the use of recycled materials."

But when all is said and done, raising garbage tipping fees is one of the most effective ways to stimulate greater competition and more recycling, composting, and reusing, Knapp believes. "It is only right and fair that tipping fees should reflect all the applicable costs of landfilling, including post-closure care and monitoring," he notes. Many communities also levy surcharges on garbage disposal to provide funds for capitalizing recycling operations. Besides raising the cost of wasting, Knapp thinks such surcharges also begin the process of converting the entrenched bureaucratic opposition to materials conservation, because they provide funds for people in government to spend on meeting or exceeding the recovery goals set in the 1980s.

"The other thing we badly need is more information on success stories. My own industry is the reuse industry, but no one even knows how many reuse businesses there are, much less how much they divert from wasting or their cost-effectiveness. The old-line reuse industry is very atomized and jealous of its information, Knapp says. He believes that one reason salvaging and scavenging efforts are unheralded and unrecognized is that they are often cash businesses with poor record-keeping and a tendency to skim profits. Not only does this cheat the tax man," Knapp suggests, "it also fails to inform the public that there is a cost-effective alternative to wasting."

Knapp believes in keeping careful records not only for the Internal Revenue Service but also to stimulate further growth of the reuse industry. "Information is what will eventually change the prevailing philosophy away from burning and burying these valuable materials," he continues. "Our society needs good information about how different discard supplies can be diverted from the landfill, what useful products can be created, and how many jobs there are at what pay levels. We need the right information and the right ideas about how to deal with unwanted materials. Then we can pick appropriate technologies to do the work we want to do," he says.

Knapp is upbeat about the future of reuse, recycling, and composting. He believes that the United States is moving toward 100 percent recycling faster than most people realize. For instance, he notes, New Jersey went from 10 percent to 46 percent recycling in six years; Fairfax, Virginia, achieved 30 percent recycling in just two years; and Florida now recycles 21 percent of its discard supply compared with 4 percent three years ago.

Reuse and recycling enterprises will eventually prove to be highly successful businesses because they employ energy-efficient practices, Knapp says. "Whether it's grinding up wood, sifting soil, handling textiles or bottles—all these products are already refined so they have a huge energy subsidy built into them. Recycling these materials also puts money into the economy—money that would otherwise literally be buried—and creates jobs without pillaging additional resources," he concludes.

9 *Walton Smith*

Return to Selective Forestry after the Failure
of Clearcutting

Walton Smith has spent a lifetime studying and practicing sustainable forestry, but no one would mistake him for a tree hugger. When the eighty-five-year-old forester stands on his land deep in the Appalachian Mountains of North Carolina and looks up at a towering poplar he sees lumber. "I'm a wood utilization man," he says simply. Smith experiences great satisfaction felling a tree that is 24 inches in diameter that he planted forty years ago, dragging it behind a tractor down to his sawmill, and sawing it up into paneling.

As he walks his 180 acres of woodland with a walking stick, and a Colt .22 pistol stuffed in the pocket of his corduroy pants (a summer drought has brought out the rattlesnakes), Smith points out a stand of white pine he planted with his eldest daughter forty-seven years ago. "We planted those trees in the snow and a month later my wife Dee gave birth to twins. It kind of dates them for me," he says.

While Smith's penchant for looking at trees in terms of board feet may give the willies to some environmentalists, he is in fact one of their best allies when it comes to moderating the impact of logging on the forests of this nation. A pioneer in sustainable forestry with 50 years' experience selectively cutting trees on his land and an extensive employment record with the Forest Service, Smith has knowledge that is invaluable to those who want to move forest management beyond the current obsession with clearcutting.

Smith reminds us is that there is nothing very new about the so-called new forestry, sustainable forestry, or low-impact forestry that is being promoted as the solution to problems caused by clearcutting. "Everything I do here on my land in terms of selective harvesting and uneven-age management I learned from my work in the Forest Service," which practiced sustainable forestry up until a few decades ago, he says. Smith's personal history as a forester tells something about the way forest management has evolved over the last fifty years.

Growing up in the Piedmont of North Carolina, Smith remembers the land as being terribly eroded. "People didn't care about forestry in this area and that bugged me a lot as a young man. I felt sure we could do better," he recalls. In an area devoted largely to the textile industry and the surrounding cotton fields, the forests were seen as entirely secondary.

Smith's first brush with forestry came as a Boy Scout in Charlotte. As a child he swam in the Catawaba River that ran blood red from year-round soil erosion. When he became director of the Boy Scout camp, Smith used the forestry merit badge as a way to get kids interested in trees. And in high school in Charlotte, when a teacher asked Smith what he was going to do after he graduated, he told her he planned to become a forester. Her response was not encouraging: "You are not going to waste your life in a firetower," she said and tried to get the principal to talk him out of it. But Smith kept his own counsel and took a degree in forestry from North Carolina State University.

At the depth of the Great Depression, in 1933, President Franklin Roosevelt launched his first attempt at unemployment relief by creating the Civilian Conservation Corps (CCC). Through a coordinated federal effort the Labor Department enrolled unemployed men; the Department of War equipped, sheltered, and supervised them; and the Department of Agriculture directed them to do fieldwork in the nation's forests. They were paid a dollar a day and provided with food, shelter, and clothing.[1]

During the summer between his junior and senior year at college, Smith worked for the CCC and remembers when the Army rolled into Durham in trucks with 200 men and a load of tents. He helped put up the first tents and worked in the Duke Forest, the Chapel Hill Forest, and the North Carolina State Hill Forest.

When he graduated from college in 1934 he was offered a job at the Nantahala National Forest. His first assignment was in timber stand improvement with the CCC near Clayton, Georgia. At that time the Nantahala Forest spread over parts of North Carolina, South Carolina, and Georgia, making it the largest forest in the East. Timber stand improvement, as it was practiced then, included uneven-age management whereby diseased or insect-infested trees were removed to allow the best trees to develop into high-quality timber. Clearcutting was not practiced. Timber was marked for sale to private bidders for saw logs or plywood using a selective process. Select mature or low-quality trees were marked for harvesting so that trees of different ages would continue to grow. This insured a steady supply of timber over time. Lower-quality trees were cut out to let the best trees grow to maturity, thus improving the quality of the standing timber.

"Back then we were entirely focused on the health of the forest," Smith recalls. "We didn't think of anything else. We would mark the timber for cutting so that twenty to thirty years down the road the harvest would be better than the one we were making."

Smith's assignments for the Forest Service gave him a broad picture of both the forest and the people who lived in it. He was assigned to develop a plan for pricing timber for sale and studied the expenses of six portable sawmills. Next, as part of a New Deal scheme dreamed up by Eleanor Roosevelt, Smith was dispatched to see if some of the poorest residents of Appalachia would sell their mountain homes in exchange for 10 acres of good bottomland. He learned that most of them would not. After that he was detailed to map the Nantahala Forest as seen from the tops of fifty mountains in an effort to find the best locations for firetowers. He was also part of the land acquisition team that acquired the Joyce Kilmer Memorial Forest and personally measured most of the big trees to develop volume tables. Subsequently, Smith was sent to help the Army Corps of Engineers figure out a way to keep the Mississippi from flooding by improving forest cover along the river. When the foresters evaluated stands of timber from the Gulf of Mexico to the Arkansas White River, Smith found himself waist-deep in the Louisiana swamp: "That was one of the roughest years I spent. We were wading in water with alligators, copperheads, and cottonmouths," he remembers.

Eventually Smith was transferred to the Southern Forestry Experiment Station in New Orleans where he worked on flood control techniques for eight years. The idea was to reduce flooding by preserving forest cover on upstream lands instead of building dams. He worked on a plan that was submitted to Congress and printed, but World War II broke out and the program was shelved. From there he was assigned to the forest products laboratory where he became a specialist in wood utilization looking for better ways to use wood and grow higher-quality wood.

After a long and distinguished career, Smith left the Forest Service in 1969. "I saw that they were switching to clearcutting and I didn't like what they were doing," he says. But don't forget that during most of its history the Forest Service practiced sustainable forestry, he emphasizes. To prove his point Smith offers a tour of a section of the Nantahala National Forest adjacent to his own land, which was cut selectively by the Forest Service 20 years ago. Today it is still in good shape and ripe for another round of selective harvesting. In contrast, just on the other side of the mountain, the Forest Service recently permitted 12 clearcuts that manifest all the problems associated with indiscriminate cutting.

For years following his retirement in 1969, Smith was reluctant to criticize the Forest Service. "I love the Forest Service. I thought it was the best agency in the world and I thought everyone in the Forest Service wore a white hat," he says. It went against the grain for him to criticize an agency that had done so much good work in the past. But in 1982—when the Forest Service management plan for the Nantahala National Forest was made public and called for 97 percent even-age management and 80 percent of that clearcutting on the lands adjacent to his property—he finally took a stand and helped organize opposition to the plan. Since then he has published through the Western North Carolina Alliance a series of "green papers" on Forest Service timber sales.

While he has fought the Forest Service over clearcutting, Smith's remains primarily focused on using his own land as a experimental forest to prove that sustainable-yield management forestry is both practical and profitable. "I didn't just want to be one of those old-timers who criticize everything," he says. "I wanted to demonstrate on my own land how it should be done right."

When Smith and his wife Dee inherited the 150 acre woodlot near Franklin, North Carolina, from her parents in 1937, it was in sorry shape, he recalls. It had previously been leased out to tenants who had overcut the land and opened large areas for pasture. As a result there was significant erosion from overcutting and overgrazing and the gullies ran red with raw earth.

To heal the land Smith and his family began an ambitious program of replanting and selective cutting. He planted 8000 white pine seedlings that caught on well on rough terrain where there were few nutrients left in the soil to stabilize the gullies. Once the land was stabilized and the worst of the erosion had been stemmed, he began planting and encouraging a diversity of species, including maple, oak, poplar, black cherry, walnut, and hickory among others. Other hardwoods regenerated naturally.

For fifty years Smith has managed his forest to maximize its sustainable yields. "Not many people can say that," he says. While many loggers will cruise a stand of timber and mark the best-quality trees for cutting, Smith takes the opposite approach: he harvests the lower-grade trees and leaves the best to grow to their maximum commercial potential so that the standing timber on his land becomes more valuable every year. Using this method he has doubled the amount of sawtimber on his land from 1 to 2 million board feet over the last ten years, he claims.

In addition to maximizing the quality and quantity of timber he can harvest from his land, Smith also aims at maintaining tree diversity, preserving habitat for wildlife, preventing erosion, and insuring that the amount of timber removed from a given area does not exceed the amount that will grow back before the next harvest.

To achieve these goals Smith practices selective, uneven-age management of his woodlot. This means always having a good reason for cutting down a tree. "I won't cut a tree unless the ground is dry and I know what I am going to do with the wood," he says. Smith will cut a tree that has reached its peak maturity and highest commercial value, he will cull out a tree that is diseased to protect others, and will fell a tree to provide more sunlight so that a more commercially desirable tree will grow faster and straighter. He also thins out trees in areas that are too dense. All these reasons for cutting, however, are aimed at maximizing the value and diversity of the timber that is left standing in the forest, as well as protecting the wildlife and ecosystem of the forest.

By selectively cutting trees, Smith limits the size of the "holes" he creates in the forest canopy. In a clearcut the forest canopy disappears and sun, wind, and rain attack the exposed soil causing erosion that sweeps sediment into the streams and rivers. By keeping the holes in the forest canopy small, erosion is minimized, the soil builds up, and a healthy habitat for forest animals is maintained.

Furthermore, in felling trees Smith takes care to drop them so that he doesn't damage younger trees. The result is that his forest is multilayered and multiaged. The land does not look as if it has undergone heavy logging. Within a year of selective cutting, an opening in the canopy looks like a nice clearing in which to picnic. There are no wide open spaces covered with a

Figure 9.1
Walton Smith, who has practiced sustainable forestry on his land near Franklin, North Carolina, for the past fifty years. Photograph by Steve Lerner.

stubble of cut tree trunks and discarded branches. "Usually if the forest looks bad, it means that it has been badly logged," Smith remarks.

Unlike clearcutting, where trees of all ages are harvested at the same time after which a single variety of fast-growing softwood trees are planted, selective cutting allows trees to grow to maturity before they are harvested. Uneven-age management permits the forester to insure that there are three or four different age groups of trees coming along so that the yield of timber remains sustainable. This method also facilitates the coexistence of a diversity of tree species on a property, Smith explains.

"The kind of continuous selection system that Walton uses on his land is a model of the kind of forestry management that we advocate," says Buzz Williams, former executive director of the American Forest Service Employees for Environmental Ethics. "The single tree and small group selection cutting method that he practices allows him to harvest timber while sustaining the forest ecosystem by not removing the whole canopy."

Leaving intact an ecosystem that supports a diversity of wildlife is also an essential part of a good forester's task, Smith asserts. He takes care to leave "den trees" that animals live in, and often leaves vines on trees that other foresters would cut because they provide berries and winter raisins for animals. Similarly, he leaves standing nut trees that bears and other animals use as a food source. In this fashion, selective harvesting allows Smith to manage his forest for the wildlife as well as for the valuable timber. As a result, his woods are alive with black bears, deer, wild turkeys, raccoons, squirrels, grouse, and a host of other animals. The bears are so comfortable here that Smith had to put up an electric fence to keep them out of the bee yard where he produces honey for sale.

Smith also encourages the growth of a diversity of trees in his forest. Rather than cull out slow-growing hardwoods in favor of the fast-growing softwoods, he considers that a diversity of species not only improves the health of the forest, it also makes sense economically over the long term. The reason for this is that species of low value today may become sought after in years to come. In an article he wrote entitled "The Fallacy of Preferred Species," Smith documents how, over time, the value of different species varies widely.[2] Years ago, for example, dogwood was a very popular wood and commanded a high value, but now the Forest Service poisons many of them in the national forests.

Selective cutting also allows Smith to use light equipment that minimizes damage to the land. Using a come-along to pull the tree in the direction he wants it to fall, he can minimize the damage a tree does when it is cut. He prevents erosion caused by heavy equipment or tracked skidders by hauling logs down to the sawmill with a small, rubber-tire tractor. By laying out his logging roads so they follow the contour of the hills and descend at a rate of less than a 10-degree slope, Smith is able to keep the soil in the forest from running off during the rains. After the logging is finished he reseeds the roads with grass to further prevent erosion. He also avoids cutting on steep slopes and leaves a corridor of trees along the banks of streams to prevent erosion and maintain water quality.

By contrast, clearcutting results in all the trees in a given area of the forest being of the same age. Clearcutting requires heavy equipment and expensive roads. As a result the Forest Service often pays more to build logging roads than it receives in return for the timber it sells. Some estimates suggest that the Forest Service lost $7.2 billion from 1978 to 1991 and $1 billion from 1992 to 1994.[3]

These "below-cost sales" remain attractive to forest districts, Smith contends, because a portion of the money for the sale goes directly to the district office where it can be used for salaries.

Clearcutting also often causes significant soil erosion as trees are felled in a large area and heavy equipment churns up the soil. The sudden loss of tree cover by clearcutting can cause flooding and destroy the habitat for a wide variety of forest animals. Furthermore, herbicides used to keep trees from sprouting or to keep down undesirable species in areas that have been clearcut can poison the water supply. And the planting of monocrops of softwoods following a clearcut vastly simplifies the forest ecosystem. "Clearcutting is great for the pulp and paper industry," Smith explains, "but the landowner gets less money over the long term because the most valuable hardwood trees on his property are not allowed to grow to maturity."

But is sustainable, selective, uneven-age, low-impact forestry management practical on a large scale? Smith argues that it is and points out that the forest yielded large volumes of timber before clearcutting became the preferred method of harvesting timber. Companies that make plywood and supply the hardwoods used for flooring and furniture, which require mature trees, often practice selective harvesting on their own lands because they know that it is the most efficient way to produce large-diameter trees for their mills.

"We are going to pay for the next ten to fifteen years for the overcutting that has been done because it will take time for the forests to recover," Smith asserts. But eventually the national forests will once again be able to supply somewhere in the vicinity of 30 billion board feet a year on a sustainable basis, he predicts.

Selective harvesting has not been practiced more widely recently because the price of lumber has been low over the last decade, Smith contends. As a result, many lumber companies have been struggling and clearcutting has allowed them to sell large quantities of small trees to the pulp and paper mills, often providing them with their profit margin.

But now that lumber prices have risen once again, private landowners are finding it worthwhile to permit their lands to be logged. Unfortunately, a lot of people sell their timber without careful planning because they are in a financial emergency and must come up with a substantial sum to pay medical bills, educational costs, or to replace their car. Then along comes a logger who says he will give them $10,000 for their trees. Too frequently landowners are taken advantage of by loggers who pay them too little for their trees, cause considerable damage when they log the land, and leave a mess behind. "After having one or two such experiences, many landowners decide that they won't mess with that again," Smith says.

Today there is an alternative to permitting loggers to run hog-wild on your land. Smith and others have been passing on their knowledge of sustainable forestry to young foresters who are now hiring out to private landowners to manage their lands on a sustainable basis. It is best to convince a number of landowners in a given area to hire a local consulting forester who can manage their lands, Smith advises. With the permission of the landowner, the consulting forester then takes bids on logging a certain area, marks the trees to be harvested, oversees the selective cutting, replants where necessary, and collects the money from the logger for the landowner. For these services he receives a percentage of the gross sale. "It's a pretty good living if you can get about 10,000 acres to manage," Smith notes. And since 75 percent of the forested land in the United States is owned privately, if consulting foresters were to practice selective forestry on these lands it would change the way logging is practiced in this country.

Other foresters are finding that they can make a good living by selectively harvesting trees. Mel Ames of Atkinson, Maine, helped send eight children through college with profits from his woodlot that he managed for fifty years without clearcutting. And Sam Brown in Cambridge, Maine, has managed his 300 acres sustainably and devised small-scale machinery—such as a radio-controlled winch mounted on a trailer—that permit him to pull trees out of the forest with a minimum of damage.[4]

While the kind of sustainable forestry that Smith practices still remains the exception, there are signs of change in the way the federal government manages its forests. For one thing, below-cost timber sales may be on the way out. In April 1993, the Forest Service decided to stop selling timber from 62 of 156 national forests because it consistently lost money on sales.[5] This was a big admission for the Forest Service, which had long argued that it was not losing money. However, environmentalists had proved conclusively that subsidizing timber sales with road building and other inducements was costing the U.S. taxpayer big bucks while at the same time destroying public forests. This case is forcefully made by Randal O'Toole in his book *Reforming the Forest Service.* O'Toole calculates that "Money-losing timber sales are costing taxpayers at least $250 to $500 million per year."[6]

Even if the Forest Service stops selling timber at below cost, we will still have to take other steps to keep our forests from being managed unsustainably. Much of the unsustainable logging on national forests could be eliminated by asking visitors to the national forests to pay a daily entry fee of $3, O'Toole maintains. In many of the national forests, the recreational use of the forest will bring in more money than logging, grazing, or mining, he says. "Marketizing" the management of the Forest Service would make below-cost sales of timber unattractive to forest managers who would naturally gravitate to finding more profitable nontimber uses of the national forests, he contends. "By giving forest managers the right—and the incentive—to charge recreational fees, marketization encourages the promotion of recreational uses," he writes. According to the Forest Service, recreation is more valuable than timber and other extractive commodities in eight out of nine administrative regions.[7]

However, getting at the root of why our forests are logged in an unsustainable fashion will require that we go beyond marketizing the Forest Service and ending below cost sales. Sustainable forestry will only become widespread when the cost of unsustainable logging is high. The proper role for government here is to stop subsidizing the destruction of our forests and start levying taxes that include the ecological costs of timber extraction on the price of lumber and other wood products. In the current economic system, a tree only has economic value when it has been cut up into lumber. None of the ecological benefits of leaving trees standing is taken into account.

A system of taxes is required that will take into account the value of trees in protecting our fisheries, watersheds, water quality, controlling erosion, acting as a buffer against climate change, providing a habitat for wildlife, maintaining the gene pool, and for recreation.

Taxes should be highest on wood products from intact forests, less on wood from secondary growth, and lowest on timber from sustainably managed cutting. Without these environmental taxes, wood products will continue to be underpriced and there will be no incentive to practice sustainable forestry. Perversely, in some states, inheritance tax laws consider trees to be a form of investment and as a result, when the owner of a woodlot dies, the children may be forced to cut the timber to pay the tax bill.

In addition to relearning the art of sustainable forest management, Americans must also learn to curb their seemingly insatiable appetite for timber products. While per capita wood consumption has declined in industrial countries over the last century, we could use a lot less by using recycled paper and recycled lumber as well as finding substitutes for wood in the construction of our homes. These economies would permit us to meet our wood products needs while at the same time managing our forests sustainably.

Today there is an urgent need to bring selection management back into the mainstream of American forestry. Photographs of the Pacific Northwest taken from satellites reveal that clearcutting has caused worse damage to the forest cover of North America than the burning of the rain forest has in the Brazilian Amazonia. Lighthawk, a group of pilots with an environmental mission, regularly takes lawmakers up over the piebald landscape of Oregon and Washington to give them a feel for the extent of the ecological devastation wrought by the buzz of the chain saw.

Despite evidence of the enormous injury caused by clearcutting, the United States has yet to initiate a sustainable forestry policy that will spare the rest of the country from a similar fate. The struggle over U.S. forestry policy is coming to a head as the pressure of logging operations on

American forests is shifting from the Pacific Northwest to the more densely populated East Coast. Having cut the majority of the commercially valuable and easily accessible timber in the West (only about 5 percent of our old-growth forest remains in the national forest), logging companies are returning to the Northeast and Southeast for their next round of clearcutting. Georgia-Pacific, for example, has already relocated in the South and the volume of lumber being cut out of the forests of Maine and other parts of New England is mounting.

The assault on the northern forest of New England, some 26 million acres that have been logged for 400 years, is once again gathering momentum. In Maine 1.2 million acres have been cleared since 1980.[8] "From the air much of northern Maine resembles a dog with terminal mange, with vast clearcuts stretching off in every direction," writes Bill McKibben.[9]

This continues a pattern the timber industry has followed for decades. In 1933 legislation was proposed that would have reformed logging practices on private land, but it failed. After logging out its own lands using unsustainable practices, the timber industry turned to public lands for its supply of timber. In 1949 some 5.6 billion board feet were cut in our national forests. By 1955 that figure had risen to 8.6 billion board feet, and by 1970 to 13.6 billion board feet.[10]

To many who work in the timber industry, clearcutting is a simple and efficient method of harvesting trees. They tend to view clearcutting as no more harmful to the forest than getting a crewcut at a barbershop: if you clip your hair off close to the skull it grows back; the same holds true for the forests, they insist.

To permit the escalating level of allowable cuts on public lands, Forest Service officials became advocates for the timber industry and developed a number of questionable rationales that excused excessive cutting of the nation's forests. Clearcutting is a beneficial practice on public lands, they argue, because it provides forage for wildlife, particularly for elk and grizzly bears; increases the diversity of vegetation; augments the flow of streams, providing additional water for downstream farmers; prevents forest fires; and reduces pest infestations.[11]

Conservationists argue with these rationalizations for increased timber sales on public lands, pointing out that far from helping farmers with their water supplies, clearcutting eliminates the forest that once held water in the ground in a reservoir, gradually releasing it and permitting streams to run year-round. Once the forests are cut, stormwater floods off the land causing massive erosion and sedimentation of the rivers and streams. Furthermore, uncut forests are more resistant to forest fires than areas that have been cut.

Clearcutting also radically reduces the variety of tree species in the forest and eliminates important habitats for wildlife, they contend. In fact, owing to the rapacious and indiscriminate nature of logging in this country, many of our forests today are little more than single-crop plantations of fast-growing softwoods, all of roughly the same age, from which much of the wildlife has fled.

In 1977 the Texas Committee on Natural Resources sued the Forest Service for permitting clearcutting on public lands and the court ruled in its favor. The judge found that clearcutting results in a fire hazard, encourages disease, accelerates erosion, reduces the amount of habitat essential to various species of wildlife, reduces the number of tree species, and destroys the recreational value of an area. Subsequent legislation, the National Forestry Management Act of 1976, permitted clearcutting on public lands under certain conditions.[12]

The debate over clearcutting remains one of the most contentious environmental issues of our day. Europeans, who ruined their forests with clearcutting and the planting of monocrops, are now turning to more flexible cutting methods that preserve the forest canopy. They see Americans making the same mistake they made. Even in the United States, large-scale clearcutting is a relatively recent phenomenon. Before the 1950s, clearcutting was practiced in relatively few of the 156 national forests, while after 1970 it was the predominant harvesting method in nearly all national forests.[13]

"Excellent forestry costs nothing more than restraint," observes Gordon Robinson, author of *The Forest and the Trees: A Guide to Excellent Forestry.*[14] Evidence of this restraint is on display in the Waldee Forest named after Walton and his wife Dee.

Smith has a different sense of time than most of us. He thinks in decades and centuries rather than in the daily-monthly-yearly perspective that most of us are locked into. His continuous selection method means that he goes into an area every twenty to twenty-five years to cut a few trees. Each tree is on its own rotation, he explains, depending on the species. Some trees, like white oaks, need 200 years to reach their peak value.

Hiking through the Waldee Forest with Smith one gets a sense not only of the history of the forest and what it is like today, but also what it will be like 50, 100, and 200 years hence. "That poplar stand regenerated naturally forty-five years ago in open pasture," he says pointing to a densely wooded area. Elsewhere, Smith cleared a 1-acre plot in 1972 where he planted 150 hickory seedlings at 10-foot intervals. "I plan to thin the hickory in 2010 when I'm 100 years old and harvest them for veneer logs in 2060," he deadpans. "When you come down to it, none of us really owns the land, we just have it for a short time," he says.

On a ribbon of land that runs through the Waldee Forest known as the Big Tree Trail, Smith has made a promise that all the majestic, old trees will never be cut. "I have a poplar that is 5 feet in diameter, a big white ash, and a big black gum on the trail," he says with obvious pride. "There are some sacred things in the wilderness that man should leave alone."

Transforming Hazardous Wastes into Useful
Industrial Materials

In a renovated helicopter factory located in an industrial park outside of Fall River, Massachusetts, Molten Metal Technology, Inc. (MMT), is developing a new way of recycling hazardous waste. If the new technology pans out, it will put this small town on the map for something other than being the place where Lizzie Borden allegedly gave her parents 81 whacks with an ax.

In the center of a huge, brightly lit room at the Fall River research facility stands a giant machine called a catalytic extraction processing (CEP) system. This Rube Goldbergesque contraption of metal boxes and pipes is advertised as being capable of digesting hazardous wastes and spitting out industrial-grade materials.

At its core the CEP is a closed-system cauldron of molten metal, usually iron, which is heated to some 3000° F. Any hazardous waste injected into the bottom of this molten metal bath is instantly dissolved. The catalytic and solvent properties of the molten metal break the molecular bonds that hold the hazardous chemical compounds together and reduce them to their elemental components. For example, polyvinyl chloride (PVC) wastes can be fed into the CEP and hydrochloric acid (HCl) and synthesis gases, or syngases (hydrogen and carbon monoxide) are recovered.

Controlling the CEP are technicians wearing blue uniforms, yellow hardhats, and safety glasses who manipulate the dissociated chemical compounds into new forms by adding materials to create desired products. The recovered materials are then partitioned into a gas phase, metal phase, or ceramic (slag) phase.

After injecting a waste into the molten metal, a portion of it is converted into a gas such as carbon monoxide or hydrogen, which can be filtered, separated out, and sold. Another part of the waste is extracted as any one

Figure 10.1
The catalytic extraction processing unit used to transform hazardous chemicals into useful industrial materials at Molten Metal Technology's (MMT) facility in Fall River, Massachusetts. Photograph by Steve Lerner.

of a number of metals such as chromium, cobalt, or nickel. The metal is concentrated into an alloy that can be sold or reused as feedstock for an industrial process. The remainder of the material, now rendered relatively benign, is partitioned into a ceramic, some forms of which can be sold as abrasives or aggregate.

Ancient alchemists would have appreciated this modern device for the way it transforms dross into valuable materials. Judged by today's standards, however, a new technology that processes hazardous waste is greeted with considerable skepticism by a public wary of technologies that typically spawn a new generation of unwanted byproducts. As a result, MMT faces an uphill struggle for acceptance. As one of a number of new environmental technology companies, it must not only prove it can recycle hazardous chemicals but also navigate a maze of environmental regulations.

Despite these hurdles, MMT has well-placed enthusiasts such as Vice President Gore, who said: "Molten Metal is a success story, a shining example of American ingenuity, hard work and business know-how, all being used to clean up our environment, and at the same time provide jobs and economic growth."[1]

Molten Metal Technology is possibly the most widely known new company in the controversial field of hazardous waste management. Informed opinion about MMT runs the gamut: there are those who think it could revolutionize hazardous waste treatment, those who think it is a sham, and those who simply consider it a promising new technology with some advantages and some problems. But whether MMT lives up to its promise or fails, its story of entrepreneurship on the frontiers of environmental science is becoming increasingly common, as inventors and businessmen rush to get a piece of the market for green goods and services.

Depending on how one defines the sector, the size of the environmental industry in the United States is anywhere from $100 billion a year to $170 billion. One estimate places the number of jobs in U.S. environmental protection in 1992 at 4 million, or about 3 percent of total employment. Moreover, the international market for environmental goods and services could well reach $4 trillion for this decade, says Michael Silverstein, editor of *The Environmental Industry Yearbook*. And while most observers agree that the United States has recently lost its lead in the green market to Germany and Japan, Silverstein believes that this country, with its 30 years of experience in environmental cleanup, is still best positioned to develop the technologies that will ultimately dominate.

William M. Haney III, president and CEO of MMT, is a young entrepreneur with substantial experience in tapping the power of green tech. Haney started young, incorporating his first environmental technology company, Fuel Tech Inc., in 1981 while only an eighteen-year-old freshman at Harvard. The company produces front-end air pollution control equipment designed to reduce nitrous oxides (NO_x) emissions. The technique, known as NOxOUT, which reduces NO_x emissions by up to 80 percent in the postcombustion phase, mixes dried urea with water and injects it into a flue gas of gas turbines, stationary diesel engines, power plants, and incinerators. "All of a sudden the pollution goes away," notes Haney, who sold his interest in Fuel Tech in 1987. It is now a company that has 75 plants operating and 200 coming on-line, he adds. Not bad for a freshman effort.

His second company, Energy BioSystems Corporation (EBC), founded in 1989, engineered a microorganism that extracts sulfur from crude oil and high-sulfur coal, permitting these fuels to be burned without generating sulfur dioxide. The microbe transforms the sulfur into sulfate, an inorganic salt which can be used as a fertilizer. EBC is currently estimated to be worth $90 million.

At this point Haney could have retired comfortably to a life of luxury, but this prospect did not appeal to him. "Walt Disney used to say 'I make money so I can make movies; I don't make movies so I can make money,'" he observes. "I want to use my money to deploy science to solve environmental problems." With this in mind Haney started another environmental technology company. Having attained the ripe old age of thirty-three, he

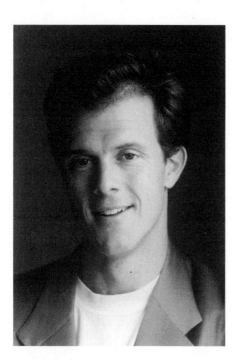

Figure 10.2
William Haney III, president of Molten Metal Technology. Courtesy MMT.

launched his third and most ambitious venture, MMT, which some think has the potential to revolutionize the $12.2 billion U.S. waste industry.

MMT is on the cusp of a scientific revolution, Haney insists. He likens the position of environmental technologies today to the stage the bio-technology industry was at in 1980 just before it took off. "If I told you I was going to analyze a cow, clone the genes in her milk, add these cloned genes to an insulin shot, and make people who are dwarves full-sized, you would probably say, 'Haney, that sounds like a bit of a stretch,'" he says with a laugh. But someone did it and there is no reason why comparable wonders cannot be achieved in the environmental field, he adds.

Haney's counterpart is the inventor of the molten metal process, thirty-eight-year-old Christopher J. Nagel, now executive vice president for Science and Technology at MMT, who sits down the hall from Haney in an office with a photo of a wild-haired Einstein staring down from the wall. Nagel's story is a classic tale of a young inventor who finds a practical application for what at first appears to be an obscure observation in an esoteric field of study.

His story begins in 1982 when, after graduating from Michigan Tech, Nagel landed his first job researching new energy conservation measures at the behemoth U.S. Steel works in Gary, Indiana. From his office—wedged between blast furnaces and torpedo cars carrying hundreds of tons of hot metal—Nagel watched the plant produce hundreds of pots of molten metal slag every day, which were simply allowed to cool. He soon came up with the idea of using the heat from the molten metal to drive catalytic reactions that would release more energy, including a reaction that used old tires in reducing iron oxide to iron.[2]

In 1986 Nagel, who is the kind of guy who carries a card with a copy of the periodic chart in his wallet, left U.S. Steel (now USX) to do graduate work in chemical engineering at the Massachusetts Institute of Technology. There he developed the concept of using molten metal to drive reactions that would recover useful materials from hazardous waste. He described the technique, which he dubbed "catalytic extraction processing" (CEP), to the head of the licensing and patent office at MIT, John T. Preston, who introduced him to Haney. From the meeting of the inventor and the entrepreneur, MMT was born.[3]

Hazardous waste treatment is a briskly moving sector among green-tech startup industries, and for good reason. According to the most recent

Figure 10.3
Christopher J. Nagel, executive vice president of science and technology at Molten Metal Technology. Photograph by Bill Gallery.

figures from the Environmental Protection Agency's toxic release inventory, there were 37.3 billion pounds of hazardous wastes generated in the United States in 1992 (and the inventory only covers production-related wastes). Some 3.4 billion pounds of this toxic material is either released to the environment or sent off-site for disposal; the rest is managed in some way through energy recovery, recycling, or treatment. What makes this sector attractive to entrepreneurs is not so much the volume of waste, but the rules governing its disposal and the public clamor for cleanup. The combination of public pressure, strict standards, and the potential for liability can persuade a company to reduce or detoxify its hazardous waste rather than dump it.

There are a number of enterprises springing up to answer this need. Thermatrix, Inc. breaks down volatile organic wastes in a heated ceramic matrix that, the company says, consumes little energy and gives off no measurable organic byproducts. ELI, Inc. uses hydrogen in a reactor to transform polychlorinated biphenyls (PCBs) and other chlorinated wastes into water, carbon dioxide, sodium chloride, and methane gas. Still other processes use plasma and molten salt oxidation.

Among these new technologies for processing hazardous wastes, MMT appears to have attracted ample investments. Investors clearly think the CEP has great potential and Wall Street initially reacted enthusiastically to the prospects for this new technology. When MMT went public, Haney found little difficulty raising $105 million on the Nasdaq exchange. And MMT, which employs more than 300 people, has in hand orders for CEP systems from a number of major corporations. (Rather than processing waste commercially at its research facility, MMT aims to set up CEP systems on site at other companies. It may either own or operate the system itself, under contract for the client's waste, or sell the system to the client.) Clients include Hoechst Celanese, a Fortune 100 chemical corporation; Scientific Ecology Group, a subsidiary of Westinghouse Electric Corporation, the world's largest processor of low-level radioactive wastes; and M4 Environmental, a partnership MMT formed with Lockheed Martin.

MMT is also one of the most environmentally ambitious of the new hazardous waste processing companies. Its officials claim that the CEP can recycle 90 percent or more of a number of hazardous waste streams the company has placed a priority on, including PVC wastes, electronic component wastes, and chlorinated solvents. They also claim that, in seven waste streams that are priorities for MMT, the CEP's destruction removal efficiency (DRE) is 99.9999 percent or greater. Haney says that the CEP, when hooked to the tail end of certain industrial processes, will create a closed-loop system, in which virtually nothing escapes and everything is reused.

Heat-based treatment of hazardous waste is far from new. Incineration also uses a chemical reaction—oxidation during burning—to transform toxic compounds into more innocuous substances. But much of the trouble with incineration stems from the fact that fire does not deliver a constant temperature. By contrast, the temperature of molten metal can be held more constant, and molecules in a molten metal bath have more intimate, more complete contact with the metal than they would with a flame, says Dr. William Moomaw, professor of international environmental policy at the Fletcher School at Tufts University. As a result, he says, toxic compounds are wholly broken down in MMT's process.

"People still ask me if this new technology for recycling hazardous waste is too good to be true," says Haney, but the question is asked of him less often now that he has a track record for developing a string of technologies that help solve environmental problems. Sitting in his modern corporate offices in Waltham, Massachusetts, dressed in chinos and a denim shirt buttoned at the throat, Haney is brimming with confidence. "If in 1950 I told you that on a piece of silica the size of your thumbnail someone would produce something that would have the thinking capacity of 400,000 people for 400,000 days, what would you have said?" he asks with a smile. "If I had told you that I could develop a missile that would fly out of a silo in Nebraska and come down within a hundred yards of where I wanted it to land half a world away, would you have believed me? Yet we now accept that as perfectly normal. I would argue that this took a much larger leap of technological fancy than what we are doing at Molten Metal."

But there is no definitive answer yet on just how closely results match theory at MMT, and independent environmental specialists are far from seeing eye to eye on it.

Moomaw is one of the more sanguine. "The CEP system really is a breakthrough technology. It is one of the few new and original ideas to come along. As a chemist, the notion of taking complex molecules that happen to be toxic and breaking them down either into their elemental forms or into simple molecules like carbon monoxide, which can be burned as a fuel, is very attractive," he observes. Silverstein, author of *The Environmental Economic Revolution: How Business Will Thrive and the Earth Survive in Years to Come* (New York: St. Martins Press, 1993), is just as bullish. He sees MMT as part of a new field of industrial ecology, which is redesigning industrial systems along the lines of natural systems where there is no such thing as waste because every waste product becomes the feedstock for a different process. "When the leaves fall off the trees in autumn you do not see squirrels picking them up and stuffing them in Hefty bags and hauling them to a landfill. The leaves become a feedstock for vegetation that grows on the forest floor," he says. "What you have with MMT is the kind of process you see with more mature ecologies where there is no such thing as waste or pollution."

Marco Kaltofen, president of Boston Data Chemical Corporation, a firm which does environmental investigations for law firms, trade unions, and nonprofit organizations, also sees MMT's process as having distinct advantages over incineration of hazardous waste. "First, MMT does not necessarily generate the huge volumes of gas that you get with incineration," he notes. With incineration, the higher the temperatures achieved to insure destruction of toxic chemicals, the greater the generation of gases that must be processed and cleaned. By contrast, the CEP process collects metals as solids instead of oxidizing them into vaporous fumes or gases, which then have to be scrubbed out in the stack.

This is a considerable advantage, says Kaltofen, who was previously director of the Citizen's Environmental Laboratory, a project of the National Toxics Campaign. But it does not make MMT a panacea for dealing with hazardous waste. It just means that some subset of wastes will be amenable to this form of treatment, particularly homogeneous streams of industrial waste.

MMT's system is superior to incineration because you can run it cleanly and you have the opportunity to recycle hazardous wastes, Kaltofen continues. "But this is a process you can screw up," he cautions: "You have to know what is in the waste stream in order to create conditions in the CEP that will allow you to avoid the creation of unwanted byproducts." And therefore the success of the CEP system will depend on the company that buys it. "As long as the people running the CEP are responsible" and have a policy of keeping emissions as limited as possible, "then they will do well with this technology," Kaltofen says.

"I think Molten Metal has to be a step forward over incineration and landfilling," agrees Paul Connett, professor of chemistry at Saint Lawrence University and editor of *Waste Not.* "It is a more rational approach than incineration where you put things through a high temperature, propel them into the air, and then do your best to catch as many of the propellants as you can. Incineration is a very unsophisticated chemistry; this sounds a bit more sophisticated." However, even Connett has reservations. "The only trouble of course is that this new technology may produce the mindset that you can continue to make toxic chemicals instead of placing the attention on front-end reduction of their use," he warns.

So is CEP a closed-loop technology? It is no academic question. To encourage recycling of hazardous waste and development of technologies that do it, EPA has set up much more lenient rules for companies that recycle toxics than for those that treat and dispose of toxics. Which classification a company receives can have a substantial effect on its profitability.

"Molten Metal Technology didn't fit into any of the established regulatory boxes very well," observes Steven DeGabriele, acting director of the Division of Hazardous Materials at the Massachusetts Department of Environmental Protection (MADEP). So state officials developed a unique research permit. Like many companies MMT is allowed to perform small-scale tests on different wastes. But Massachusetts gave MMT a special permit to perform tests on quantities over 250 kg, provided its data show that the material can be effectively recycled. When MMT sells a CEP system its client must apply for a recycling approval in the appropriate state and for the specific waste stream involved.

At its Massachusetts facility, MMT has performed some 15,000 hours of tests. And, at least in some cases, the waste loop is far from closed. True, in 26 months of operation, MMT has reclaimed 140,000 pounds of iron and 15,000 pounds of nickel that was sold to local scrap metal dealers.

But some of the waste streams that have passed through MMT's nine operating CEP units at its Fall River facility have left residues behind—twenty-four 55-gallon drums of hazardous solid waste and 23,255 gallons of liquid hazardous wastes were generated per year in 1994 and 1995. The solid wastes are principally particulate matter collected in a Gore-Tex fabric filter in the baghouse that comes off the molten metal bath, which is subsequently shipped away for disposal at a hazardous waste treatment facility, notes Gene Berman, vice president for regulatory and community affairs. But most of the liquid wastes would be recovered and reused at a commercial-scale facility, he asserts.

For example, thousands of gallons of hydrochloric acid is scrubbed out of the gaseous stream to make aqueous-grade HCl. "At a commercial facility this is a desired and valuable product, but at our Fall River facility it is not made in sufficient volume for commercial purposes," he observes, so it is shipped away for treatment at a hazardous waste treatment facility. Similarly, syngases are burned off at the Fall River facility because it has proved uneconomical to collect and compress them for sale in small quantities. In large-scale operations, however, these gases would be used, Berman emphasizes.

Another byproduct of the CEP's operation, some slag or ceramic material, was reincorporated into a metal alloy which was subsequently sold. In a large-scale operation, such as the one being set up at Hoechst Celanese in Texas, ceramic or slag materials will be separated out from the metals and sold as aggregate or abrasives, Berman says.

But there is no guarantee that the slag produced at MMT's client facilities will be marketable. The only case study of MMT done by a regulatory agency was authored by Paul Boerst who wrote in a report to Congress that "in certain applications, a CEP unit may produce not only valuable products, but also non-salable residue that must be disposed."[4] MMT officials contend that any waste it produces will at least be benign because toxic organic compounds will have been reduced to usable components, heavy metals will be recovered, and what remains will be sealed in a vitreous lattice.

Some of the uncertainty about how to classify MMT stems from lack of information. "I think the process is very exciting if it can do even a part of what it claims it can do. What I like about it is the concept of integrating this equipment into industrial processes. To the degree to which they can recover valuable materials and not create waste streams certainly moves us in the right direction," says Catherine Koshland, associate professor of environmental health sciences at the University of California at Berkeley. But Koshland, who holds a Ph.D. in mechanical engineering, was troubled by the fact that MMT had published nothing in the peer review literature.

For Joel Hirschhorn, MMT has yet to prove itself. Hirschhorn, who holds a Ph.D. in material engineering, worked for twelve years at the Office of Technology Assessment and now heads his own consulting company. He was hired to assess MMT for an investment company he declines to identify. "Bill Haney is a most effective person when it comes to public relations and marketing. The man is a genius. He is constantly pumping up the story," Hirschhorn says. One of the reasons MMT officials have been so successful at hyping their technology is that a theoretical description of how it works sounds great, he continues. But once you get past the "Science 101 level" with this technology, the problems begin to become apparent, he says. And Hirschhorn asserts that MMT is not making public the data that would allow for an independent assessment of its effectiveness.

MMT officials respond that they have submitted thousands of pages of data to the EPA and the state of Massachusetts on its emissions testing and general performance since 1993. Furthermore, Chris Nagel wrote an article entitled "Catalytic Extraction Processing: An Elemental Recycling Technology" that was published in *Environmental Science and Technology*, a prestigious peer review journal.[5]

While some have been watching to see if the CEP will pass muster with a peer review journal, rapid changes in the technology and in the market have taken MMT for a roller-coaster ride that is not for the fainthearted. The development at MMT of a Quantum CEP that can handle radioactive wastes was intriguing enough that the Department of Energy ponied up a total of $30 million over four years to demonstrate that this technology can decontaminate or reduce and stabilize radioactive wastes.

In partnership with the Scientific Ecology Group, a subsidiary of Westinghouse, MMT has built a plant at Oak Ridge, Tennessee, for the processing of ion exchange resins, an organic material used to soak up radiation in the cooling waters of nuclear power plants. Once the resins are fully loaded with radioactivity, they are taken out of the cooling water and packed into high-integrity containers (HICs), each filled with 200 cubic feet of resins, and then trucked to the Quantum CEP in Oak Ridge. By partitioning the radioactive materials out of the resin into a ceramic phase, the Quantum-CEP vastly reduces the volume of radioactive wastes that must be disposed of by more than 25 to 1. It also puts the rad-wastes into a more stable ceramic form that can be more easily handled and safely stored, Gene Berman says. Some of the radioactive isotopes in the wastes, such as cesium-137, can actually be isolated and recovered for use in food irradiation equipment, he adds.

Nothing, however, runs entirely according to plan, and after having processed 150,000 pounds of ion exchange resins, Westinghouse decided to sell all of its environmental businesses, including its 50 percent interest in the Quantum CEP and the facility on which it had been built. To protect its interest in the business of treating ion exchange resins, MMT bought back its interest in the equipment and now fully owns and operates the Quantum CEP at Oak Ridge, which is now fully operational, says Berman.

MMT also joined with Lockheed Martin to form a limited partnership called M4 Environmental. With a $50 million capital investment from Lockheed Martin, M4 Environmental has gotten off to a fast start, building a 150,000-square foot plant at Oak Ridge to treat hazardous, radioactive, and mixed waste problems. One of the waste streams targeted for processing in 1996 is depleted uranium hexafluoride, known as tails, a byproduct waste generated from the enrichment of uranium for nuclear weapons by the U.S. Enrichment Company (USEC). There are mountains of these wastes awaiting treatment: USEC produces 15 million kilograms of tails each year at two plants and there is over 550 million kilograms of DOE tails in storage.

In addition, M4 Environmental is processing radioactively contaminated metal equipment. The Quantum-CEP separates the radioactive contaminants from the metal, permitting the metal to be reused. The process can completely partition radioactive elements such as uranium, plutonium, and thorium out of a waste stream into a volume-reduced final form, but if the wastes contain certain other radioactive elements such as cobalt-60, trace amounts of this contaminant may remain in the recovered metal, Berman reports. Nevertheless, this recovered metal containing trace radioactive contaminants can be made into metal shielding blocks used in nuclear facilities or shaped into metal containers for radioactive wastes. This makes it unnecessary to use virgin metal, which will be exposed to radiation in these applications.

Currently M4 Environmental has three Quantum CEP units (which they call "radiologic processing units") operating and a fourth will soon be activated. RPU 3 has already successfully processed radioactive-contaminated mixed wastes from the Duke Nuclear Power Plant in South Carolina, says Berman. When RPU 4 goes on line, it will be able to process larger chunks of equipment contaminated with radioactivity, he adds.

M4 Environmental is also pursuing contracts with the Department of Defense (DOD) to process its huge stockpiles of nerve gas and mustard gas. And M4 Environmental is among three companies being considered to do this work for the Army's chemical demilitarization program and has presented a proposal to DOD to convert 100 percent of nerve gas into useful products. The nerve gas—a highly complex and toxic molecule made up of carbon, hydrogen, sulfur, and chlorine—can be broken down and recovered as syngas and commercial-quality sulfur, Berman says.

With a technology that had a potential not only to treat hazardous chemical waste, but also get a piece of the lucrative market to treat radioactive wastes, MMT became a darling of Wall Street, and its stock rose dramatically. That love affair ended, however, on what is known in-house as "black Monday": October 21, 1996. On that day MMT stock plummeted 49 percent.

One of the reasons for this precipitous drop was reports that MMT was losing its DOE Funding. The *Wall Street Journal* reported that DOE had "declined to renew a research contract with the company."[6] These reports were inaccurate because MMT continues to receive DOE research money, and may receive more in the future, says Berman. What really happened was that MMT had overestimated the amount of money it would be receiving from DOE. Furthermore, the confusion came about because MMT was moving out of the research and development phase and into the marketplace, where it has to compete with other technologies to win contracts, he adds.

To clear up any misunderstanding, two days after MMT's stock plummeted, DOE Deputy Assistant Secretary Clyde Frank issued a statement saying: "The department is enthusiastic about Molten Metals Technology for use in cleaning up mixed waste throughout the DOE complex."[7]

Further evidence that MMT technology has not lost favor at DOE came when the agency announced that a team led by Lockheed Martin Corporation and using MMT's patented Quantum-CEP technology had been selected as one of two teams chosen for the first phase of the cleanup of radioactive wastes at DOE's Tank Waste Remediation Program at Hanford, Washington. "The DOE doesn't say, 'Your technology is bad' and then go give you (millions) of dollars to clean up waste at the Hanford site," Douglas Augenthaler, an analyst with Oppenheimer & Company, is quoted as saying by the Bromberg Business News.[8]

But how does one explain the drop in MMT's stock given all this good news. Berman speculates that one factor that spooked investors was that MMT was not bringing facilities on line to profitably process commercial volumes of waste as quickly as investment analysts had anticipated. While MMT had started out looking like a promising technology with a number of applications, analysts are now taking a cold, hard look at the bottom line. "What Wall Street is looking for is for us to crank up the commercial volumes of waste we treat to the level where we will make some real money," Berman notes. Once the two facilities in Oak Ridge begin to process large quantities of radioactive wastes and the new plant designed to handle chemical wastes in Texas gets going, revenues should rise and Wall Street's faith will be renewed, Berman predicts.

MMT is close to opening a $25 million, commercial-scale CEP facility at Bay City, Texas, for the Hoechst Celanese Chemical Group, one of the largest U.S.-based chemical companies. The plant will be operational by mid-1997, Berman says, and will recycle up to 24,000 tons of biosolids from wastewater treatment plants into syngas, which Hoechst will use onsite to manufacture oxoalchohol products.

While it has been difficult to convince the chemical industry to use the CEP technology, Berman is confident that MMT will ultimately capture some of the huge market to treat chlorinated hydrocarbon wastes. That effort was given a boost recently when the EPA gave CEP technology a Best Demonstrated Available Technology (BDAT) determination of equivalency treatment (DET), which states that MMT's equipment is at least as good as incinerating or combusting wastes from PVC and other chlorinated plastics. In the past, chemical companies were required to incinerate these wastes, but now they have the option to use the CEP, Berman says. MMT also received a BDAT DET for processing toluene diisocyanate (TDI) wastes.

The Electric Power Research Institute (EPRI) is also interested in CEP, and plans, in conjunction with member utilities, to test the new technology. A three-way synergistic arrangement is being considered whereby a chemical company would install a CEP to treat a waste stream. Out of the waste stream the CEP would recover syngas that would either be used by the chemical company or sold to the nearby utility. The utility, in turn, would provide the power to run the CEP unit, Berman explains.

Knowledgeable observers of the CEP remain divided on whether or not they think this new technology will prove widely applicable. Pat Costner, senior scientist with the toxics campaign for Greenpeace, is surveying innovative technologies for processing hazardous wastes. Where MMT is concerned she is on the side of the skeptics. "I personally think that every omnivorous technology for treating hazardous wastes, such as MMT, has a common array of disadvantages," Costner says, adding that none of these innovative technologies will work on all waste streams. "This technology has been configured specifically for circumventing Resource Conservation and Recovery Act (RCRA) regulations," Costner asserts.[9] She notes that for years the incineration industry claimed that nothing came out of its stacks except water, steam, and carbon monoxide. "How many years did we listen to that?" she asks. "Is this the same kind of ballgame?"

But Marco Kaltofen argues that while MMT should not be regulated as a recycling operation, it is nevertheless a valuable new technology. Fundamentally, he says, the business of this company is "making your waste problem go away without creating any environmental problems—and that alone makes it a technology worth encouraging."

"What MMT's story shows is how sophisticated the public has become environmentally," adds NRDC senior scientist Linda Greer. Any new technology will have to open up its data and undergo serious scrutiny, because "people now realize how important it is to get this stuff right." And Greer believes that "in a way that is the most hopeful sign of all. Whether or not it's MMT that helps us get there, public pressure is moving toward a sustainable society. Innovative technology has a big role to play in that transformation."

11   *Paul Mankiewicz*

Urban Rooftop Agriculture

Statistics about the number of people moving into cities around the world are enough to make your jaw drop. Consider the following disturbing projections:

• By the year 2005, the number of people living in cities is expected to outnumber those in the countryside for the first time in human history.
• Fifteen years later, in 2020, the number of urban residents will increase to 3.6 billion, more than twice the 1.4 billion who lived in cities in 1990.[1] By another calculation, the number of people who live in cities will reach 5 billion people by 2025.
• The number of "megacities" with populations over 8 million is also mushrooming. In 1950 New York and London were the only two megacities on the planet, but by 1990 there were 21 of them; by 2015, there will be 33 megacities, 27 of them in the developing world.[2]

With this phenomenal increase in the urban population on the horizon, many of the most farsighted eco-pioneers are working on ways to make the cities of tomorrow more ecologically efficient. And while some of their visions may sound far-fetched today, in the future these practices may be considered quite ordinary.

Two of those who have been looking for ways to turn urban waste products into resources are Paul and Julie Mankiewicz, co-directors of the Gaia Institute in New York. Paul Mankiewicz, a biologist, is developing new on-site systems for composting urban and suburban food waste in order to turn it into a rich nutrient base for growing food on the rooftops of city buildings and suburban supermarkets. Julie Mankiewicz, who received her Ph.D. on the biochemistry of plant cell wall components, is testing these innovative compost systems and the qualities of the compost they produce.

By turning urban food waste into compost for agriculture, Paul Mankiewicz hopes to simultaneously reduce the urban waste stream and provide
farmers with a substitute for chemical fertilizers. To make this possible, he
developed a system for collecting and composting 5 to 25 tons a day of food
waste in both urban and suburban settings. Under a grant from the Environmental Protection Agency and in conjunction with the New York City
Department of Sanitation, Bureau of Reuse, Recycling, and Waste Reduction, Mankiewicz looked at a variety of existing composting systems, including the agitated bay, tunnel system, silo system, and rotating drum. He
concluded they were all too expensive or were not designed to "provide optimal conditions for biological performance."

In the course of doing research to determine what conditions would
cause decomposition to take place most efficiently, Mankiewicz found that
when food waste is chopped into pieces about 1 centimeter square, it provides the best surface area and air pockets for bacteria to thrive. He also

Figure 11.1
*Julie and Paul Mankiewicz, director of the Gaia Institute at the Cathedral of Saint
John the Divine with their children, Phoebe and Tighe, sitting atop an old prototype
of the proprietary composting technology they developed with funding from the
EPA and the New York Department of Sanitation. Photograph by Richard DeWitt.*

discovered that decomposition took place most rapidly if the compost was not agitated or turned, as long as there was enough air being pumped through the pile to keep an anaerobic state from developing.

With these lessons in mind he set about building his own prototype, a 5-ton-per-day composting vessel that would be odorless and efficient enough to be accepted as a "good neighbor." What he came up with was a large container made of reinforced polyvinyl chloride (PVC) plastic with two air vents, one near the top and the other near the bottom. He cut the PVC in a cylindrical pattern on the floor, as one would cut a clothing pattern, and then sealed it with an adhesive which he found to be stronger than heat-welded seams. Into the bottom of the composter he installed a coiled, perforated hose covered by some screening material. The hose was then attached to a one-eighth horsepower blower so that air could be introduced into the bottom of the composter to control temperatures and keep an anaerobic state from occurring.

Into this vessel he placed shredded food waste along with some leaf mold as a bulking agent. On top of these organic materials he spread 8 inches of finished compost to suppress odors. In eight to twelve hours the whole system ran at 130° F. In four days he was able to achieve a 50 percent mass reduction and within two weeks the compost was ready to use. "Basically, all we did was figure out how to treat microbes well," Mankiewicz says modestly. "It is just more efficient than most people's compost."

In effect, Mankiewicz found a way to reduce by half the volume of New York City's food waste in four days. This is a significant achievement considering that organic, compostable material comprises anywhere from 12 percent to 30 percent of the municipal waste stream, depending on whether or not yard waste and some compostable paper are collected along with the food waste. Using Mankiewicz's on-site, in-vessel composting system, by the end of two weeks the compost could be sold to farmers as a substitute for chemical fertilizers instead of being hauled to the dump.

Not only had Mankiewicz come up with a workable way to compost municipal organic matter, it also made sense economically. The cost of the composting vessels was relatively cheap: $650 to $800 for a 1- to 5-ton-per-day vessel. A further advantage of these collapsible composters was that a dozen of them could be rolled up, transported in a small van, set up at the compost site, and connected to the air delivery system by two people in a morning.

While some inventors would have stopped there and claimed victory, Mankiewicz realized that unless he designed a start-to-finish, practical system that would save the Department of Sanitation money, the project had little chance of being adopted. To this end he began to look at what a 10-ton-per-day modular system for collecting, shredding, composting, and disposing of food waste would look like.

In scouting for space to compost food waste in Manhattan, Mankiewicz targeted the basements of large commercial and residential buildings. He calculated that if each household generated 1.5 pounds of compostable solid waste per day and captured 60 percent of it, in order to collect 10 tons per day he would have to service a three-by-five-square-block area, encompassing a little more than one quarter of a mile square, in which an average of 800 households per block were located.

By dividing the city into these one-quarter square mile zones, each generating 10 tons per day of compostable waste, Mankiewicz was able to avoid the use of the city's standard, large, noisy, garbage trucks, which cost about $110,000 each. As a substitute, Mankiewicz adopted the suggestion of William Kinsinger, an industrial engineer, who recommended the use of an $18,000 electric cart that could quietly and efficiently pick up the food waste at night when there was little traffic to interfere with collection. Mankiewicz figures that the electric collection vehicles' capital cost is between one-sixth and one-half that of the large garbage trucks on a tons-per-day basis. The annual cost of running these electric carts a couple of hours a day would be about $50 to $100. And the average distance from the composting facility in their zone would be about a two-minute drive.

These little electric carts would zip around the neighborhood sidewalks late at night and early in the morning collecting food waste. When they had a full load they would dump it into a chute that emptied into a shredding bin located in a 5000 square foot area of basement in a commercial building. There, the food waste would be shredded, weighed, mixed with leaf mold, and loaded onto pallet trucks for delivery to a "material placement gantry" equipped with a spreader. The material placement gantry, which would run on a set of elevated tracks, would spread the shredded food waste into one of twelve composting vessels. Subsequently, 8 inches of finished compost would be spread onto the food waste to suppress odors. Then the container would be sealed for a two-week composting period dur-

10 TPD Basement Compost facility

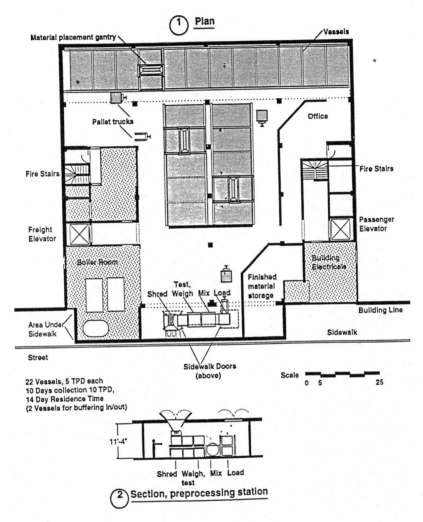

① Plan

Material placement gantry

Vessels

Pallet trucks

Office

Fire Stairs

Fire Stairs

Freight Elevator

Passenger Elevator

Boiler Room

Building Electricals

Test, Shred Weigh Mix Load

Finished material storage

Building Line

Area Under Sidewalk

Sidewalk

Street

Sidewalk Doors (above)

22 Vessels, 5 TPD each
10 Days collection 10 TPD,
14 Day Residence Time
(2 Vessels for buffering in/out)

Scale   0   5   25

11'-4"

Shred Weigh, Mix Load test

② Section, preprocessing station

Figure 11.2
Schematic of 10-tons-per-day basement compost facility. Courtesy the Gaia Institute.

ing which air would be pumped through the compost to speed decomposition. At the end of the process the compost would be loaded into 44-cubic yard roll-on–roll-off containers that would be picked up every week.

But what would the city do with all this nutrient-rich compost? One solution would be to sell it to farmers at a cheap price as a substitute for chemical fertilizers. This would certainly beat dumping it in a landfill. But significant transportation costs are involved in trucking it out of the city. Instead, Mankiewicz has a vision of using this material to create urban farms.

In the past, urban agriculture has been restricted because it was prohibitively expensive to grow food on valuable city real estate. But recently planners have begun to look at city rooftops as a potential large expanse of sunlit territory on which to grow food. "Once you look around, you realize that the only open space in the city that is available for agriculture is roof space," Mankiewicz observes.

The vision of a city covered by vegetable gardens is seductive. Think what it would be like if everywhere you looked buildings had greenhouses on top of them where vegetables, fruits, and flowers were growing in profusion. Not just a few potted geraniums here and there, but rather full-fledged urban farms. It would add a whole new green layer to the urban landscape. In their book *Bioshelters, Ocean Arks, City Farming: Ecology as the Basis of Design*, Nancy Jack Todd and John Todd have written about the potential of rooftops to be "a combination greenhouse and hotwater collector." Their vision went further to suggest that solar collectors could be mounted on rooftops to power grow-lights for the floors below. "Old warehouses could be converted into ecologically inspired agricultural enterprises floor by floor where fish, poultry, mushrooms, greens, vegetables, and flowers could be grown in linked, integrated cycles."[3]

A pipe dream you say? But why couldn't it work? What are the constraints that prevent us from turning the vast unused rooftop acreage in Manhattan and other cities into productive agricultural plots?

Mankiewicz asked himself these questions and came up with answers. There was no way that you could pile tons of soil on top of Manhattan buildings because the additional weight would cause the roofs to cave in and crush the residents below. Most city apartment building rooftops are designed to hold 30 to 40 pounds per square foot while soil that has been watered weighs more than 100 pounds per cubic foot. To support this additional weight would require major retrofits, making an investment in an urban greenhouse unprofitable.

Rather than reinforce rooftops at great expense, Mankiewicz decided that the only way around this problem was to invent a superlightweight soil. The solution was improbable enough. He found that by shredding post-consumer expanded polystyrene (a.k.a. EPS or Styrofoam) he could make a soil substitute that weighs only 20 pounds per square foot, a mere fifth as much as dirt. When watered with diluted compost, this shredded Styrofoam acts like a sponge and holds nutrients until plant roots can make use of them.

Critics argue that mixing soil and Styrofoam is not a good idea because it will be difficult at a later date to separate them into organic and technological loops for composting and recycling. But Mankiewicz says that separating the Styrofoam out of the compost is not a problem. The Styrofoam can be blown into a bag placed on a screen over a low-powered blower. Or, if water is cycled through the mixture in a container, the Styrofoam floats to the top and can be skimmed off.

Growing food in the cities not only provides a local source of nutritious sustenance, it also reduces energy use and pollution by radically decreasing the distance over which food must be transported. It is estimated that the average unit of food now travels 1300 miles from field to table.[4] By reducing transportation and middleman costs, locally grown, rooftop food can also be sold below prevailing prices. Urban rooftop farming also holds out the possibility of providing some jobs for underemployed people with minimal education precisely in the inner-city areas where this population is concentrated.

Mankiewicz made urban farming even more attractive by inventing specialized equipment that eliminates many of the back-breaking, stoop-labor aspects of farming. This further opens the work up to both the elderly and handicapped population. In order to make urban gardens efficient and profitable, Mankiewicz and colleagues designed a "pulsed nutrient system" that calls for an array of perforated tubes to be laid out under the superlightweight soil. The perforated tubes can deliver measured doses of "water, nutrients in solution, microbes, carbon dioxide, and high concentrations of oxygen to enhance plant growth, control predators, and regulate soil temperatures," he argues.[5]

For large-scale operations, Gaia Institute colleague William Kinsinger designed a motorized gantry system that permits crops to be planted, tended, and harvested from above. The advantage of this device is that precious growing space is not taken up with footpaths, allowing large uninterrupted beds to be intensively sown with crops. The gantry itself is a moving steel platform that spans the width of the roof and rolls along tracks set on the parapet walls. It is designed so that urban agricultural workers can lie on their stomachs with their hands reaching down into the soil (like someone floating on an air mattress in a swimming pool) to tend the plants. For those who want to invest in more high-tech equipment, Kinsinger also designed an automated harvesting and planting mechanism (using existing technology), which can be fitted onto the gantry, further reducing the need for manual labor and increasing the profit margin.

Stretching over these urban gardens would be greenhouse structures made of materials much lighter and less expensive than conventional glass greenhouses. Mankiewicz favors a greenhouse built by Advanced Greenhouse Systems of Burlington, Vermont, which is constructed out of lightweight steel framing and a film-glazing technology. Working with Mankiewicz, structural engineers William Kinsinger Associates and Peter J. Galdi Associates have modified these greenhouses to make them adaptable to a number of building types.

Mankiewicz calculates that an 8000 square foot rooftop greenhouse garden would cost $360,000 to build and about $100,000 a year to operate. He figures that annual profits would be in the range of $100,000 to $180,000, permitting the operator to recoup the initial capital investment in two to five years. But these estimates are conservative: "Winter season prices for tomatoes and specialty crops could make it possible to recover the entire cost in one to two years," he claims.

While it is impossible to say whether these calculations will hold up in the real world, the exercise does suggest that some kind of reasonable profit can be realized with rooftop gardening. But beyond the direct profits involved in growing food and flowers on urban rooftops, there are other substantial advantages to Mankiewicz's plan. For one thing he projects that an 8000 square foot garden on a rooftop could supply fresh vegetables for some 2000 people for a year.

A garden on the roof of a building also adds a significant insulating factor, which keeps the building from heating up in the summer and losing heat in the winter, thus providing real energy savings. Furthermore, rooftop gardens will help moderate temperature extremes in the city and purify the air, as well as reduce the amount of waste that must be hauled to landfills by providing decentralized locations for composting organic wastes.

For suburban areas the collection of food wastes and the growing of suburban crops would be configured in a somewhat different fashion, Mankiewicz's colleague William Kinsinger suggests. Unlike the high-density urban area with a quarter square mile collection area and 3 miles of street on the collection route, the low-density suburban area with only sixteen households per block would comprise 5 square miles in order to be able to collect enough food waste to supply a 10-ton-per-day composting facility. The collection route in the suburban area would be an estimated 120 miles of street compared with a route of only 3 miles of street in the city. This longer street route might justify the use of large garbage trucks, Mankiewicz allows.

But where would composting facilities and gardening space be found in the suburbs? In most suburban areas there are shopping malls covered with huge, unused rooftops where composting food waste facilities and rooftop gardens could be sited, Mankiewicz points out. He envisions drawing a circle 2 miles in diameter around a typical shopping mall and establishing this as the collection zone. Food waste would be picked up in this area and delivered to a composting building that would be erected adjacent to one of the walls of a supermarket located within the mall. Within this structure, composting of the food waste would proceed in a fashion similar to that described above for the urban basement facility.

The only difference would be that instead of shipping the finished compost out to urban rooftops, the suburban compost would be taken to a greenhouse built on the roof of the supermarket where it would be used to grow produce. Once the compost is taken up to the roof by elevator it would be mixed with shredded post-consumer Styrofoam collected in the mall or on the shopping strip close to it. The Styrofoam would be shredded to make superlightweight soil. To maximize the growing surface, Mankiewicz would transplant seedlings into crop trays and place them in three-tiered hammocks suspended from the steel trusses that support the greenhouse roof. Supplemental, high-intensity, discharge lighting would be used to increase plant growth rates at off-peak hours when electricity could be purchased at reduced rates. Once the produce had grown to maturity, the trays could be taken down by elevator to the produce section of the supermarket where customers could literally pick their own produce, insuring its freshness and avoiding spoilage and refrigeration costs.

Growing fresh produce on the roof of a supermarket makes a lot of sense financially. Since much of the fresh produce sold in eastern cities in the United States comes from Florida, California, or Central America, significant savings can be realized on transportation costs alone. By growing food on the roof of a supermarket one eliminates interstate hauling charges, crating at the farm, local wholesale distribution facilities, sorting for spoilage, loading on trucks for local delivery, uncrating at the retail store, and a host of other expenses, Mankiewicz points out. Further savings are realized because intermediate profits are taken at each stage of this process. At the end of the chain Mankiewicz calculates that there is a 70 percent cost reduction on food that is grown on the roof compared with that shipped in over long distances. In addition, it is undeniably fresher when sold.

Although no one has yet attempted to implement this urban/suburban food waste to food production system on a large scale, it remains one of the more practical and well-thought-out options for making cities more ecologically sustainable. To prove that his composters will perform as advertised, Mankiewicz is starting small by installing a Department of Sanitation–funded, superefficient composting system for food wastes at the Brooklyn Botanic Gardens (BBG), the fourth most visited cultural institution in New York City. Food waste from BBG's palm house and cafeteria is shredded and then composted in one of Mankiewicz's composters. The composter, located in one of the BBG greenhouses, permits the 100,000 schoolchildren who visit the greenhouses every year to become familiar with this sophisticated composting process.

The in-vessel system at BBG, which reduces the volume of food waste by half within three or four days, has attracted the interest of restaurant owners and hospital food service operators who can see that virtually all the processing steps in food waste composting can be mechanized. Negotiations are underway with officials at one hospital who are interested in a combination food waste composting system and rooftop greenhouse, Mankiewicz reports.

In addition to providing highly nutritious food, grown without the use of pesticides or chemical fertilizers, Mankiewicz's urban rooftop gardens also provide urban dwellers with a whole new green environment within the city that they can visit and use as a refuge from the harsh reality of the streets below.

Imagine living or working in a building that has a farm on the top floor! The rooftop farm would be a pleasant place to visit where one could relax in a green open space. It would enable residents to show their children how food is grown. Some residents might even want to volunteer to work in the garden a few hours a week. Having a productive agricultural plot on a roof-top might also prove to be an excellent selling point for owners.

Transforming Chattanooga into an Environmental City

City Councilman David Crockett is cruising around downtown Chatta-nooga in his black Chevy Suburban whistling the theme song from *The Dukes of Hazzard,* a television series that featured good ol' boys in souped-up cars racing over the back roads of southern Appalachia. But it is with a wry smile that Crockett evokes this Southern stereotype: "Many people still think of us as hillbillies down here in eastern Tennessee and not as folks on the cutting edge of urban design and sustainable development."

Six feet 5 inches tall, flamboyant and outspoken, and a great-nephew of the famous frontiersman, Crockett is one of the better-known exponents of an environmentally based urban recovery movement taking root in a num-ber of communities around the country. What participants in this "sustain-able communities" movement share is a commitment to broad-based local participation and an ideal of environmental and economic health within the community.

Building on environmental breakthroughs of recent years in urban plan-ning and architecture, committed to the concept of "local empowerment," these community activists are engaged in efforts to preserve the environ-ment while keeping and creating jobs. The definition of local sustainability may vary from place to place and from person to person, but the ideal of taking a combined approach to solving environmental, economic, and so-cial problems has led to a catch phrase: a sustainable community is a place "where the pieces fit together."

Chattanooga is widely regarded as the granddaddy of this movement and Crockett is one of a determined group of residents who are seeking to re-make Chattanooga as America's premier "environmental city".

Already, Chattanoogans have made some real progress in pursuit of this title. Over the past two decades the combined efforts of thousands of

Chattanoogans have cleaned up much of the city's air pollution problem, manufactured one of the nation's largest fleets of electric buses, erected the nation's largest freshwater aquarium, constructed an extensive network of riverwalks and greenways, recycled old factories and houses to bring people back to a formerly abandoned downtown, and made plans for a zero-emissions industrial park. To realize these goals, Chattanooga launched a number of public-private partnerships for sustainable development that generated a total investment of $739 million, 1300 permanent new jobs, and 7300 temporary construction jobs.[1]

At a meeting of the President's Council on Sustainable Development held in Chattanooga in January 1995, Vice President Gore praised the city as a place that "has undergone the kind of transformation that needs to happen in our country as a whole." Extolling the green virtues of the city, Gore said that if the CEOs of Fortune 500 companies are looking for a community "deeply committed to sustainable development and a high quality of life, then Chattanooga is the place to come to."

How much of this hoopla about Chattanooga being the environmental city turns out to be real and not a lot of happy talk remains to be seen. To be sure, Chattanooga is no Ecopolis. *Money Magazine* ranked it 274th out of 300 in a 1994 "Best Places to Live" survey. Its industrial past has left a legacy of serious pollution problems, many of them borne by people of color; in 1991, one researcher found that a daycare center in a predominantly African American neighborhood was within 300 yards of six different toxic waste sites.[2]

Crockett himself is quick to point out Chattanooga's flaws. "Look at that," he says driving past a filthy creek that runs through a concrete culvert. "Do you call that sustainable?"

Yet Chattanooga's unresolved environmental problems make it the perfect "living laboratory" where new environmental initiatives, technologies, and designs can be tested, Crockett asserts. "Calling ourselves an environmental city does not mean that we are pristine: it is not a statement that we have arrived, but rather a goal, a journey," he explains. "We don't have all the answers, but at least we are asking the right questions. That is why we now talk about Chattanooga as a defining place for sustainability."

Chattanooga has not always been touted as the nation's premier environmental city. Twenty-five years ago the idea would have been laughable. At that time air pollution in the city's downtown district was so bad that residents sometimes had to drive with their headlights on in the middle of the day. The air was so thick with soot that it made smudge marks around the noses and mouths of pedestrians and turned white shirts gray after a brief walk. Some folks went so far as to store extra shirts at work to keep up appearances. Unfortunately, there were no equally simple solutions to the health damage the air pollution caused. Mortality rates from tuberculosis were triple the national average and in 1969 a report published by the federal Department of Health, Education, and Welfare awarded Chattanooga the distinction of having the highest level of suspended particulates in the air in the country.[3]

The source of this filthy air was not hard to trace. All along the banks of the Tennessee River, factories, foundries, and coke plants burned soft coal that was mined locally. Chattanooga, which is nestled in a bend in the Tennessee River at the southern foot of the Appalachian Mountains, has a pocket-like topography that creates thermal inversions, trapping the foul air above the city for days at a time. "From Lookout Mountain you are supposed to be able to see seven states, but there were days when we couldn't see seven blocks," Crockett recalls. "We were like a drug addict who hits rock bottom, OD's, and needs some rehab."

The cleanup of Chattanooga's air was crisis driven: everyone recognized that the status quo was intolerable, Crockett recalls. The 1970 Clean Air Act gave shape to the cleanup activities. To meet the health standards of the new law, the Chattanooga–Hamilton County Air Pollution Control Board required that air pollution control equipment be installed by local industries. To drum up support for this drive, both of Chattanooga's major newspapers educated readers about the need for stringent air pollution control measures. Owners of factories spent millions of dollars on pollution control equipment and as a result, by 1988, the city was "in attainment" for all pollutants regulated under the Clean Air Act in those areas that were monitored—a considerable accomplishment given that some 100 communities have yet to meet these same standards.[4]

The realization that Chattanooga could help revive its flagging industry by making a business out of cleaning up its environment was sparked by the fact that local industrialists spent $40 million on air pollution control equipment. Impressed by these figures, one local firm quickly geared up to capture some of this business by manufacturing the air pollution "scrubbers" that are installed in the smokestacks of factories.

The impetus to clean up Chattanooga also arose from a civic effort to pull the community out of the steep economic recession it suffered in the mid-1970s. During that period the outlook for Chattanooga's industrial sector was bleak as many of the city's older industries—particularly companies that manufactured carpets, appliances, and auto parts—laid people off when the home building industry slumped. Some of these industries relocated out of state or overseas, leaving factories rusting and abandoned along the banks of the Tennessee River. An exodus to the suburbs followed and large parts of the inner city were afflicted with unemployment and crime. "The city had a coronary and needed a quadruple bypass," Crockett recalls.

Recognizing that the city they loved was in critical condition, groups of Chattanoogans began to plan initiatives to revitalize the inner city. Crockett worked with a series of grassroots environmental organizations during the 1970s and 1980s, including Streams, Trails, and Rivers Trust (START) and Hixson Environmental Land Protection (HELP). It was during this period that Crockett, community activist Jerri Spring from the Guardians of Chickamauga Creek, and a wide spectrum of other organizers began work on the first integrated planning process in a high-growth suburban area near the city where development was spiraling out of control.

"We got together environmental groups, civic groups, schools, builders, realtors, mortgage bankers, hospital administrators, carpenters, plumbers, farmers, hunters—such an ungodly coalition of folks that the local government eventually had to listen to us," Crockett recalls. The planning group looked at road carrying capacity, the siting of schools, transportation patterns, and zoning issues. Ultimately, the integrated plan was adopted, a large farm near the city was preserved as open space, wetlands were saved, and plans to shave off the top of a mountain for a development were halted.

But this was just the beginning of a long campaign to revitalize the city, preserve natural resources, and reverse the stream of people moving to the suburbs that was causing the inner city to wither. Instead of hiring expensive consultants to advise them what to do, Crockett and others did the obvious thing: they questioned those who had moved to the suburbs about why they had abandoned the city and what they liked about their new neighborhood. Almost universally they were told that the suburbs were cleaner, greener, and safer than the inner city. So a decision was made to try to put those values back in the city, says Crockett. He even made up a slogan to capture the gist of what they wanted to achieve in Chattanooga: Put the Country Back in the City and Save the Country.

Putting the country back in the city required reconnecting people with nature and specifically with the Tennessee River. For too long residents had been cut off from the river by abandoned factories and private land enclosed by chainlink fences. "At that time most people looked at city creeks as urban sewers and the closest they got to them was driving down the highway at 60 miles an hour with their windows rolled up," Crockett says. To change all this the city began an ambitious program of downtown revitalization that involved building a riverwalk where people could stroll, jog, bike, fish, or picnic. While this was no revolutionary environmental project, it did seem to propel Chattanooga into making a series of incremental changes that have improved the quality of urban life and brought residents closer to nature.

The vision of "bringing the country to the city" and positioning Chattanooga to attract environmentally related industry was not fashioned by Councilman Crockett and a cabal of backroom politicians. Rather it was modeled on the experiences of Indianapolis, Dallas, St. Louis, and Minneapolis, all of which had invigorated inner-city redevelopment initiatives by giving residents a voice in deciding what their city should look like. Building on this prototype, in 1984, a nonprofit organization called Chattanooga Venture, initially funded by the Lyndhurst Foundation, launched a project known as Vision 2000. The project brought together thousands of Chattanoogans from all walks of life to build a consensus about what the city could be like at the turn of the twenty-first century.

"We invited the whole community to bring their ideas about how to change the city," says Eleanor McCallie Cooper, executive director of Chattanooga Venture from 1990 to 1992. "Before that people felt as if decisions about the city were made by a small elite. So when we said everyone can contribute to shaping the city's future, suddenly a whole lot of people with energy and good ideas came into the process." What was most remarkable about this exercise in participatory democracy was that it made residents realize that if they wanted a better urban environment they would have to do the heavy lifting themselves.

No idea was dismissed as too trivial to consider. For example, in discussions about how an area of the city should be reconfigured, one local resident asked where she was going to be able to buy diapers under the new plan. This question prompted a productive discussion about how to create a mixed-use neighborhood with a good balance between commercial and residential elements.

"Chattanooga Venture provided a safe place for people to brainstorm about how to change the city and everybody parked their turf battles at the door," says Geri Spring, now coordinator of the Chattanooga/Hamilton County Neighborhood Network, one of a number of neighborhood groups that grew out of the visioning process. The brainstorming drew some 1700 people into a twenty-week series of visioning sessions. These meetings generated thousands of ideas and produced a consensus for a cleaner, greener, safer city with rehabilitated housing and nonpolluting jobs.

The raw material from these visioning sessions was then boiled down into thirty-four concrete goals. City officials, chamber of commerce members, and participants in Chattanooga Venture then recruited investors for the public-private partnerships that were to turn these goals into reality. In all, 223 projects arose from this exercise, including support for the construction of the Tennessee River Park, the Tennessee Aquarium, the Bessie Smith Hall, renovation of the Walnut Street Bridge and Tivoli Theatre, and a commitment to upgrade all substandard housing in the city.[5] Then Governor Lamar Alexander helped channel $9 million in state funds to help pay for these projects. Vision 2000 proved so successful—85 percent of the goals were met by 1992—that in 1993 the community went through the process again in Revision 2000 during which another 2500 ideas about how Chattanooga could change were solicited from residents.

"If anyone is looking for a model of how to build a sustainable city and they say let's get the mayor together with eight other people, decide on a plan, and execute it on the count of three like a football play—they will find it won't work," Crockett warns. Nor will simply organizing a visioning process in which you gather the residents of the city together. "If you just call everybody to a meeting to put together a plan, people will be running out of the building in ten minutes and there will be fist fights in the alley."

Building a movement to fashion Chattanooga into "a defining place for sustainability" took several decades, Crockett says. Everyone had a role and everyone had to do his or her part to make it work; leadership came from all quarters. For example, there was an old fisherman named Hubert

Fry who wanted kids to be able to get down to the river to fish. People worked on that idea for a long time and eventually the concept was captured in the visioning process and mechanisms were put in place to turn it into a reality. In a similar fashion, a couple of design students thought up the idea for the aquarium and then it was picked up by other people who made it happen. "The process works like a big ecosystem: a general theme is set and then all kinds of groups go off on their own to save a gorge or lobby for a greenway," he observes.

One of the most important lessons that can be drawn from Chattanooga's experience is that haranguing people about the need to act in a more ecologically sensible manner will not work by itself. Advocates for a riverwalk or preserving a creek can talk endlessly about the need for green space, but unless they start from the perspective of the people they are trying to convince they will never get anywhere, Crockett asserts.

"You have to start where people are," he explains. "A high-school superintendent doesn't wake up in the morning thinking about sustainability and the environment. He is thinking about how to improve the school system and get test scores up in eight subjects. So you have to start where he is at and ask him what *his* priorities are. Then, if you are a good salesman, you may be able to convince him that using environmental themes throughout the curriculum and getting the kids out of the classroom applying principles of sustainability in the community has the potential to make learning more exciting and result in higher test scores."

In the evolution of his own thinking, Crockett recalls a moment on a flight coming home from San Jose, California, when it came to him how Chattanooga might be able to position itself to snag some of the billions of dollars spent on environmental remediation and environmental goods and services. From his work with IBM he recalled how Research Triangle Park in North Carolina had positioned itself to capture much of the computer business by developing graduate programs in computer technology at a number of local universities. The city then carved out an industrial zone that could take advantage of the trained, computer-literate labor pool. The strategy worked famously and everybody who was anybody in the field of computers opened up shop there. Why not position Chattanooga to follow a similar pattern in the field of environmental sciences?, Crockett asked.

Figure 12.1
Councilman David Crockett, a persuasive advocate for the greening of Chatta-nooga. Tennessee. Photograph courtesy of David Crockett.

Armed with a strategy that called for using the environmental city concept as a common theme for Chattanooga, Crockett and others brainstormed about how to sell the idea. It would not be enough just to get a bunch of environmental groups to champion the idea, Crockett realized, because then it would be seen as a soft social issue. "It was no accident that we concentrated on the hard-core business community first," Crockett says. He set about selling the business community on the idea that the next megatrend will be industries that use energy and resources efficiently and that eastern Tennessee could become the Silicon Valley for cutting-edge, environmental technologies. Gradually, people in business began to see that this was more than a do-good social issue and that the environmental city theme could lead to economic revitalization.

To make the strategy work, the academic community also had to be convinced to train local students to fill environmental technology positions. Chattanooga State College, a two-year program, rose to the challenge by establishing an Environmental Science and Technology Division where some 350 students are being given hands-on experience learning how to control chemical spills and leaks at their environmental test facility. The school offers students a course in hazardous materials technology that prepares them to work as environmental technicians in cleanup actions or doing compliance work. The possibility is being explored of establishing a training area at the old federal Volunteer Army Ammunition site, one of the most polluted sites in Chattanooga, where both toxic and explosive materials are commingled in the soil. The University of Tennessee at Chattanooga is also starting a Ph.D. program in environmental engineering in its Department of Environmental Studies.

Improving the environmental quality of life in Chattanooga will give it a reputation as an environmental city, Crockett argues, which in turn will attract businesses that want to be known for cleaning up the environment and their commitment to operating in an environmentally responsible manner. There are a number of reasons why advanced environmental technologies might choose to establish a plant in Chattanooga. For example, the city is home to the Tennessee Valley Authority's sophisticated water quality laboratory and already there are efforts to organize a freshwater research consortium. The Electric Vehicle Test Track facility in Chattanooga might also become the hub for a number of electric car companies. An agreement between the city of Chattanooga and the National Laboratories to test some of their advanced technologies in an urban setting may also spawn energy and resource–efficient technologies.

A strategy has also been devised to bring advanced environmental technologies to the city through a series of conferences. For example, a conference on zero-waste and zero-emissions technologies took place in May 1996. Among those attending were representatives from Dow, DuPont, Monsanto, General Motors, Chevrolet, Cannon, BMW, Volvo, a number of Japanese firms, and other "wild-eyed idealists," Crockett observes. "Some of these folks have already figured out how to recover every single material out of a car seat. That's the kind of discussion we want going on here in Chattanooga."

But it has not always been easy to convince business leaders that the city should climb on the environmental bandwagon. Some of them initially feared that doing business in an "environmental city" would mean complying with more rules and regulations. One official in the city administration is reported to have said, "When Crockett first started talking about Chattanooga as the environmental city I thought he had been smoking dope."

The president of the local chamber of commerce, James G. Vaughan Jr., however, argued that the environmental city is one with new jobs and new business opportunities. In the contest to attract business, he says, Chattanooga's interest in "the business of the environment" gives it a niche—"a way to market our community for business relocations, for investments, and as a site for conferences."

Convincing teachers, business owners, politicians, landlords, municipal employees, and other local residents that an improved city environment could benefit them and should be made a top priority required a lot of work and endless jawboning, but these efforts have not gone unnoticed or unrewarded. Chattanooga was the first North American city selected as a "best practices" case study for the international Habitat II conference held in Istanbul in 1996. "Chattanooga, Tennessee, may boast the greatest political outcomes of a citizen involvement process," write Francis Moore Lappe and Paul Martin Du Bois.[6]

On the ground, the tone for the downtown development effort was set in 1985 with the construction of the Riverwalk. This 5-mile-long riverfront park was built on the site of an industrial wasteland. Planned ultimately to extend 22 miles along both banks of the river, the Riverwalk is a key element in a more ambitious plan to weave a 75-mile network of greenways (long narrow parks) throughout the city.

Another hugely successful downtown initiative was the Tennessee Aquarium, which houses the largest collection of freshwater fish in the nation.[7] Opened in 1992, the aquarium was built at a cost of $45 million from private investors; $10 million more was raised from the city and state for the plaza and river park that surrounds it. Chattanooga is an appropriate place for this facility because Tennessee is host to a greater diversity of freshwater species than any other state in the union. The aquarium, heralded as "Chattanooga's cathedral of conservation," celebrates this rich biological endowment while explaining the rationale for protecting freshwater resources from pollution.

While early critics argued that exhibitions of freshwater fish without exotic species from abroad would prove too boring to attract much of an audience, some 1.5 million visitors in the first year proved them wrong. The aquarium also contributed an estimated $133 million to the local economy.

Figure 12.2
Overview of Riverpark, which provides residents of Chattanooga with access to the Tennessee River. Photograph by Bob Boyer.

"Five years ago you would have driven through this area where the aquarium is now located with the windows on your car rolled up and the doors locked," recalls Crockett. "Now it is a tourist attraction." In fact, the aquarium became a catalyst for revitalizing a significant stretch of the downtown waterfront.

Nearby, the 104-year-old Walnut Street Bridge, once scheduled for demolition, was recycled into the nation's longest pedestrian bridge over water. Spanning the Tennessee River, the 2370-foot suspension bridge provides a vital link to the north side of Chattanooga.[8] The pedestrian bridge not only offers a spectacular overlook of the river, it also demonstrates a commitment to shield parts of the city from the negative impact of vehicular traffic. Other cities might have scrapped the old bridge and built a modern replacement carrying a major highway into the heart of downtown, but designers at Chattanooga's Riverfront/Downtown Planning & Design Center had seen other downtowns "blown out" and turned into concrete wastelands by such construction projects.

Instead of blasting out the downtown area with a major freeway, a slower-paced, inner-city atmosphere was enhanced as fountains were in-

Figure 12.3
The Walnut Street Bridge was converted from vehicular to pedestrian and bicycle traffic.

stalled, pocket parks planted, and the railway station renovated. Rather than tearing down dilapidated buildings, great pains were taken to renovate the old Tivoli Theater and other historic structures. Old warehouses were made into an attractive mall that is surrounded by outdoor cafes and restaurants. Instead of looking for a big hotel chain to build some huge, incongruous edifice, city planners encouraged the renovation of small hotels, inns, bed-and-breakfasts, and restaurants around the train station and convention center. To entice people back to the city's heart, attractive streetlamps and benches were installed, banners and new signs were hung, street vendors and musicians encouraged, and outdoor musical events held on weekends.

Moreover, Chattanooga's planners sought to reduce the environmental damage done by downtown development. For example, public and private funds were raised to rebuild the sidewalks with beveled paving bricks that permit rainwater to be absorbed into the ground, thus limiting the amount of stormwater runoff the city's sewer system must handle. Trees were also planted along the streets and around surface parking lots to provide shade, clean the air, and moderate temperatures.

Another of the more significant environmental achievements in Chattanooga to date has been the purchase by the city of a fleet of electric buses and the nurturing of a local electric bus manufacturing company. To reduce vehicular traffic, eliminate sources of air pollution, and make the downtown district easily accessible by public transit, city officials at the Chattanooga Area Regional Transit Authority (CARTA) put out a bid for twelve battery-powered electric buses. The bid was won by a local startup firm called Advanced Vehicle Systems Inc. (AVS), which built the six 22-foot electric buses that now ferry passengers for free around a 4-mile loop through downtown Chattanooga. The buses, which arrive quietly and without exhaust fumes every five minutes, are financed with the revenues from a multistory garage located near the Chattanooga train station. In 1993 the electric shuttle ferried 1 million passengers around the city.

A prototype of a second model electric bus, measuring 31 feet, is now being tested on Chattanooga streets, says AVS president Joe Ferguson. He expects to build a total of sixteen electric buses for Chattanooga by the end of this contract and looks forward to building more. CARTA just received federal financing for an additional ten buses, he notes. Chattanooga and Santa Barbara are currently the cities with the largest fleet of electric buses

Figure 12.4
View of Broad Street, which has been landscaped with shade trees and beveled paving stones that permit rain water to soak into the ground. Photograph by Bob Boyer.

in the nation and once all the contracted AVS buses have been built, Chattanooga will have bragging rights as home to the largest fleet.

Ferguson distinguishes his electric buses from those of other companies that convert diesel buses to battery power. The converted buses are too heavy to be efficient, argues Ferguson, who believes in building electric buses from the ground up. The curb weight (weight when empty) of an AVS electric bus is about 14,000 pounds compared with a comparable 20,000 to 24,000 pounds for a converted diesel bus. "We are building an antelope to be an antelope, not taking an elephant and trying to turn it into an antelope," he says.

Furthermore, the economics of electric buses are attractive. While an electric bus sells for about the same amount as its diesel equivalent, Ferguson explains, the maintenance costs on electric buses are roughly half those of comparable diesel buses; and they can run for 6 cents per mile compared with 16 cents a mile for diesel, he adds. In addition to supplying Chattanooga's needs, AVS is now shipping electric buses to the historic districts of Savannah, Georgia, Charlotte, North Carolina, and Anderson, Indiana. Ferguson also sent some of his electric buses to the Olympics in Atlanta and to the Habitat II conference in Istanbul.

This double-win scenario of solving urban environmental problems (vehicular congestion and pollution) while providing inner-city jobs (building electric buses) is not accidental. During the period that Chattanooga was cleaning up its air, local entrepreneurs found they could make money manufacturing air pollution control equipment. Now AVS has hired thirty-five employees to make electric buses. This is quite deliberate. Chattanooga wants to be a city that gains a reputation for both practicing sustainable development and attracting environmental business.

To capture more eco-dollars, city officials and urban planners now have ambitious plans on the drawing boards to transform Chattanooga's south central business district (a blighted industrial area) into an environmentally advanced commercial development. This blueprint, prepared for River Valley Partners by William McDonough & Partners and Calthorpe Associates, is probably the most ambitious set of plans in the country for transforming a polluted inner-city industrial area into an environmentally friendly neighborhood.[9] The plans were developed through a consensus process of some 130 "stakeholders," including local residents, commercial property owners, city officials, urban planners, and architects.

The design envisions a mixed-use neighborhood to include a sports stadium, conference center, eco-industrial park, business incubator, residential housing, greenway, and electric shuttle buses. It calls for gradually piecing together a neighborhood in which a residential area would be placed adjacent to an eco-industrial zone with the intent to create a zero-emissions industrial park that would permit people to live near their place of work. Such an arrangement would also permit a mix of residents with varying incomes to live next to one another.

Nancy Jack Todd and John Todd have anticipated this mixed-use approach to urban planning: "Building and architectural forms can be created in which living, manufacturing, food growing and processing, selling, banking, schooling, waste purification, energy production, religious activity, art guilds, governance, and recreation are woven together on a neighborhood scale." With this mixed-use design "a neighborhood could function in a manner analogous to an organism."[10]

In most major urban centers today, the core of the city is deteriorating as more and more people flee to the suburbs. Instead of investing in the renovation of the inner-city infrastructure, new roads, sewer lines, transportation systems, and schools are built out into these suburbs at enormous expense. "The growth patterns in many cities today look like the rings on a tree," Crockett notes. "People who live next to each other look pretty much the same, make the same amount of money, drive the same cars, and hold the same values. This stratification of society makes it hard to forge a political consensus on anything."

Many of these problems could be solved through a deliberate attempt to design mixed-use and mixed-income neighborhoods, Crockett believes. However, convincing local businessmen to buy into the idea of a zero-emissions industrial zone, which makes a mixed-use neighborhood possible, is a hard sell. So Crockett invited eco-architect William McDonough to make the pitch, but warned him that the idea would not go over well at first: "I told him that it would be like teaching a mule to play the violin. It will sound terrible and the mule won't like it," Crockett recalls.

But McDonough made a strong case for zero emissions. He argued that while the zero-emissions requirement sounds radical at first, when looked at in the context of other industry goals it is a natural evolution. After all, industrialists adopted the goal of "zero accidents" with the motto Safety First; "zero defects" was popularized when companies like Ford came out with the slogan Quality is Job One; and more recently companies are requiring "just-in-time" delivery of components to their assembly line, which is essentially a "zero-warehousing" policy. Zero emissions is simply the next step and companies that adopt it first will be the big winners, McDonough contends.

To reach zero emissions, McDonough has worked successfully with a number of industries redesigning their product and process so that they don't produce anything that is not compostable or recyclable. For example, the carpet manufacturers Collins and Aikman now print an 800 telephone number on the back of their carpets guaranteeing that they will take them back after their useful life and recycle them into parking stops, highway guardrail offset blocks, and soundwall barriers. Officials at Collins and Aikman say they are willing to give the zero-emissions industrial park a try, McDonough says. The next step will be to convince local authorities to pass zoning laws for the eco-industrial zone that require zero emissions.

Companies targeted for this eco-industrial zone or zero-emissions industrial park would agree that either their wastes become the feedstock or fuel for another manufacturing process or be piped to an ecology center where they would be detoxified using natural processes. John Todd at Ocean Arks International has demonstrated that water tanks filled with plants, algae, microorganisms, and fish can be used to detoxify municipal sewage. A similar system is being considered to treat the wastes generated in the eco-industrial zone.

If companies can be enticed to locate in the eco-industrial zone it will permit residents to live in adjacent areas without suffering the ill effects of pollution. This will allow people to walk or bicycle to work instead of enduring a lengthy and polluting commute. They will also be able to catch an electric shuttle bus to other parts of the city. And a greenway, planned to intersect the south central business district, will provide a natural area for recreation and a nonvehicular pathway to other parts of the city.

To jump-start this urban renewal project, there are a number of catalyst projects planned. One is an expansion of Chattanooga's trade center using a green design. McDonough would like to see a trade center with grass and trees growing on its roof and other environmental features that would attract conferences on environmental issues.

While plans for the ecology center are still being formulated, Todd's organization, Ocean Arks International, would like the center to serve a number of purposes. [11] Todd wants to begin by detoxifying some of the contaminated soil that surrounds factories in the south central business district. To do this he would construct a sealed yet movable greenhouse on an eighth of an acre of contaminated soil. Inside the greenhouse he would rototill the soil repeatedly and add microorganisms and minerals to it to help speed nature's ability to purify itself. Off-gases from the contaminated soil would pass through an "earth filter" that would trap contaminants. After the soil was decontaminated using this process, the greenhouse would be moved onto the next area of contaminated soil where the process would begin again.

Another catalyst project in the south central business district is a plan to build a 30,000- to 35,000-seat stadium. In other cities the building of a stadium often overwhelms neighborhoods by occupying huge areas with surface parking lots. To avoid this, planners have proposed parking lanes on either side of broad, tree-lined boulevards leading up to the stadium, as well as special "parking streets" where cars can pull off the road onto a meridian paved with a honeycombed "grasscrete"—a concrete-like material with holes in it that grass can grow on.

There are also plans to plant an urban forest along the now vacant highway right of way that borders the south central business district, an area that could be a nursery for appropriate species of trees that could be transplanted on city streets, greenways, and pocket parks. One of the main streets could be placed off-limits to vehicular traffic and planted densely with trees to give people a sense of living in an urban forest, suggests McDonough.

Finally, there would be an area designated as a business incubator facility where startup enterprises would be provided with cheap rent, shared office support facilities, and business counseling.

If Chattanooga urban planners pull off their green plan for the south central business district, as they did the revitalization of the riverfront area of downtown, it will secure Chattanooga's reputation as the nation's premier environmental city. Already the renovation of some old hotels has started, low- and medium-income housing is in the works, funding for the stadium and greenways has been secured, parking lots are to be made of grasscrete, the Ross-Mehan factory is to be retrofitted as an environmental center, and a sales tax will help fund some of the other projects.

Yet, while plans for the south central business district are well advanced, finding the political will to carry them out remains an uphill struggle. It is easy to get carried away by plans to turn a city into an environmental Ecotopia without acknowledging the real fiscal and political constraints. Certainly, the meeting of the President's Council on Sustainable Development gave city leaders a sense that it was possible to garner considerable attention for the city by making it a model of urban environmentalism. But whether or not the enthusiasm will last over the years it will take to build the urban revitalization project sketched above remains to be seen.

Not surprisingly, the Chattanooga environmental city concept has its critics. "I think Chattanooga wants the image of being an environmental success story, but I have my doubts about how serious city officials are about doing it. They are talking this really good talk about making an environmental city, but it will be up to citizens and environmentalists to make them walk the walk," says John Johnson, a twenty-four-year-old part-time construction worker who belongs to the local chapter of Earth First!

Denny Haldeman, age forty-two, an environmental activist, carpenter, and organic farmer who was a founding member of Earthworks ("Chattanooga's All-Purpose Cleanser"), says it was the grassroots activist community that pushed civic and business leaders to put Chattanooga on a green trajectory. Had local greens not blocked construction of an incinerator, the curbside recycling program Chattanooga is so proud of today would not have been possible, he asserts.

By making room for community-based recycling in Chattanooga, local environmentalists made it possible for Orange Grove, a nonprofit that provides training for mentally handicapped adults, to process Chattanooga's recyclables. Instead of building a $10 to $20 million mechanized system that would have employed ten people at the landfill to process recyclables, Orange Grove built a $2.5 million plant in the center of the city that is operated by 110 mentally handicapped persons. This deliberately labor-intensive approach provides residents at Orange Grove with dignified work processing 400 tons a month of plastic, glass, aluminum, and paper. Since the recyclables are separated carefully by hand, the recycled product is of high quality and can be sold at higher rates than less well-sorted materials.

It was also grassroots environmental groups that initiated and provided much of the people power to carry out an annual cleanup of the Tennessee River. "The TVA people acted like God by lowering the level of the river, but we were the ones in the muck hauling 89,000 pounds of trash in an afternoon," says Haldeman.

Local and regional environmental groups also led the fight against the chip mills that would have wiped out the hardwood forests surrounding Chattanooga if they had been granted a permit to operate. A coalition called Tennesseans, Alabamans, and Georgians for Environmental Responsibility (TAGER) organized a network of people opposed to the chip mills and only subsequently did the business and civic leaders in Chattanooga climb on the bandwagon, Haldeman says.

Crockett has no quarrel with this reading of history. As a former grassroots activist, who later ran for the city council on an environmental ticket, Crockett knows well the essential role the small nonprofits played in the greening of Chattanooga. "Denny Haldeman sounded the klaxon on the chip mills; he was the guy who stepped up and said these chip mills are a dumb idea, and he woke me up to the importance of this issue," Crockett says. Another local activist, Bob Mitchell, a high-school teacher, then began a campaign against the chip mills, Crockett recalls.

"But what is unique about this city is that Haldeman and Mitchell's campaign was embraced by the whole community. That is something I will always be proud of our community for," Crockett says. After the initial wake-up call, the business community saw that creating a few jobs while despoiling a significant portion of the local environment did not make sense. The tourism board, chamber of commerce, city council, county commission, Kiwanis club, hunting and fishing clubs, and environmental groups all joined in opposing the chip mills. Other cities that had not spent as much energy debating how to improve the city's environmental profile would not have had such unanimity on this issue, Crockett maintains.

Of course, not all environmental groups are thrilled by the "environmental city" theme. Some dismiss it as a "greenwash" in which city officials and business groups are talking a good game about the environment while doing very little about it. "But I tell them, listen, not everyone involved may be a candidate for the John Muir Award, but I'm trying to find a downside to everyone in town talking about sustainability and the environment. Somehow I think this approach is going to work out," Crockett remarks.

There remain other parts of Chattanooga, however, particularly some of the old industrial districts, where residents face a Herculean task in terms of environmental cleanup. Best known is the case of Chattanooga Creek, which runs through Alton Park and Piney Woods, two predominantly African American communities. A 2-mile segment of the creek has been designated a federal Superfund site and it is estimated that there are some 2500 cubic yards of sediments contaminated with coal tars and other toxic substances in the creekbed piled 8 feet deep in some places. In addition to the contaminated sediments, some forty-two sites in Alton Park and Piney Woods have been identified as potentially contaminated.

It took residents decades of activism simply to attract attention to this problem. It was only when students at the University of Tennessee at Chattanooga (UTC), members of environmental groups, and community activists banded together to form Stop Toxic Pollution (STOP) that they began to get a hearing, says Milton Jackson, one of the early local organizers. By then, word that Chattanooga Creek was heavily contaminated had spread beyond the borders of the city. "I remember when I drove down to Houston, Texas, and a man saw my license plates were from Tennessee, he asked if I lived in the city with the bad creek running through it," Jackson says.

To date, few practical steps have been taken to remedy the problem. The state has started to clean up twelve of the forty-two contaminated sites in the area, but federal officials expect at least two more years of studies before cleanup begins at Chattanooga Creek. "I tell people they spent 75 years messing the place up, give us at least five years to clean it up," says Wayne Everett of the state division of Superfund.

But to Peter Montague, director of the Environmental Research Foundation and coauthor of a 1991 Greenpeace report on environmental problems in South Chattanooga, the state of Chattanooga Creek is an indictment.

"It is a bad joke that Chattanooga is being held up as a model environmental city," says Montague. He argues that the city's behavior is consistent with a pattern of environmental racism documented elsewhere in the country. "It may be true that they are cleaning up the air where white people live in the city, but I find no evidence it is true for the African American population of Chattanooga," Montague asserts.

"In some ways this is a case of environmental racism," agrees Tommie Ashford, president of the Alton Park and Piney Woods Community Coalition. "In the beginning, city officials turned a deaf ear to our problems. But then when UTC students got involved—and they were all white—suddenly we began to get some attention and city officials started coming to our meetings."

As environmentalism has become the keynote of the city's public image, Chattanooga Creek has attracted still more attention. New warning signs have been posted along the creek. Mayor Gene Roberts, Crockett, and others have donned protective clothing and held a cleanup session. The city arranged a seminar on the creek for the meeting of the President's Council on Sustainable Development with speakers from Alton Park and Piney Woods. Denny Haldeman suggests a still more public approach: while residents wait for the cleanup to begin, the city could establish an educational "toxic tour" of the creek, with signs describing what is in the sediment, how it got there, and what can be done to clean it up.

Ultimately, South Chattanooga may prove to be the most serious test of the city's environmental commitment. Publicity is a start, but far more is needed. "We don't want another Love Canal, but that is what it might lead up to," says Tommie Ashford. "We live in a beautiful community—but now South Chattanooga is an area where people are sick and dying and the doctors don't know, I think, what is really wrong with them."

While Chattanooga clearly has a long way to go to clean up its toxic legacy, there are a number of signs that the city government is taking steps to keep the city from being further polluted. Tom Scott, manager of Chattanooga's stormwater system, has already installed over a hundred oil skimming devices on parking lots and commercial properties around the city and plans to require the installation of another 2000 of these devices. The stormwater skimmers capture oil derivatives and other floatable matter in a sludge which is periodically pumped into trucks for removal to the landfill. In addition to capturing oil sludge that otherwise would have gone into the city sewer system, Scott has created a business opportunity in Chattanooga. When he first started installing the skimmers, the hardware alone cost $3600 each, he explains, but now a local sailboat manufacturer produces them for $300 each.

If Scott encounters recalcitrant property owners unwilling to meet local stormwater abatement regulations he can always send them to the environmental court run by Judge Walter Williams at the Chattanooga City Court. Seeing that many environmental crimes and code violations were going unpunished because they were crowded out by crimes against people and property, Judge Williams instituted an environmental docket where environmental cases are heard every Thursday. Williams also organized an "E-team" made up of representatives from the police, building inspectors, health inspectors, stormwater office, and utility companies that meets regularly to discuss problem properties. Building code inspectors and others find that infractions are being remedied more quickly now that property owners know they will be taken to a court that takes these cases seriously, Williams reports.

City planners also recognized that in addition to improving the urban environment there had to be a simultaneous effort to meet human needs. One of the most significant outcomes of the public visioning process is a public-private partnership called the Chattanooga Neighborhood Enterprise (CNE), which has invested $60 million in affordable housing since 1986. Working with federal funding agencies, the city public works department, and local banks, CNE has helped 1000 first-time home buyers, made affordable loans to 1300 low-income families for household repairs, and produced 300 new rental units. CNE also purchases dilapidated rental units from absentee landlords, rehabilitates the buildings, and sells them back to renters for a mortgage payment no higher than the rental rate.

CNE's investment in inner-city housing is visible in Lincoln Park, a traditionally African American neighborhood where the Negro National League's Nashville Elite Giants once played exhibition games before the neighborhood infrastructure began to crumble. Residents have now banded together to restore the quality of community life and CNE has helped. "Not only did CNE rehabilitate many of the houses in Lincoln Park, it also helped repave our streets and got us sidewalks, which we had never had before," observes Bessie Smith, a resident of Lincoln Park since 1934 who has been part of a resident-led movement to improve the quality of community life.

"Neighborhoods can be kept alive and well—and I guess that is what sustainability is all about—if the people who live there are willing to fight for them," Smith says. "But people cannot always do it on their own. They need an organization like CNE to help them repair older homes, build new homes, and make homes affordable for people without a lot of money. . . . It is a lot easier to feel proud and work hard to make our neighborhood clean and safe if you have decent housing that people are proud to call home."

City officials are also using every tool at their disposal to educate residents about sustainability. Some of John Todd's "living machines" that purify sewage using plants and microorganisms are in the high schools. And along the Riverwalk there are educational displays such as that built by the Tennessee American Water Company that show people where their drinking water comes from and how it is purified.

On a midweek morning in October at one of the educational pavilions located on the Riverwalk, about a hundred school children romp in a playground despite a light drizzle. Teachers siphon off small, manageable groups of students to visit an educational resource center where they learn about the different varieties of trees, fish, and birds that live in and along the river. Amid this youthful cacophony, adult inner-city residents can be found baiting their hooks with grasshoppers in hopes of catching dinner off a fishing pier built near a huge TVA dam.

Surveying this oasis of urban sanity Crockett stands above the fray smiling: "I can't define sustainability, but I know it when I see it. When you hear laughter in a public place like this you know something sustainable is going on. People having a good time in the city while surrounded by nature: that's sustainability."

Crockett has a theory about the behavioral importance of reconnecting urban residents with nature: "We all get a little weird when we are cut off from nature too long," he says. "Take an inner-city kid off the asphalt and put him in a more natural setting and you won't recognize him in two weeks," says Crockett who is proud of an urban farm the city established in one of its greenways where kids from public housing projects can enjoy and learn about nature.

While friends kid him that he always has an "idea de jour" about how to make Chattanooga the nation's leading environmental city, Crockett is unapologetic about the schemes he floats. For example, he wants to convince city and TVA officials to dig ponds around Chattanooga and stock them with fish. At these ponds only kids would be allowed to fish; adults would have to bring a kid with them before being permitted to drop a hook in the water. It would be part of our "Get Hooked on Fishing, Not on Drugs Campaign," Crockett says. He also wants to get a local seed company involved in selling the seed of native plants so that landscaping can be done with plants that come from the region. But while Crockett dreams of urban fishing and native plants, there are already enough accomplishments on the ground in Chattanooga to make it a remarkable example, despite its flaws, of urban sustainable development.

It is deeply moving to watch residents of the city enjoy a great river that runs through their community after having been cut off from it for decades. One cannot help but feel that this is the way a city is supposed to be after wandering along the Riverwalk and seeing schoolchildren, joggers, bicyclists, lovers, people fishing, and birdwatchers. Many of us carry a carefully concealed burden of grief over the incremental destruction of nature that we have witnessed in our lifetime. So when one comes across an example of how a city such as Chattanooga can be designed to

Figure 12.5
Chattanoogans making use of improved access to the Tennessee River. Photograph Bob Boyer.

work with nature, suddenly a sense of hope bubbles up. If there is any lesson to draw from the Chattanooga experiment it is that the phrase "environmentally friendly city" is not necessarily an oxymoron. As a member of the President's Council on Sustainable Development said as he left the meetings there, "Let a thousand Chattanoogas bloom across the nation."

Redesigning Buildings and Building Materials for
Environmentally Intelligent Architecture

William McDonough, dean of the School of Architecture at the University
of Virginia, likes to contrast two building technologies: the Bedouin tent
and the urban skyscraper.

Consider the Bedouin tent, says McDonough, age forty-five, a delicate-
boned man with a fondness for bow ties who is one of the nation's foremost
eco-architects. At first glance the Bedouin tent appears primitive, but closer
examination reveals it to be an elegantly designed, sophisticated piece of
equipment that achieves its purpose with a minimum of materials. The tent
provides shade from the sun, while the loose weave of the fabric permits
desert breezes to penetrate the interior and cool its occupants. At the same
time, the material allows filtered sunlight to illuminate the interior with a
hundred thousand points of light. When it rains the fibers swell up until the
tent fabric is tight as a drum. Furthermore, the tent is portable: it can be
rolled up, packed onto camels, and re-erected elsewhere. Finally, when the
tent wears out, it can be patched or left to compost without doing harm to
the environment.[1]

By contrast, consider the modern skyscrapers that dominate the urban
landscape today, continues McDonough. Built of concrete, steel, and glass,
these towering buildings are typically gas-guzzlers of the first magnitude.
The manufacture of both concrete and steel requires large amounts of energy, and the expanse of glass called for in the construction of these high-
rises can condemn the buildings to excessive rates of heat loss in winter and
heat gain in summer, requiring massive amounts of energy to keep interior
temperatures comfortable.

While the skyscrapers look impressive, their design is less sophisticated
than the Bedouin tent, McDonough says. They fight the environment rather

than take advantage of natural forces. "Our culture has adopted a design stratagem that essentially says that if brute force or massive amounts of energy don't work, then you are not using enough of them," says McDonough, who is founder and principal partner of William McDonough & Partners in Charlottesville, Virginia.

Most architects use three basic design criteria: aesthetics, performance, and cost, he explains. But given the significant harm to the environment that modern building techniques wreak, these criteria are no longer adequate, he argues. To deal with the threat to nature posed by the modern design of the built environment, McDonough offers two additional criteria: buildings must be designed to be environmentally intelligent and built of materials that are produced in a socially just or nonexploitative fashion.

The phrase "environmentally intelligent design," which McDonough coined, follows three basic principles. His first principle is that "waste equals food," a concept that comes from a close study of nature where the waste of one species provides nutrients for another. For example, using this principle, the graywater from a building's plumbing system should be used to water landscaping plants. His second principle is that buildings should take advantage of "current solar income" to power the lighting, heating, and cooling of a building's interior, rather than depending on fossil fuel or nuclear generated energy. His third principle is that all building techniques should respect and protect biological and cultural diversity. Making his work even more challenging, McDonough seeks out building materials that are free of carcinogens, mutagens, bioaccumulative and persistent toxins, heavy metals, and endocrine disrupters.

While these building principles sound impossibly idealistic to some, McDonough argues that it is possible to protect the environment not only by regulating toxic chemicals and ecologically unsustainable practices but also by improving the design of buildings, products, and manufacturing processes. In a sense, we have not been asking enough of our designers because customers have settled for inferior products that are energy intensive and contain toxic materials. Requiring that designs meet strict environmental standards should not be seen as an obstacle. Instead, these environmental standards should be viewed as a challenge, which, when met by the designer, have the potential to create a new aesthetic, McDonough maintains.[2]

Figure 13.1
William McDonough, founder and principal partner of William McDonough &
Partners in Charlottesville, Virginia. Photograph by Jeremy Green.

Designers must widen their agenda so that they are not simply interested
in the way a house looks, how much it costs, and whether it performs well.
While all of these criteria are important, designers must also be concerned
with the materials of which a house is made. They must ask themselves
whether their building materials are toxic, whether the harvesting of the
building materials destroys other species, and whether the building requires disproportionately large quantities of energy to construct and maintain. In brief, McDonough wants to "design out" some of the toxins and
"design in" ecologically intelligent features that fit in with the cycles of
nature.

Ecologically intelligent design requires the architect to be very selective
about the characteristics of building materials used in construction. "As I
began my architectural practice I became aware of toxic chemicals in some
wood products, paints, and carpets. I realized I couldn't put that stuff into

my buildings because they just didn't meet my standards," McDonough recalls.

"I take a somewhat different approach than many other architects because I go out to the factories and talk with manufacturers about how things are made," he says. For example, when it came to painting one of the offices he designed, McDonough didn't just buy off-the-shelf paint. Instead, he visited a local paint manufacturer and convinced the owner to supply him with a special batch that contained no fungicides or biocides. McDonough discovered that paint manufacturers mix these chemicals into their products to prevent a green mold from growing on the surface of their paint when it sits too long on the shelf. He eliminated the need for these toxins by buying fresh paint from the manufacturer and using it immediately.

Each time McDonough starts a new project, he identifies local suppliers who can provide him with the most "ecologically intelligent" building materials. For example, he will go to a brick manufacturer and ask him how his bricks are made, whether he gets his fly ash from a municipal waste incinerator, and if the bricks contain heavy metals or dioxins. "I like to spread this kind of thinking like a virus," says McDonough, the Tokyo-born son of a Seagram's executive, who attended nineteen schools in different countries as he followed his father from post to post.

McDonough had an opportunity to put his theories about environmentally intelligent design to the test when he was asked by Wal-Mart Stores officials to consult with their architects (BSW Architects of Tulsa, Oklahoma) on the design of a one-story, 121,267 square foot "Eco-Mart" store, which was to be constructed in 1993 in Lawrence, Kansas.[3] The idea for the store is reported to have come from Hillary Clinton, who was then on the Wal-Mart board of directors.

Working with energy-efficiency expert Amory Lovins from the Rocky Mountain Institute, McDonough decided that one of the environmental features with the biggest payback that he could introduce into Wal-Mart stores would be the use of daylight to supplement their lighting needs. "It just amazes me when I go into a big building and the sun is shining outside and all these lights are on inside burning up the kilowatt-hours," McDonough says.

It quickly became apparent, however, that using standard skylights to bring daylight into the Wal-Mart store would not work. The light from an off-the-shelf skylight is uncontrolled and produces a high-contrast rectangle of light that travels like a spotlight across the floor of the retail space as the sun moves through the sky. These bright splotches of light cause glare that makes it difficult for customers to see merchandise on walls and shelves. The unfiltered light can also cause ultraviolet degradation or fading of some materials displayed for sale.

Instead of giving up on using skylights to "daylight" the cavernous interior of a one-story retail store, McDonough took his design problem to the Andersen window and door company in Bayport, Minnesota, where he worked with engineers to come up with a new generation of skylights that use a holographic film to disperse and filter daylight. With the holographic film, it is as if the light goes through millions of tiny prisms, McDonough explains. Andersen engineers also devised a sensor that monitors how much daylight passes through the skylight and controls a dimmer switch that automatically adjusts the store's electric lights depending on how much natural light is available. The redesigned skylights produce twice as much illumination as previous versions while using half the space, thus reducing the amount of heat loss in the winter and heat gain in the summer. Using daylight in this fashion cuts Wal-Mart's electric bill by 40 percent, McDonough claims.

Another environmental feature that McDonough insisted upon was that the retail space be designed in such a way that it could be converted into a residential living complex. One of the most wasteful aspects of the construction sector today is that we tend to abandon or knock our buildings down every forty years or so as needs shift.

To make the one-story commercial space adaptable into a two-story residential complex, McDonough laid out the concrete block that formed the walls in such a way that blocks could be easily removed and additional windows and doors installed at a later date. In addition, McDonough raised the store's roof slightly to make room for an extra floor inside should it ever be required. He also eliminated the deep steel-truss roof construction with a drop ceiling for a number of reasons. First, steel is more energy intensive than wood: the amount of energy embodied in steel construction can require 300,000 BTUs per square foot compared with 40,000 BTUs for wood. Overall, by building with wood and concrete block and avoiding the use of steel where possible, construction materials will require 33 percent less energy to produce. Second, McDonough wanted to eliminate the drop ceiling because it would block the daylight streaming in through the skylights. Third, he wanted to get rid of the 6-foot-deep steel trusses because they would cut into the interior living space were the store ever converted to residential use. In place of the steel truss–supported roof, he designed a wooden roof that provides an interior ceiling appropriate for future residential use. This design also permits light from the skylights to illuminate the retail space during the period of time that the structure is used for that purpose.

But where was McDonough to buy the wood for his roof? Most architects would have left that decision up to the building contractor. But McDonough required that the $500,000 worth of wood products—construction lumber, plywood, and engineered timbers—come from sustainably harvested lumber. "One of our design principles is 'adore diversity,'" McDonough explains, and if you buy lumber that is harvested using clearcutting techniques that are followed by the replanting of a single species, then the architect and builder are complicitous in a drastic simplification of an ecosystem that leads to a loss of biodiversity.

Launched on another material search, McDonough began to look for companies that supply sustainably harvested lumber. Unfortunately, there were no lumber supply businesses that could provide the volume he needed nor were the companies that claimed to provide sustainably harvested lumber monitored by outside auditors capable of verifying their claims. So McDonough had to organize his own source of supply from scratch. To this end, in 1991 he helped launch a company called Timber Source, now renamed Forest Partnership, which works with wood producers, wood users, and environmental groups to promote sustainable forestry. [4]

Forest Partnership oversaw the search for sustainable lumber and helped set up a number of harvesting operations. One was in Eugene, Oregon, where 200,000 board feet of second-growth Douglas fir were harvested. On the East Coast some 600,000 board feet of longleaf pine came from the James Madison and Zachary Taylor forests in Virginia. Two outside auditors were hired to certify that the lumber was sustainably harvested: the Rogue Institute in Ashland, Oregon, and Global Resource Consultants in Manassas, Virginia.

Since then the Forest Stewardship Council has set up a program to accredit certifiers of sustainably harvested lumber and the American Society of Testing Materials (ASTM) is developing protocols for sustainably harvested wood. The Forest Stewardship Council will likely play a role in this process, McDonough says. Furthermore, the Forest Partnership is spinning off a for-profit private company that will supply sustainably harvested lumber under the trademark Forest Friendly Products. Demand for sustainably harvested wood from chains such as Wal-Mart will help create a market for these products, McDonough says.

But just setting up a system that could supply him with sustainably harvested lumber was not the end of it. In order to build and support the roof, McDonough needed laminated (glue-lam) beams of I-joists and plywood. To produce these McDonough worked with Trus-Joist International to insure that the glues were not toxic and that the top and bottom members of the glue-lam beams were not clear, long boards that came from trees harvested out of old-growth forests. "You have to get right down to that level of detail," McDonough says.

On the Wal-Mart project, McDonough's design specifications also required a cooling system that would run on hydrofluorocarbon (HFC) 134-A instead of ozone-destroying chlorofluorocarbons (CFCs). Furthermore, the design team took care to recycle materials from a previous structure on the building site. They called for the recycling of steel from a preexisting structure and crushing concrete block for use in the construction of the parking lot. In addition, fifty-four of sixty trees located on the site before construction were replanted elsewhere. And stormwater runoff is captured in a pond dug to provide a reservoir for water that will be used to irrigate landscaping plants. The store is also equipped with its own recycling center with bins for fifteen different kinds of materials. Construction costs were slightly higher than usual, but McDonough claims that savings on utility bills and greater productivity will more than compensate for this within a short period of time.

Despite these eco-friendly construction practices, some argue that if McDonough is interested in sustainable development, he should have refused to work for Wal-Mart, the largest retailer in the nation, and a company that is "malling America," promoting rampant consumerism, and destroying the independent retail business base of our small towns. Data from the National Trust for Historic Preservation reveal that in Iowa the average Wal-Mart increases area sales by $9 million a year, while the company grosses $20 million a year, stripping some $11 million in sales from existing stores. Another study reveals that a typical discount store (such as Wal-Mart) creates 149 jobs while destroying 230 higher-paying jobs.[5]

But McDonough looks on the opportunity to work with Wal-Mart from a different perspective. "Unless we engage in a dialogue with these retail companies, all we are doing is setting ourselves up against them," he argues, and opportunities are missed to design more environmentally intelligent buildings and products.

When Wal-Mart officials asked McDonough if he would design a store for them, he responded: "I'm interested in 'waste equals food.' I'm interested in people with lives rather than consumers with lifestyles. And we need to talk about Wal-Mart's impact on small towns." After the initial contact, McDonough thought it unlikely that he would ever hear from Wal-Mart officials again. But a few days later they called back and asked him: "Are you willing to discuss the fact that people have the right to buy the finest quality products at the lowest possible price?"

Thus began a collaboration and exchange of ideas in which Wal-Mart officials argued that providing people with quality goods at the lowest possible price frees up disposable income for them and improves their lives. McDonough countered that a time will come when the highest-quality, lowest-price goods will be produced locally because they will require less costly packaging and shipping. Such a shift would provide more local employment and use less fossil fuel. But these kinds of changes will only be possible "if we can use the momentum of the current delivery system to achieve our ends. Let our retailers solve the problem they have created by doing better business."

Wal-Mart is going to build the stores whether or not he makes their design more environmentally intelligent or not, he argues. By taking the job designing their eco-store, he would be in a position to convince them to reduce their energy demand, reduce the use of chemicals that destroy stratospheric ozone, and help create a demand for sustainably harvested lumber and other green building materials. Were large retailers to build all their stores with the new energy-saving Andersen skylights, for example, it would mean the purchase of a half mile of skylights every day, McDonough calculates. This kind of volume can kick-start a whole new industry, he notes.

Pleased with the store McDonough helped design for them, Wal-Mart now plans to build fifty more outlets using a number of the environmentally friendly features from McDonough's prototype design.

"Design is the first signal of our intent," observes McDonough in his most pedagogical style. If, for example, a house is designed that uses lumber harvested by clearcutting a forest, it indicates what kind of world we want to live in. If an office is built with materials that emit toxic chemicals that cause illness, again, it signals a set of values. In a sense, design reflects our level of maturity as a civilization because an analysis of our manufacturing processes and architecture reveals whether or not the built environment is in harmony with nature or seeks to overwhelm it by brute force. This is more than an aesthetic issue because those civilizations that seek to overpower nature are doomed to destroy the life-support system upon which they depend.

Having earned his architecture degree from Yale and built Ireland's first solar-heated house, McDonough was afforded the opportunity to address these issues in 1985 when he designed the Environmental Defense Fund's (EDF) 10,000 square foot offices in New York with 14-foot ceilings and large windows that can be opened. To improve the indoor air quality in these offices, McDonough increased fresh-air circulation significantly, used beeswax instead of polyurethane on the floors, employed jute rather than synthetic material under the carpeting, and bought desks made of wood and granite instead of plastic. He also created a boulevard that runs between the offices, complete with ficus trees and streetlamps.

In 1988, McDonough won an international competition for his conception of a 50-story trade center to be built in Warsaw. Caught up in Poland's political and economic turmoil in the late 1980s, the project was never built, but one of the features of McDonough's design is interesting. As part of the plans, McDonough calculated the energy costs to build and maintain this high-rise and concluded that 6400 acres of forest would be needed to offset the effects on climate from its energy requirements. Before plans for the project were shelved, McDonough says he had convinced developers to plant and maintain some 10 square miles of forest somewhere in Poland at a cost of $150,000.[6]

Planting trees to compensate for the ecological cost of construction is something McDonough has done before, albeit on a much more modest scale. When he was remodeling the premises of Paul Stuart clothiers on Madison Avenue in New York, McDonough called for the use of two trees' worth of English Oak paneling. To more than make up for this withdrawal from nature's account, he and friends planted more than a thousand acorns around the country. Even taking into account a low survival rate, McDonough figures he more than replaced the English Oaks.

Planting trees as an adjunct to building a structure has ancient roots in the practice of architecture. McDonough likes to recount a story told by Gregory Bateson about the renovation of the roof of the main hall at New College, Oxford. As the story goes, building inspectors discovered that the roof's brown oak beams (40 feet long, 2 feet thick, and dating from the 1600s) were suffering from dry rot. A committee formed to look into their replacement found that it would cost $50 to $60 million. Faced with this impossible expense, the committee inquired of the college forester if there were trees on the grounds that could supply the timber. The forester replied: "We have been wondering when you would ask this question. When the building was constructed 350 years ago, the architects specified that a grove of trees should be planted and maintained to replace the beams in the ceiling when they would suffer from dry rot."[7] Bateson is reported to have observed: "That is the way to run a culture."

One of McDonough's furthest forays into the redesign of a manufacturing process came when he was asked to design a portfolio line of furniture fabrics for DesignTex, a subsidiary of Steelcase Furniture, the largest manufacturer of office furniture in the United States.

Knowing that McDonough went beyond the surface in design, DesignTex officials asked him to design for them some environmentally friendly materials that would be used on "upper-end" couches and chairs. When McDonough told them that he had to design what the material was made out of as well as what it looked like, they were unfazed. In fact, they suggested that the material be made out of natural cotton and recycled PET (polyethylene terephthalate) plastic. What could be more ecologically chic?

To their surprise, McDonough nixed the idea. First, he pointed out, most cotton, although a natural fiber, is produced in a manner that is ecologically disastrous: a large percentage of the pesticides used worldwide are sprayed on cotton and many dyes and fixatives (mordants) are also considered toxic. Furthermore, cotton has often been grown in a manner that uses excessive amounts of water and pesticides; it also has a history of being produced in a manner that exploits farm workers.

But beyond the question of whether or not a supply of cotton could be found that is not sprayed with pesticides and is grown in an ecologically intelligent manner that is also socially just, McDonough had a problem with the idea of mixing organic cotton with plastic. "I see materials as nutrients," he explains. "They are either part of the organic nutrient cycle or they are part of the technical nutrient cycle." Mixing the two cycles is a mistake because what are you going to do with it after its useful life is over? he asks. The mixed materials cannot be composted because they have plastic in them, and they cannot be recycled because they have organic matter in them. "So it is an unintelligent product and we should not make it," he concludes.

Instead, McDonough decided to design an organic fabric that could be composted. The design challenge was to find a fabric that was naturally flame retardant, stood up well to wear and tear, and permitted air to move through it. Chairs that do not breathe make you hot and sweaty, he explains, requiring more air conditioning for comfort. Furthermore, if a fabric is used in wheelchairs, it is important that it permit good air circulation so that sweat does not build up.

Ultimately, the design team came up with a fabric made of wool and ramie, a nettle-like Asian shrub with fibers that resemble those of cotton. The combination of New Zealand wool and Filipino ramie was perfect for chair fabrics because while wool absorbs 35 percent of its weight in water, ramie wicks moisture away, keeping the person sitting in the chair dry and cool. Furthermore, this fiber combination is naturally fire retardant, reducing the need for additives.

But what would be used to dye the fabrics and fix the dyes? After selecting a Swiss fabric manufacturer who was diligent about meeting environmental standards, the team began looking for a source of safe dyes. They began to collaborate with Michael Braungart, an ecological chemist from Hannover, Germany, whose focus was on redesigning manufacturing processes to eliminate the generation of toxins and waste, and who has been an inspiration for McDonough's design protocol. The search led them to many chemical companies where they asked for permission to investigate their deep chemistry. At Ciba Geigy they were initially told

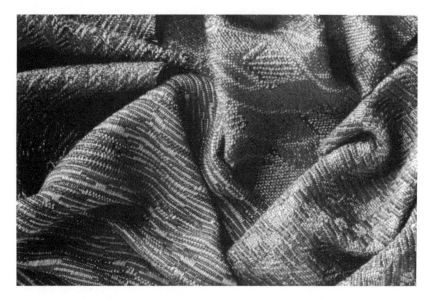

Figure 13.2
This nontoxic fabric is used in the William McDonough Collection by DesignTex.
Courtesy William McDonough.

such information was proprietary, but after reiceiving permission from top officials, Braungart worked with company chemists to examine their dyes and mordants to see if they were carcinogenic, mutagenic, persistent, or bioaccumulative. "In three weeks we completely redesigned the chemical protocol so that it was as safe as we could make it," McDonough claims. When he asked the chemists why they hadn't done this before, their answer was startlingly simple: "No one ever asked us to."

McDonough is struck by the absurdity of the industrial situation that has evolved. "If you go to some of these fabric mills and look at the inch of trim on either side of the bolts of material that are cut off at the end of the manufacturing process, you learn that this trim material is classified as hazardous waste. Then you ask yourself, if the trim is hazardous, what is the product itself?" McDonough concludes that we have reached the point where we are deliberately producing products with which we have intimate contact that are made of hazardous materials. This is what the new breed of designers must "design out" of the manufacturing process, he argues.

"We are not going to solve these problems by making slightly better engines every year, or by reducing the number of toxins in a product. The only way we are going to get there is to redesign our products. All of our products. Everything. Completely," he writes.[8] Redesign of products should be the first line of defense of the environment, followed by the more familiar principles of reducing, reusing, and recycling of materials.

McDonough's redesign of manufacturing processes to eliminate toxins and use energy and resources more efficiently is not novel. Industries have not been entirely blind to the fact that their manufacturing process can be manipulated so they can recover, as useful byproducts, materials that previously were defined and treated as wastes. Taking the next step, a number of major American companies are engineering their products so that they are "designed for disassembly" (DFD) in such a way that many components can be reused, refurbished, or recycled, writes Gene Bylinsky in an article that details some of the following examples.[9]

In Highland Park, Michigan, the Big Three automakers have established a Vehicle Recycling Development Center where efficiency experts watch how long it takes to disassemble a vehicle and come up with ways to simplify the design of cars so that they can be dismantled more rapidly. For example, instead of gluing and welding parts together, components are redesigned with snap fasteners or screws that permit quick disassembly.

The personal computer business is also undergoing DFD redesign. With some 70 million obsolete computers in the country on their way to the landfill, computer company officials are concerned that they will be held legally responsible for groundwater contamination associated with their disposal. As a result, IBM, Hewlett-Packard, and Siemens Nixdorf have redesigned some of their personal computers so that they are easier to disassemble for recycling.

Xerox redesigned its copiers so that potentially reusable parts were easy to remove with snap fasteners, saving the company $200 million a year through reuse of parts. After its fiasco with disposable cameras, which won the company a congressional "Wastemaker of the Year" award, Kodak developed recyclable cameras. Telephones are probably next. Already the Canadians are disassembling old telephones, replacing the plastic housing, and sending them out again for lease. In this country, AT&T is experimenting with a new generation of green product designs.

But the question remains as to what to do with unmarketable byproducts of manufacturing. Michael Braungart, the German ecological chemist who works with McDonough, suggests substituting the word "product" for what we previously defined as a "waste." Through this semantic device, all wastes are redefined as products, albeit of greater or lesser value. Braungart and McDonough further categorize all products as being one of three different kinds: consumable products, products of service, or unmarketable products.[10]

The first category, consumable products, is the easiest to deal with because they are all compostable. For example, food is a consumable product. If properly treated, a tortilla, either eaten or thrown away, can be turned into compost and returned to the soil. Similarly, a cotton or wool garment or a product made of wood can be safely composted if it is not treated with flame retardants, dyes, paints, or varnishes that contain bioaccumulative toxins.

The second category, products of service, includes cars, television sets, and a host of other noncompostable products. Disposing of these products should ultimately be the responsibility of the manufacturer. Instead of buying an industrial "technical" product from a manufacturer, the user should lease it, McDonough and Braungart argue. Were this the case, manufacturers would retain responsibility for the disposition of the product once the user no longer wanted it. And the manufacturer would continue to accrue "technical" materials. With mountains of technical materials coming back to manufacturers (and high fees for waste disposal) there would be a powerful incentive for manufacturers to design products that could be composted, reused, or fully recycled.

This brings us to the third category: unmarketable products. It is no secret that many products today contain (or generate in their manufacture) a variety of byproducts—such as dioxins, polychlorinated biphenyls (PCBs), or radioactive materials—which no one will buy. Under Braungart's system, once the manufacturer of these products is identified, the toxins would be stored in government operated "parking lots." The government would then bill the manufacturer for safely containing these toxins until the technology was available to neutralize them and reintroduce them into the technological or organic cycle. These "parking fees" would not only pay for the safekeeping of toxic materials, they would also discourage their generation.

McDonough sees a great future in the power of design to solve some of our environmental problems. He is currently working with a major carpet manufacturing company on a design that he and Braungart have devised that will "revolutionize" the carpet industry. This company, Interface, the world's largest manufacturer of carpet tiles, has adopted a "product of service" protocol and its complete redesign implications, McDonough says. But he is not divulging what this new product will be made of until the intellectual property is secured and the deal is struck. The gleam in his eye, however, suggests that not only is this good business for everyone, including the designers, but that he is also having great fun redesigning manufacturing processes along green lines.

Moving Out of the Flood Plain and Designing an
Environmentally Sustainable Community

When the rains first started in earnest in June 1993, everyone in Pattonsburg, Missouri, began to keep an eye on the river. The Grand River jumped its banks but that was not all that unusual; the water had escaped the riverbed before without flooding the town. But by July, town residents, like the people in hundreds of other communities up and down the Missouri and Mississippi rivers, were fighting to keep the water out of their homes. Their efforts with earthmoving equipment and sandbags failed. Defeated by the power of rising rivers, they packed their favorite possessions into cars and pickup trucks and moved uphill.

Not everyone, however, was in a hurry to get out of town. One farmer, who had seen floods come and go, just took his armchair upstairs and rode out the flood on the second floor. The same kind of "flood macho" was demonstrated by a number of local patrons at the Double Eagle bar, a friendly establishment on Main Street with a pool table and tap beer drawn into frosted mugs pulled from a freezer. There, a stalwart group of regulars refused to be chased out by the flood and sat drinking beer and playing dominoes as the waters inched up over their boot tops.

As the flood waters continued to rise, a local farmer, who was helping with the rescue effort, rode his horse into the bar in search of some refreshment. Shirley, the barkeep and owner, kept the bar open as long as she could but abandoned ship when the tables began to float by with the bottles still on them.

At Bettie's Cafe on Main Street the flood turned killer. Bettie Valenti's husband, Robert "Banny" Gardner, a skilled coon hunter and restaurateur, was salvaging what he could from the flood when he was electrocuted by a wave of current that arced through the waters lapping around his legs.

Bettie reached out to grab him but was warned off touching him by a neighbor.

On the outskirts of town, the flood waters were also rising on Route 69 where David Warford, elected mayor only a few months earlier, was scrambling to save as many of his shiny-new, all-terrain vehicles (ATVs) and motorcycles as he could in the Honda dealership he owns and operates. A good deal of his merchandise was ruined. When the flood waters finally receded, he and the other residents of Pattonsburg shoveled the muck out of their homes, threw away dumpster loads of ruined belongings, and helped one another fix their houses up so they could move back in. But no sooner had they buried their dead and scraped the mud from the floors when a second flood rolled into town. This time the waters rose 8 inches higher than the previous flood.

"After the second flood people were saying: 'I don't think I can go through this again,'" recalls thirty-five-year-old Mayor Warford. "It wasn't hard convincing folks that we needed to move the town to higher ground."

The flood of 1993 was hardly the first major inundation that Pattonsburg residents had experienced. This is a town where the biblical Job would have felt right at home. Since it was founded in 1845, Pattonsburg has been flooded at least thirty times, suffered a devastating fire, and been partially destroyed by a tornado. At one point the U.S. Army Corps of Engineers planned to build a dam that would have turned Pattonsburg into a lake.

While the dam was never built, in the 1970s the people of Pattonsburg were hit by an economic catastrophe more powerful than any natural disaster: an interstate highway, I-35, was built a few miles east of them. Until then, Pattonsburg had drawn much of its commerce from Route 69, a two-lane, north-south blacktop that truckers dubbed the Ho Chi Minh trail because it was narrow and dangerous.

When the interstate bypassed Pattonsburg, businesses shriveled up. The intermittent flooding didn't help either. Once a bustling town of 2000 residents, Pattonsburg's population dwindled to 316. Evidence of the town's abandonment can be seen everywhere: many homes are boarded up or stand gape-jawed with missing doors and windows. Along Main Street the hardware store, grocery store, pizzeria, and a host of other enterprises have gone out of business. Yet it wasn't until the great flood of 1993—the nation's costliest flood, which did $12 to $16 billion in damages in nine states—that Pattonsburg residents finally decided to move to higher ground.[1]

While the flood was having its way with the residents of Pattonsburg, Nancy Skinner was warm and dry in her apartment in Chicago watching television coverage of the massive Midwestern flooding. At the time Skinner was a 30-year-old self-employed entrepreneur and environmentalist who sold environmentally safe paints free of volatile organic compounds (VOCs). She also moonlighted as a comedienne.

As the extent of flood devastation unfolded, Skinner had an idea: since the Government was poised to spend $6 billion on flood relief in the Midwest, why not use the funds to relocate communities out of the flood zone so that in the future federal dollars would not be needed to bail them out again? And why not rebuild these communities on higher ground using the best available environmental technologies and energy-efficient practices so that the new communities would become models of sustainable development in the Midwest?

These were reasonable questions, but for most people this kind of brilliant idea about how things ought to be would have faded within an hour. Skinner, however, is a persistent woman whose bouts of enthusiasm have staying power. Needing someone to bounce the idea off of, she called her father who heard her out and then said with more than a trace of sarcasm: "Great idea, Nancy. What are you going to do? Call the White House?"

In fact that is exactly what she did; she telephoned Washington the next day. Over the following weeks she ran up an impressive phone bill calling the Environmental Protection Agency, the Federal Emergency Management Administration (FEMA), the Department of Interior, the Department of Energy, the Small Business Administration, the White House Office on Environmental Policy, and a host of other bureaucracies. Her suggestion, that flood relief funds be used to rebuild communities along more energy-efficient and environmentally sound lines, was generally received as a good idea. But the federal employees she spoke with just pushed her along the line to some other bureaucrat who might be interested.

Skinner's telephone marathon continued without rest until she spoke with Bill Becker at the Department of Energy who had unique experience that permitted him to appreciate the potential of Skinner's suggestion. As the former editor of the *Kickapoo Scout,* a newspaper in Soldiers Grove,

Wisconsin, he had been instrumental in lobbying to see his town moved out of the flood zone to higher ground.

As with many other communities located along rivers, Soldiers Grove had no problems with flooding for the first 50 years of its existence. It was only after extensive logging in the area caused the river to silt up that the river jumped its banks. The first flood was in 1907, but by the 1970s there was a major flood every decade. In 1975 the U.S. Army Corps of Engineers proposed building a $3.5 million levee to protect approximately $1 million of property from flooding, but the expense of maintaining the levee would cost more than double the amount the town raised from annual property taxes. As an alternative, town officials proposed rebuilding the business district on higher ground. Federal officials considered this a radical proposal until the flood of 1978 destroyed Soldiers Grove's business district. At that point, a trifle late, the Department of Housing and Urban Development (HUD) came up with $1 million to fund the relocation.

Instead of just rebuilding the business district along conventional lines, the money was spent constructing passive solar, superinsulated, energy-efficient buildings that were cost effective. The most advanced of these were built with translucent plastic panels in the roof, which permitted winter sunlight to heat up the attic spaces; fans then circulated the heat throughout the building, reducing heating bills by 75 percent.

The town also passed the first solar ordinance in the country requiring that newly constructed commercial buildings derive at least half of their heating from the sun. Planners also conducted a microclimate analysis of building sites that allowed them to plant trees in a pattern that blocked winter winds while channeling summer breezes. From the Soldiers Grove experience, Becker learned that the practical and frugal residents of small Midwestern towns could be convinced to relocate and rebuild along ecologically sustainable lines.

When Skinner and Becker connected over the telephone, two of the pieces fell into place that were needed to convince federal officials to use disaster relief as an opportunity to rebuild communities using an ecologically sustainable design. Skinner served the function of a gadfly, willing to pester the federal bureaucracy into action, while Becker was a federal official with personal experience in the Midwestern flood plain. What they needed was someone who could pull together specialists from around the nation who knew how to design and build an environmentally friendly and energy-efficient community. Becker knew just the man for the job.

Robert Berkebile is a prominent Kansas City architect who was spurred into an interest in ecologically intelligent architecture as a result of a tragedy. Berkebile was the lead architect on the Hyatt Hotel in Kansas City where a skywalk collapsed killing dozens of people. When the skywalk collapsed, Berkebile sped to the hotel and joined efforts to rescue survivors from the rubble. For the next year and a half he was engaged in litigation over responsibility for the tragedy and while he was ultimately exonerated the experience left a scar.

In the wake of the collapse of the skywalk, Berkebile began to rethink all the buildings he had ever designed. He went back to his drawings to see in greater detail how they were put together. In the process, he learned a great deal about where building materials come from, what they are made out of, what toxins they contain, and what energy is required to manufacture and transport them.

For example, he learned that not only does the manufacture of aluminum require large amounts of energy but that one of the key ingredients in aluminum is bauxite, an ore frequently mined from beneath tropical rain forests. On another building, where his design called for the use of marble from Minnesota, he learned that the marble had been shipped to Mexico to be cut and then to Italy to be polished before it was transported to the construction site in the Midwest.

In the process of reviewing all the materials used in his buildings, Berkebile became an expert on the environmental costs of building materials and various construction methods. He subsequently founded and became chairman of the American Institute of Architects' (AIA) Committee on the Environment (COTE).

By the time DOE's Bill Becker called him in 1994, Berkebile had become one of the focal points for a loose network of green architects and experts in a number of fields, including the use of daylight to illuminate the interior of buildings to reduce energy consumption, passive solar design, the construction of wetlands to handle stormwater runoff, and the use of local building materials to avoid transportation costs. This network permitted

Figure 14.1
Robert Berkebile, president of BNIM in Kansas City, Missouri, is founder and chairman of the American Institute of Architects Committee on the Environment and helped coordinate the effort to provide an ecologically sustainable design for New Pattonsburg, Missouri. Courtesy BNIM.

him to assemble a team of ecologically sophisticated professionals who could travel to flood-devastated towns and help the residents with cost-effective plans to relocate and build ecologically sustainable communities.

The design team assembled by Berkebile first met in January 1994 at the Johnson & Johnson Wingspread conference center in Racine, Wisconsin. To keep the forty experts grounded, Dennis Knobloch, the mayor of Valmeyer, Illinois, was invited. Valmeyer (pop. 900) was a town, largely destroyed by the flood of 1993, whose residents were living in trailers provided by FEMA. The government was providing $30 million in disaster relief to relocate the town to higher ground.

While he had been reluctant initially to attend the conference, by the end of the presentations Knobloch had heard enough useful ideas that were relevant to the rebuilding of his town that he invited the design team to use Valmeyer as a site for their first demonstration project. "He really got religion about building an ecologically sustainable town," Skinner recalls, "and he wanted us to jump in with both feet."

Unfortunately, the design process in Valmeyer was so far advanced that the design team had only marginal influence, recalls Christopher Kelsey, an architect who is the Pattonsburg project manager for BNIM, Berkebile's architectural firm in Kansas City. The regional planners had already designed a suburban cul-de-sac community—the streets were laid out in the wrong direction to take advantage of the sun—and they refused to change their plans, he recalls.

Becker finds this entirely understandable. "In a post-flood situation, where people are living in trailers and their kids are going to school in temporary structures at a local fairground, people want their lives to get back to normal as quickly as possible. They want to spend their next Christmas at home and not in a trailer." So when a design team of eco-experts comes flying in and tries to revise plans that have already been made, it provokes resistance because of the delay it entails, he notes. Nevertheless, some modifications were made in the plans for the new town, including the use of ground-source heat pumps in a couple of buildings and improved energy-efficiency measures. "It was a very good learning experience for us, even though the results were not all what they might have been," Becker concludes.

Fortunately, Skinner (who now runs Daybreak International, a nonprofit organization that consults on sustainability projects) had invited David Warford, the mayor of flood-damaged Pattonsburg to attend the design workshop in Valmeyer. She invited him because she had heard from a FEMA official that the Pattonsburg residents' idea of how to relocate out of the flood zone was to move their town to an off-ramp of an interstate and throw up some truck stops to capture the passing vehicular business. "See what you can do for them," the official suggested.

It turned out to be a good match. Pattonsburg's Mayor Warford was looking for all the help he could find in terms of funding and expertise to move his town. He saw that Valmeyer was not taking full advantage of what the design team had to offer because of the unwillingness of residents to abandon the traditional suburban design. At the end of the workshop, Warford offered Pattonsburg as the next demonstration project for the design team. He pointed out that the residents of Pattonsburg were ready to relocate and that they were not wedded to a car-oriented suburban design.

Figure 14.2
Nancy Skinner asked why flood-damaged towns could not be relocated and rebuilt using ecologically sustainable techniques. Photograph by Suzanne Plunkett.

But not everyone is anxious to move to higher ground. "Why should we move? I'm used to the flooding. I've lived with flooding all of my life. It's just part of living here," shrugs a man wearing knee-high rubber boots. "It's a hell of a lot better than living in Los Angeles with those earthquakes. What is the government going to do about that? Move LA?" he asks.

While the federal government can't move Los Angeles, it can move a small town the size of Pattonsburg out of the flood zone. Furthermore, it makes sense economically. In the last eight years federal officials paid out some $26 billion in disaster relief and much of this money was spent to build communities back to predisaster status. A study directed by Brig. Gen. Gerald E. Galloway, an Army engineer, found that federal flood control efforts would be less expensive and more effective if people were moved out of flood plains.[2] Experimenting with this flood relief strategy, government officials financed the largest post-flood relocation in the nation's history and spent 15 percent of disaster relief funds earmarked for the flood of 1993 on relocating 7000 people off 35,000 acres of flood plain.[3] Pattonsburg was an obvious candidate for one of these relocation efforts and $12 million was spent to move the town 2 miles to a higher elevation.

Pattonsburg officials recognized that this large infusion of federal funds into a small town like theirs was an event unlikely to reoccur any time soon. To help them invest the money wisely, Berkebile's design team of environmental architects and alternative technology experts arrived in Pattonsburg in September 1994 for a three-day planning session. The designers were scheduled to listen to what Pattonsburg residents wanted their new town to look like at a "visioning session," then present the residents with an expanded menu of options about what was possible, and finally draw up a town plan.

Convincing skeptical Midwesterners to move their town was a task only slightly less daunting than negotiating an international nuclear arms reduction treaty. The first meeting, held in the Pattonsburg school gymnasium, was led by Milenko Matanovic, a consultant from the Pomegranate Center for Community Innovation based in Issaquah, Washington. Matanovic loosened up the crowd with a story from the town where he grew up in what used to be Yugoslavia where people had no say about how their town was designed and could only complain about it after it was built. By contrast, Pattonsburg residents had a rare opportunity to decide what kind of town they wanted to live in.[4]

Residents were then shown a documentary about how the people of Soldiers Grove, Wisconsin, moved to higher ground and built an energy-efficient business district. After the screening, Matanovic pinned a large map of Pattonsburg to the wall and began to solicit ideas about what residents treasured in their community and wanted to incorporate in New Pattonsburg. Ideas began to trickle in: some people liked the feel of walking down Main Street; others had grown accustomed to the faded red brick façade of the stores; some liked having separate entrances to each of the shops and didn't want to replace them with a shopping mall; someone spoke lyrically of a neighbor's garden; another resident didn't want to lose the sound of the town whistle that blows morning, noon, and night. The list of town treasures grew.

Residents and members of the visiting team of experts then visited the site of the new town, 640 acres purchased from several local farmers. The land stood on a hill, several miles from the old town and about a quarter of a mile from the interstate highway. It was good agricultural land with no structures on it save a dog kennel slated for demolition.

Figure 14.3
Pattonsburg residents provide input to eco-experts assembled to help them with plans for their relocated town. Courtesy: BNIM.

Standing at the edge of the cornfield, the designers took note of a line of trees they wanted to save. They brought with them large topographical maps of the site that permitted them to plot precisely how stormwater drained from the land. From this inspection of the site, they recommended that the existing contours of the land be preserved as far as possible so that the land would continue to drain naturally. This would save the town the considerable expense of building and maintaining a conventional stormwater system. Instead, a constructed wetland that took advantage of the natural contours of the land could be used to drain stormwater out of the new town. Existing ponds could be augmented to work as detention ponds that would permit much of the stormwater to go back into the ground instead of running off the land. If the right plants were grown in this constructed wetland, contaminants in the stormwater could be filtered out so that the runoff from the town would actually be cleaner than the water currently draining from the cornfield.

Local farmers and rural residents easily grasped the advantage of using the lay of the land to drain stormwater from their new town. "The people of this community have embraced the ideas of sustainable development wholeheartedly," says Mayor Warford. "We come from a farming community and a lot of the these ideas are not so different from things farmers do regularly. People in farming communities are very aware of the cycles of nature and they tend to take the long view. When you make your livelihood from the soil you realize that you have to protect it or you won't have an income any more."

The design team also pointed out that the land left open for natural stormwater drainage would also provide an uninterrupted greenway weaving through town that could serve as a bicycle path and nature walk as well as a habitat for a variety of plants and animals.

After plotting out the stormwater drainage–greenway system, planners suggested that the streets of the town be oriented along an east-west axis so the houses could take advantage of passive solar gain during the cold Missouri winters. Planting deciduous trees on the sunny side of the house would shade them in the summer while allowing the sun to shine through the bare branches of the trees to warm the house in the winter. Furthermore, creating a tree line of windbreaks would protect the houses during the winter, channel summer breezes, and muffle noise from the highway.

The town was also designed to be pedestrian-friendly. It was laid out in such a way that everyone was no more than a five-minute walk from downtown. Housing for the elderly was sited in the center of town so that older residents would not be isolated from the town's daily activities. The town was also planned so that its commercial and industrial zone was closest to the interstate, while its Main Street was set farther back from the highway where it would have a quieter, slower-paced, feel to it. Following a neo-traditional town design, planners also attempted to recreate some of the best aspects of the old Pattonsburg's Main Street. The retail section of the new town, for example, will be a single-walled structure, but each business will have its own entrance and the height of the facade will vary as it did in the old town.

To keep their new town on a sustainable trajectory, Pattonsburg residents appear to be willing to put their environmentally friendly principles into a written code. Pattonsburg's town council is poised to adopt a number of "covenants and restrictions" drawn up with the help of Dan Slone, an environmental attorney in the Richmond, Virginia, office of McGuire, Woods, Battle & Boothe. In the prologue to these new town regulations, the town of New Pattonsburg commits itself to "encourage, and in some instances require, the use of energy-efficient designs and sustainable construction techniques within its limits."

The new regulations require that all buildings be designed and located "so as not to interfere with the reasonable use of adjoining properties for solar applications."[5] This means that your house cannot infringe on the solar rights of your neighbors, Slone explains. Furthermore, all residential lots must have trees planted to maximize summer cooling, and existing trees and natural drainage patterns must be preserved wherever possible. In addition, impervious groundcover must be minimized, and roof runoff must be diverted for yard irrigation or into infiltration trenches to promote groundwater recharge. The use of certain pesticides and chemical fertilizers will also be restricted and the planting of drought-resistant native plants and perennial groundcovers is recommended.

The energy standards for new construction, which became effective 1 January 1997, will also be higher than state standards. They call for R-38 insulation in the ceilings of single-family houses and R-45 in the ceilings of multifamily dwellings. Windows must be double glazed and with low-E coating, the number of doors and windows on the north wall of the house must be minimized, hot and cold water pipes must be insulated, and automatic setback thermostats on gas heat are required. The new regulations also call for water-saving measures such as the installation of low-flow showerheads, low-flush toilets, and aerated kitchen and bathroom faucets. Storage areas must also be set aside for recycling.

At first the idea of instituting a building code met resistance, Slone recalls. Rural Missouri has no building codes and people told Slone: "You don't understand. We don't like all these laws in Missouri." But when Slone talked with Pattonsburg residents about shared rules, he found that people did want a community covenant that everyone would live by. He was surprised, for example, that there was no opposition to his recommendation that the town adopt a building code requiring that all new construction be energy efficient. Instead, Pattonsburg residents objected to his recommendation that clotheslines be screened from the view of neighbors by a vine-covered arbor. That sounded like more work than was necessary to a resident accustomed to putting up a clothesline in full view of the neighbors.

Slone explained that many communities now prohibit clotheslines because of complaints from residents who object to "looking out the window at their neighbor's underwear flapping in the breeze." Some communities have passed ordinances prohibiting clotheslines and there are even those who detect a conspiracy in the banning of clotheslines, he said. According to this theory, following World War II, when there was a surplus of power, the utility companies pushed for local regulations prohibiting the use of clotheslines in order to force more people to buy electric or gas dryers. While Slone finds this a bit paranoid, he does believe that it makes sense to use the sun to dry clothes whenever possible because of the energy savings. In this context, Slone's solution, which involves screening clotheslines on three sides, proved acceptable to the people of Pattonsburg.

These covenants and restrictions insure that while New Pattonsburg will be not a perfect model of sustainability, at least it is headed in the right direction, Becker notes. "We could try to turn Pattonsburg into a utopian community but it wouldn't work. In the end it has to be what the community can afford to live with. What we hope is that New Pattonsburg will be far more sustainable than it would have been without input from the design team. We hope that it will be a model of an enlightened way to build a town. It won't be perfect, but it will be far better than the conventional approach."

Beyond these new building standards, the effort to design New Pattonsburg as an ecologically sustainable community is beginning to stimulate a regional review of energy and resources efficiency. The most visible sign of this is an ongoing analysis of whether it is cost efficient to capture methane gas from pig manure and use it to generate electricity.

Mayor Warford would like to see a sign at the edge of town saying, New Pattonsburg: Powered by Pigs. The idea is not entirely far-fetched. Continental Grain is building facilities for some 200,000 swine and 80,000 of them will be at a site about 5 miles from New Pattonsburg. Initially there was local opposition to Continental Grain building such massive facilities because of the odor such a large concentration of swine produces and because other hog farmers worried that they would be driven out of business. To improve community relations and to help abate their potential odor problem, officials at Continental Grain are reported to be considering the possibility of trapping methane gas from their swine waste.

In a strange twist, some nuclear engineers are helping with this project. It happens that Allied Signal Corporation has a mammoth facility outside Kansas City where the non-nuclear components of nuclear weapons are manufactured for the Department of Energy. While this would appear to have nothing to do with pig waste, the fact that there is a reduced demand for nuclear bombs means that DOE is looking for ways to keep the team of engineers at Allied Signal intact so that a nuclear deterrent can be maintained. This, in turn, requires finding civilian projects for Allied's engineers.

For decades the nuclear weapons manufacturing community has been very close-knit, explains Jack Quint, who works on agricultural conversion projects at Allied Signal. "We have been locked behind razor wire for almost 50 years, so the possibility of using our expertise to help our neighbors is very exciting." Engineers from Allied Signal and experts on energy efficiency from another DOE facility in Golden, Colorado, are working with officials at Continental Grain to see if methane can be captured by covering the manure ponds with plastic sheathing and then burning the methane to turn a turbine to generate electricity. In addition, there is a possibility that the solids from pig waste could be dried out, turned into pellets, and burned (along with abundantly available switchgrass) to generate electricity. Or through a process called pyrolysis, the pig manure could be turned into a combustible oil. Phil Tate, a local state representative, has organized a regional coalition called the Heartland Bio-energy Partnership to determine which of these methods is the most appropriate technology for transforming pig waste into fuel. The group hopes to come up with a model process that could be replicated elsewhere around the nation.

"Pig power is a great theory," Gene Walker, superintendent of the Pattonsburg school opines, "but I don't know if it will ever happen." While Walker was on a trip to Washington, D.C., he met an expert who told him that it takes eight pigs to generate a kilowatt-hour of electricity. "I just had to walk away when I heard him say that. It just struck me as a strange statement," he recalls.

Both the Pattonsburg and the Valmeyer planning and relocation efforts demonstrate that sustainable development is more than just an attractive theory. The multidisciplinary design team brought together to help these communities build more energy-efficient and environmentally sustainable communities found that it could "sell" Midwesterners on a variety of cost-efficient strategies.

Nancy Skinner recalls an interview that she did with a resident in Valmeyer who said: "When members of the design team first talked about passive solar energy, I thought this was just some weird liberal concept that had nothing to do with me. But when they explained what it meant in practical terms—that it was just how you orient your house in relation to the sun and the type of materials you build with—then it made perfect sense to me."

Similarly, when surveyed, residents of Pattonsburg voted overwhelmingly to build a resource-efficient town largely for economic reasons, Skinner notes. When they saw the film of what people did in Soldiers Grove, they realized they could save money building energy-efficient homes. "That message played very well in Pattonsburg. People said that they didn't care how freaky some of these newfangled ideas sounded as long as they worked and could save them money," Skinner observes.

However, just making the economic argument for environmentally friendly technologies is often not enough, Skinner points out. "You can show people the economic advantages of environmentally friendly technologies until you are blue in the face, but the status quo is powerful and people tend to want to use the technologies they are familiar with."

Gene Walker agrees: "Most of the people we deal with out here are first-class, original hardheads, for want of a better term. You can tell them about a way to save 50 percent on their utility bills by building in a different way, but they won't believe you until you show them." The best way to convince them, Walker says, is to get five people to build energy-efficient homes and let the coffee shop talk do the rest of the work. "If they can go see Joe's house and it looks like a nice setup and Joe paid very little for his heat last winter, then it will be an easy sell to convince others to use the same building techniques." They don't call Missouri the Show Me State for nothing.

The progressive design of New Pattonsburg did not come without a struggle. There were heated debates over some aspects of the plan. For example, early plans called for houses set back from the street, whereas the design team suggested a "build to line," which would require all houses to be constructed closer to the sidewalk. When residents objected, members of the design team took them for a walk through their town and pointed out that in the old days people all built their houses in a row close to the street and that this contributed to the "front porch culture" that residents all liked so much. Similarly, residents had to be convinced by members of the design team that in a cul-de-sac community people tend to drive more and that means less pedestrian traffic where folks can stop and chat when they meet in the street.

Then there was the issue of the size of the road. Small towns have narrow streets that slow down traffic and allow people to walk across the street and chat with a neighbor. "If you build a road where two huge tractors can pass each other, then it will be so wide it will take people ten minutes to cross it. What we wanted to do was get people out of their cars and build a place where they can work and shop and live, all within walking distance," Dan Slone says.

In the end, Pattonsburg residents agreed on an energy-efficient and resource-efficient design for their new town. "There are almost no towns that are incorporating the sustainability aspects that are being built into Pattonsburg," observes Slone. In fact, Pattonsburg officials adopted a "charter of sustainability" in which they "recognize our responsibility to plan for the needs of present generations without compromising the ability of future generations to meet theirs." The charter goes on to call for "preserving the character and health of our natural environment, using and reusing materials, energy and water we need as efficiently as possible and eliminating waste."

Pattonsburg officials also adopted some "immediate development objectives" that call for mixed-use development that will place residents in proximity to the goods and services they need. They called for "pedestrian-oriented development and the adoption of energy performance standards for new construction that are 40 percent more efficient than the current average per capita consumption of energy among Pattonsburg residents.

"In the end it comes down to a matter of heart," Nancy Skinner says. "When people learn about the depth of our environmental problems, they have a motivation that is greater than just saving money. That is when they say to themselves: 'Aha, this really makes sense.' That is when they have the motivation to change the way they do business," she concludes.

Mayor Warford says that the design team helped change his thinking. "It sensitized me to environmental issues. I was aware of some of these issues before, but now I am seeing solutions to problems instead of just problems."

While looking for funds to move Pattonsburg to higher ground, Mayor Warford visited Washington, D.C., to meet with federal officials. While he was in Washington, he was struck by the number of homeless people. "Being from a small, Midwestern town we don't have a lot of homeless people. And I was told I would go broke in six months if I lived in Washington because I bought sandwiches for some of the homeless," he says.

Warford has a particularly strong recollection of standing outside the Smithsonian when a teacher, leading a group of students, advised his charges to look the other way and pretend the homeless people did not exist. "It struck me that perhaps part of the problem in this country is that we ignore our problems and hope they will go away. With our environmental problems we can all put our heads in the sand and say there is no problem with the ozone layer. But I am sure not going to tell my 13-year-old boy to pretend that these homeless people don't exist. I am going to say, Aren't we fortunate we are not in the same situation and what can we do to help? Of course one person can't help all the homeless people, but maybe as a nation we can. And maybe as a nation we can help solve the environmental problems that we face today."

Promoting Ecologically Sustainable Businesses in
West Coast Temperate Rain Forests

When Alana Probst argues that it is possible to make a livelihood that does
not overtax nature, she cannot be accused of being an armchair environmentalist who doesn't know what it is like making a living off the land. Her
father was the general manager of a large sawmill operation that used clearcut techniques on some of the best timber and forests in Oregon. A tall,
blond woman with an easy smile, Probst grew up in a number of small
towns, moving from one logging community to another as the supplies of
standing timber were exhausted. She knows what it is like to see people lose
their jobs and be forced to move out of town after an area is logged out.

It is this background, as a member of a fourth-generation logging family,
as well as her training as an economist that equips Probst to talk with
loggers, farmers, and fishermen about changing the way they harvest the
bounty of nature in the Pacific Northwest.

Probst is one of a number of economists who are pioneering a new field
of conservation-based development in which the emphasis is on creating
jobs that are ecologically sustainable. The central thesis of conservation-
based development is that passing laws and regulations that protect nature,
without providing for the needs of local people, is a strategy doomed to failure, and that in order to protect an ecosystem, the people who live in it must
have access to the information, tools, and capital necessary to build enterprises that do not destroy the vitality of the ecosystem on which they depend. Probst and other conservation-based development economists are
asking the question: Can economic development restore balance to ecosystems while bringing reasonable prosperity to local residents?

This conservation and job-generation approach is being tested by Probst
in Willapa Bay, Washington, a watershed about the size of Rhode Island

that is host to one of the last pockets of temperate coastal rain forest remaining in the lower forty-eight states. A century ago an extensive temperate rain forest ran up the coast through parts of northern California, Oregon, Washington, British Columbia, and Alaska. Since then, much of the old-growth timber, including giant red cedar and Douglas fir, has been cut down and replaced with plantations of softwoods. As a result the abundant biodiversity in many of these temperate coastal rain forests has been reduced.

In an effort to save part of what is left of the temperate coastal rain forest, Spencer Beebe, president of Ecotrust, a small conservation group based in Portland, Oregon, began to map where they still existed.[1] He focused on the pockets of temperate, coastal rain forest in North America that had survived, relatively intact, the juggernaut of development. Cartographers at Ecotrust discovered that outside of Alaska and British Columbia, Willapa Bay, located north of the Columbia River delta, was the last sizable stretch of temperate coastal rain forest in North America where biodiversity is concentrated. In fact, the Willapa estuary and forested uplands, where anywhere from 85 to 200 inches of rain fall annually, "is the most productive coastal ecosystem (estuary) remaining in the continental United States," Beebe reports.

While the tropical rain forests have a well-deserved reputation as being "hot spots" for large concentrations of species, temperate coastal rain forests such as Willapa Bay are also rich in biodiversity. Willapa Bay has twenty-one birds, nine plants, two salamanders, a butterfly, and a snail on the threatened or endangered species list.[2] The Douglas fir and western hemlock also provide a habitat for elk, black bear, mule deer, mountain lion, bobcat, coyote, raccoon, beaver, river otter, northern flying squirrel, and many small rodents and insect eaters. Willapa's eelgrass beds and marshland provide critical habitat for seventy species of migratory birds, and visitors are sometimes able to see up to 35,000 shorebirds at one time or 80,000 migrating waterfowl.

Willapa's rich ecosystem also provides a living for many of its 20,000 year-round residents and as a result much of Willapa is essentially a farmed ecosystem. One out of every six oysters consumed in the United States is cultivated in 10,000 acres of Willapa tidelands. Some 25,000 to 200,000 chum, coho, and Chinook salmon are caught in these waters every year, although, as the salmon catch has declined, hatcheries are replacing the wild run of these fish. Some 14,000 acres of cranberry bogs form the agricultural base of the community. And over the years loggers harvested Douglas fir, western hemlock, western red cedar, and Sitka spruce from the surrounding forest.

Not surprisingly, local harvests have put tremendous stress on the ecosystem. By 1977, half, or 12,469 acres, of shoreline wetlands had been diked. Since then, road building and industrial development have further diminished tidal marshland, reducing its filtering capacity. Runoff from housing developments caused fecal contamination of the oyster beds one year and the harvest had to be stopped and a water treatment plant installed to solve that problem. More threatening to the oyster beds, however, are two unmarketable varieties of shrimp—mud shrimp and ghost shrimp—that have wiped out 20,000 of the 30,000 acres of cultivated mud flats. As the ghost shrimp burrow into the sediment they compete for food with the oysters and bury them in mud.

The epidemic of ghost shrimp may have been caused by the reduction in the population of chum salmon, which are believed to have feasted on ghost shrimp larvae and kept their population in check. Efforts are now underway to enhance the salmon population and smoke them locally to add more value to the product before they are shipped out for sale. If the salmon raising, fishing, and smoking business becomes more lucrative, it is argued, there will be an economic incentive to help restore the salmon population and the resulting rise in the salmon population will keep the ghost shrimp in check. Another threat to the estuary comes from non-native cord grass (*Spartina*) which is displacing eelgrass and filling in parts of the bay, placing three-quarters of the tidal lands at risk.

Willapa's forests have also taken a big hit. From 1949 to 1986, 14 billion board feet of lumber were pulled out of the Willapa hills. Once home to some of the largest trees in the Pacific Northwest, the upland forests of the Willapa ecosystem are now mostly second growth. Currently nine out of ten trees are even-age plantation trees less than seventy years old and four in ten acres of trees have been planted in the last twenty years.

While this human activity, harvesting the fruits of the Willapa ecosystem, puts pressure on it, it also has the potential to create a significant constituency within the Willapa community that has an interest in maintaining the vitality of the ecosystem. Oystermen, for example, know that pollution from new development can contaminate the bay and destroy their

Figure 15.1
Alana Probst, economist at EcoTrust in Portland, Oregon, visiting Dave Nisbet, president of Goose Point Oysters in Willapa Bay, Washington. Courtesy Alana Probst.

cultivated oysters, wiping out a season's harvest at a single stroke. As a result, they have become some of the leading advocates for responsible stewardship of local resources.

It is just this kind of symbiosis between humans and nature that Beebe and Probst at Ecotrust hope to channel so that those who make a living from the land become its chief stewards. By generating more ecologically sensitive enterprises within the watershed, they are attempting to galvanize a local constituency that will protect the integrity of the ecosystem.

Having identified the Willapa watershed as an ecosystem worth working to preserve, Beebe assembled a team of conservation-based development specialists to provide financial support, business planning assistance, and help with marketing to ecologically sustainable enterprises within the watershed. Alana Probst joined this team and has since moved to Willapa Bay where she is working with a select group of local entrepreneurs. "We want to find ways for small businesses to become the engines for environmental restoration of ecosystems damaged by humans," Probst says.

But from the outset, any project such as this is fraught with pitfalls. Many rural communities are sharply divided over conservation and resource extraction issues. It is not unusual for local farmers, loggers, fishermen, and ranchers to be suspicious of big city environmentalists who blow into town for a few days to lecture them on how they are not treating Mother Nature with the respect that is her due and then proceed to tell them how they should run their businesses.

Fortunately, Probst and others at Ecotrust were savvy enough to recognize that this kind of approach was doomed to failure. Instead, they moved to town quietly and began searching out local entrepreneurs with whom they could work. In concert with the Nature Conservancy, they organized a meeting in 1992 out of which the Willapa Alliance was formed, a group of local entrepreneurs that includes oyster growers, fishermen, farmers, small business operators, landowners, and members of the Shoalwater Bay Native American tribe. Describing itself as "dedicated to developing and implementing strategies for sustainable, conservation-based economic development in the Willapa ecosystem," the Alliance was formed out of a sense that previous efforts to keep the Willapa Bay ecosystem healthy had failed to see the ecosystem as a whole.

The need for a systems-wide approach was made painfully clear to many in Willapa Bay when the oyster harvest was shut down because of a high fecal coliform count due to runoff from increased housing around the estuary. At that point it became obvious that development around the bay would have to proceed carefully if the oyster operations were to be kept viable. The Alliance then commissioned a series of studies on the vitality of the Willapa ecosystem including detailed reports on the salmon fishery and its prospects for recovery, as well as a broader study of local resources and assets.

Probst describes this study, the Willapa Indicators for a Sustainable Community, as a monitoring system that provides a feedback loop for residents to assess the health of their community and of the surrounding environment. Fifteen different indicators are followed including the average birth weight of babies, the percentage of young people leaving the region, the percentage of students graduating from high school and college, as well as water quality and the number of birds in the Christmas bird count. This kind of information, provided on a regular basis, stimulates residents to think about how whole systems work and has made them more receptive to innovative approaches such as natural systems for treating sewage, Probst notes.

But building a sustainable economy in Willapa means more than just talking people into installing efficient septic systems so that sewage does not contaminate the bay. It also involves creating jobs that add value to sustainably harvested resources. During the first year she lived in Willapa Bay, Probst watched thirteen small businesses and noticed that none could get a bank loan to expand its operations. To meet this need for capital, Probst went looking for money to invest in fledgling eco-sensitive enterprises. Her search led to the South Shore Bank in Chicago, a progressive bank that has invested $345 million in low- and moderate-income minority communities in Chicago using innovative yet rigorous banking practices. By 1994 Ecotrust and South Shore's holding company, Shorebank Corporation, began to work together with a small number of local entrepreneurs. The idea was to focus on enterprises that rely on a healthy, productive, and intact ecosystem for successful production.[3]

One of their first loans went to Tim and Sharon Schmitz, owners of Skamokawa Creek Enterprises, a family-owned sawmill that is specializing in ecologically sustainable forestry in the Willapa watershed. In a pasture not far from their home, the Schmitzes set up a small sawmill under an open-air shed. In a light winter rain, Tim can be found there feeding red alder logs into his Woodmiser saw, and then stacking the reddish hardwood planks.[4]

Alder has long been considered by conventional sawmill operators as little better than an overgrown weed that crowds out more commercially desirable species. But Tim Schmitz thinks that, for a number of reasons, alder has potential as a marketable hardwood for people who care about whether or not the wood they buy is sustainably harvested. First, if alder proves viable in the marketplace, it will keep woodlot managers from killing it off with a herbicidal spray that pollutes the groundwater. In addition, a market for alder will encourage woodlot managers to let another tree species grow in their forests, which in turn will provide habitat for a wider diversity of plant and animal species. Furthermore, alder has the advantage of growing quickly and maturing in thirty-five years, and it should be attractive to builders because it is a great imitator wood that can be stained to look like mahogany or walnut, Tim Schmitz says.

In the Schmitzes' modest sawmill operation, Probst sees the beginnings of a successful, environmentally sustainable enterprise that most banks would have refused to lend to. She recognized that Tim and Sharon Schmitz needed help with their business plan, marketing their product, and finding a loan to buy equipment. As a first step, through Ecotrust, she loaned them $600 to allow Tim Schmitz to take time off from his carpentry work and travel to northern California to look at sawmills that practiced sustainable forestry. She subsequently helped them put together a business plan so they could apply for loans.

In the process of formulating a business plan, the Schmitzes split with their business partner, an event which set them back because it meant buying him out. At this point most bankers would have abandoned the couple, but Probst saw the split as felicitous. "It was a lot better that they recognized they had different goals early on rather than further down the road when more was at stake," Probst comments philosophically. She stuck with the Schmitzes and her group provided them with their first loan of some $20,000 to buy alder from a neighbor's land. And they also introduced them to people at EcoTimber in San Francisco, a lumber supply company working with small sawmill operators who harvest wood sustainably. EcoTimber subsequently signed up to buy the Schmitzes' alder boards.

By granting a loan to the Schmitzes to mill red alder for the emerging green market for sustainably harvested lumber, Probst had a number of goals in mind. First, she wanted to reduce the damage done to the ecosystem by unsustainable logging. To this end she introduced Tim Schmitz to a forester at Ecotrust who oversaw his logging plan to insure that it was sustainably drawn up and executed. Second, she wanted to generate environmentally sustainable jobs in an area that had been hard hit by unsustainable logging.

Central to these kinds of loans is the idea of supporting enterprises that add value to a product locally, a practice which generates more local jobs than if round logs are shipped off to some distant point to be milled. While other logging trucks thunder down the highway loaded with uncut logs destined for distant regions, Schmitz adds value to the alder trees he fells by sawing them up in his pasture and he creates local jobs by hiring people to help him run the mill.

Further value is added to the wood when it is kiln-dried locally. Schmitz struck on an inventive solution to his need for heat to run a kiln. He made a deal with a local glass blower who was willing to divert some of the unused heat from his furnace into an empty portion of his barn where the lumber could be kiln-dried.

The Schmitzes also plan to add more value to their lumber by milling it into wainscoting that will command a higher price than plain cut boards. In addition, Tim Schmitz discovered that if he leaves some of the alder logs lying on the wet ground until they begin to rot on the outside, once he mills them they retain a beautiful color and patina that wood craftsmen and furniture builders value highly. Little is wasted in this operation because the off-cuts from the mill are sold as firewood.

What remains to be seen is whether the marketplace will take to alder. Architects and builders are conservative in their choice of building materials. They are wary of materials they have never used before because they do not know how they will age. To allay some of these concerns and interest customers in the environmentally friendly manner in which the alder is harvested, Probst helped the Schmitzes develop a brochure about the origins and properties of their wood for consumers. By documenting the harvesting of the alder, Tim and Sharon Schmitz hope that customers will see that by purchasing alder they not only buy a useful wood but also help maintain a healthy forest.

Since moving to Willapa Bay Probst has opened the ShoreTrust trading group—"the first environmental banking corporation"—a small nonprofit bank backed by Ecotrust and Shorebank Corporation, which provides loans and business expertise to environmentally sensitive local businesses. ShoreTrust now has a $5 million revolving, high-risk loan fund dedicated to ecologically sustainable enterprises.

One of the local entrepreneurs Probst supported is Karen Snyder, owner of Anna Lena's, a manufacturer of specialty foods made from cranberries. For decades cranberries have been the most valuable agricultural crop in Willapa Bay because the bogs along the coast provide ideal conditions for them. Transplanted in 1883 from the cranberry bogs of Massachusetts, cranberry plants thrived and there are now 14,000 acres of cultivated bogs in Willapa, making cranberries the most valuable food crop in the region. Most of Willapa's cranberries, however, are shipped off to large juice manufacturers such as Ocean Spray instead of being processed locally.

Karen Snyder decided to buck this trend. "It was sad when people couldn't buy cranberries in an area where they are surrounded by them," observes Snyder, a former daycare director who also worked for eleven years as a cranberry grower for Ocean Spray. Using a recipe for cranberry relish passed down to her by her great-grandmother, Anna Lena, Snyder cooked it up, bottled it, and sold it locally. She now sells forty specialty items made from cranberries, including cranberry catsup, cranberry mustard, cranberry scone mix, cranberry raspberry jelly, cranberry raspberry vinegar, cranberry curd, crannies (dried cranberries), hearty cranola granola, and, of course, cranberry sauce.

As the business grew, Snyder needed help with marketing. Through the ShoreTrust trading group, Snyder was introduced to some new marketing opportunities. She also redesigned her labels and promotional material to emphasize the firm's environmental values, the pristine environment in which the cranberries are harvested, and the natural processing techniques used. Now in its seventh year of operation, Anna Lena sells about a half-million dollars of products annually through its mail-order business. It employs three full-time, year-round workers and up to eight workers at the peak of the cranberry harvesting season. Recently, Snyder expanded her operation by planting her own cranberry bog and opening a local retail store.

ShoreTrust also made a loan to a local crabber that included some environmental standards he had to meet in order to qualify for the loan. For example, crabbing during molting season was prohibited because the crab's shell is soft and can be easily damaged. When the crabber became a major buyer of crabs in the area, these environmental standards were applied to all those who sold to him, multiplying the environmental benefits of the loan.

A similar pattern was followed when ShoreTrust made a loan to a salmon fisherman and helped him become the supplier for a chain of natural food stores and quality urban restaurants. The natural food stores and restaurants set high-quality standards for the fish they buy: the salmon had to be bled, iced immediately, transported within 12 hours, and of a minimum size. By meeting these standards the fisherman tripled his profits and subsequently started his own distribution company buying fish that met these criteria from other local fishermen. The fishermen also donate a percentage of their profits to salmon habitat restoration and volunteer their time to clear streams so that the salmon can swim upstream to spawn, Probst says.

Another loan aimed at shifting someone from a resource extraction job into a value-added enterprise went to a fisherman who started a fish-processing operation and is selling his product by mail order. Loans are also going to someone who will make crab pots out of recycled materials; and to a low-income housing project where people can invest sweat equity in their own homes, which will be designed to be resource and energy efficient. ShoreTrust is focusing its loans on farming, fishing, forestry, and tourism because these are all enterprises that are resource-based and depend on good stewardship of the environment to survive, Probst explains.

In the future Probst hopes to help people who make a living gleaning products out of the forest such as ferns, mushrooms, and seeds. These family enterprises usually sell their products to commodity buyers who pay very low prices. Gleaners need help putting together an institution, in which they will have an interest, that will make the rural-urban link and take their goods directly to market, thus providing them with access to better prices, Probst says.

While these enterprises are relatively small, they nevertheless provide good jobs in an ecosystem that is critical to protect. Already the husbanding of these environmentally sensitive enterprises is having a ripple effect and other businesses are beginning to take similar approaches. As similar businesses start up and gain strength it is hoped that there will be enough people in the Willapa watershed who are so dependent on the vitality of nature that they will fight to protect it.

More loans for ecologically sensitive enterprises will be made now that Probst has obtained a charter for the ShoreTrust Bank for which $6.5 million in EcoDeposits has already been raised. Probst expects that sum to increase to $12 million before opening the bank in mid-1997. While the bank's headquarters will remain in Willapa, loan offices will likely open in Portland, Seattle, Vancouver, and northern California, making it possible to replicate this model of development in a wider area.

Prior to her work in Willapa Bay, Probst worked to make local communities more economically viable by keeping money from leaving town. When the Oregon economy hit the skids in the 1970s, she decided to see if she could keep money circulating in the community instead of going out of state. To this end she was involved in creating Oregon Marketplace, an import-replacement service that found local suppliers for local manufacturers.[5]

In essence, Probst acted as a commercial matchmaker, convincing local manufacturers to buy (whenever possible and not at a higher cost) from suppliers within their community instead of importing resources from out of state. The rationale for this was that buying locally keeps money circulating in the community, thus stimulating the economy and creating new jobs. It also makes environmental sense in that it reduces pollution and energy consumption by eliminating the need for long-distance hauling.

Probst learned that, not infrequently, a local company would purchase goods from out of state that they could have bought more cheaply just down the street. For example, she discovered that purchasing agents for a local television station in Eugene, Oregon, were buying a multipage, news form from out of state. "There is nothing we are buying out of state that we can get locally," the general manager told her confidently. But Probst proved her wrong by finding a printer within sight of the television station that produced an identical form and saved the purchaser $1500 on the first order.

Probst made a list of local manufacturers and called them up and asked them what large-ticket items they would be purchasing over the next year. She then sat down with the local yellow pages and found local producers who could provide the manufacturers with goods of comparable quality at a better price. In many cases she discovered that the matching system worked. Calling her service Oregon Marketplace, she brought together a frozen food processor and a local poultry operation. The match transferred between $250,000 to $500,000 a year from out-of-state sources to the local economy. The poultry company added a $1.5 million plant and ninety workers to accommodate the newly identified market. Probst also matched a regional general goods chain with a local dairy company that provided it with distilled bottled water. The dairy made $1 million in sales on the bottled water in the first year and added three to five jobs.

In its first year Oregon Marketplace produced $1,858,500 in new contract agreements with eighty new jobs and $1.5 million in capital investment and financed its operations with a 5 percent commission on matched sales from the purchaser. Purchasing firms experienced up to 50 percent savings on some items. "Making a local connection also reduced the order-delivery time lag, communications costs, inventory costs, and warehousing needs," Probst reports. Buying locally saved local firms money, generated jobs, and reduced environmental damage caused by unnecessary transportation, she adds.

The program proved so successful that the state of Oregon opened thirteen computer-linked offices. Since then, twelve states and a number of other countries (including Canada and Micronesia) have initiated import-replacement programs based on the Oregon Marketplace model.

Students Swap Protests for Practical Work Building
an Ecologically Sustainable Campus

Daniel Einstein can trace his family tree to his famous fifth cousin twice removed, Albert. But the relative he admires most is a cousin who invented a
rotating lawn sprinkler. This preference for the practical over the theoretical is typical of the thirty-eight-year-old Einstein, who is Environmental
Management Coordinator at the University of Wisconsin at Madison, a
key position in one of the nation's most innovative campus environmental
programs. A lanky postgraduate student, Einstein, who is also director of
the Campus Ecology Research Project, channels the energies of undergrad-
uate and graduate students into finding practical ways to make the campus
more ecologically sustainable.

Convincing students to be practical, however, is not always easy. "Students want to stop pollution, stop global climate change, and save the rain
forest," Einstein says with a smile. This description fits twenty-four-
year-old Suzanne Tegen, a student of German literature, who says she
was "struck by save-the-worldness" during a trip to East Germany where
she was appalled by the degraded condition of the environment. "I wanted
to rush back here and solve all the environmental problems," she says,
laughing at her idealism.

But when she arrived back at college Einstein was there to bring her
down to earth to work on projects she could complete in one semester. "I
want students to get their hands on something doable and feel at the end of
a term that their work has made a difference," says Einstein, who engineered a job for himself as the university's environmental management coordinator, brokering projects that bring students, faculty, and staff together
to work on solving campus-based environmental problems.

Scaling down her ambitions, Tegen joined a team of students Einstein su-
pervises who are looking for ways to reduce the campus waste stream. She

zeroed in on one small piece of the problem—the large number of student newspapers going to waste. After two months monitoring thirty sites around campus, she estimated that every day approximately 11,500 copies of *The Badger Herald* and *The Daily Cardinal* are never taken out of the newspaper racks. To help other students visualize the monumental waste of resources involved, during Earth Week Tegen built a 21-foot tower of cardboard boxes covered with newspaper to represent half of the newspapers wasted on campus every day. For a while the student newspapers cut their press runs, but have since increased them again. As a result, the glut of newspapers remains a problem on campus. Einstein has directed another student to continue the campaign begun by Tegen and hopes that the new initiative may eventually serve as a national model. After she graduated in 1994, Tegen's college work on recycling landed her a job at the South Pole, recycling waste at the McMurdo research station.

Another former student of Einstein's, twenty-four-year-old Neil Peters-Michaud, now works for him on the Solid Waste Alternatives Project (SWAP) which is attempting to put in place a comprehensive materials management system for the university that will deal with materials from purchase to disposal.

As part of this effort Peters-Michaud set up a SWAP Shop on campus that over the last year has successfully diverted some 114,000 pounds of materials from the university waste stream. With grants from the Wisconsin Department of Natural Resources and the university, Peters-Michaud has arranged for university trucks to pick up materials that otherwise would have entered the university waste stream. The materials are then transported to a 1000-square foot university storage area where they are entered by students into a computerized inventory database. Pictures of the objects stored at the warehouse are taken with a digital camera and included in the inventory, which is circulated by the World Wide Web network. Around campus and elsewhere around the region, clients can cruise the inventory and see if there is anything they want.

One of the secrets of the Swap Shop's success is its marketing system. Peters-Michaud contacted 450 charities, nonprofit organizations, salvage operations, and remanufacturers and elicited from them a "wish list" of materials they were looking for. When items come into the SWAP's inventory, they are matched with these wish lists and calls are placed to alert clients that something they were looking for is now in stock.

Figure 16.1
Neil Peters-Michaud with computer equipment that previously would have been dumped in landfills but now is collected and resold to government agencies and nonprofit organizations through the Solid Waste Alternatives Project (SWAP). Photograph by Daniel Einstein.

On campus, SWAP saves the university money as materials that are no longer needed by one department are picked up by another at a cost of no more than 5 cents a pound. For example, surplus centrifuges, ovens, electric cabinets, and growth chambers that regulate temperature, no longer needed in the zoology department, were snapped up by the plasma physics department. Similarly, opal white glass was picked up by the art department and reblown. Last spring, when students moved out of their dorms, SWAP set up areas where departing students could leave their furniture. "We ended up finding someone to reuse every disgusting and foul-smelling carpet and sofa," Peters-Michaud says.

To date, SWAP has been able to move 95 percent of the materials it collects. It has been so successful, in fact, that the university will adopt it for wider use by hiring Peters-Michaud to run its surplus operation. A number of other universities have requested the SWAP's database so that they can copy this program.

In addition to collecting and redistributing reusable items, Peters-Michaud also nudged the university procurement program into buying more materials with recycled content. Through SWAP he put together a catalogue of 1200 office supplies that could be ordered through the state system and distributed the catalogue to forty-five departmental representatives at a workshop he gave to promote the use of recycled materials on campus. To reduce packaging waste on campus, Peters-Michaud convinced the campus state consolidated store, which buys truckloads of goods and repackages them for sale to different university departments, to experiment with the use of plastic reusable "tote" boxes instead of repackaging goods in corrugated cardboard.

This greening of campus procurement offices is happening at a number of universities around the country, notes Julian Keniry, author of *Ecodemia: Campus Environmental Stewardship at the Turn of the 21st Century*, the most up-to-date and comprehensive overview of practical environmental projects being carried out at colleges around the nation.[1]

For example, at Rutgers University, Keniry reports, the procurement and contracting department uses the competition inherent in the bidding process to push vendors into providing environmentally sensitive packaging and the purchase of recycled construction materials such as ceiling tiles, wallboards, insulation, plastic lumber, roofing products, snow fences, and parking bumpers. Purchasing agents are buying non–chlorine-bleached, post-consumer paper products; coordinating printer cartridge take-back programs; and buying recycled plastic garbage bags, among a host of other items. The switch to recycled garbage bags alone saved the university $18,000 a year, Keniry says. University bookstores and university-affiliated stores are also being persuaded to stock earth-friendly wares and buy merchandise in a more ecologically sensible fashion. Some colleges are giving a discount on the purchase of new batteries if old batteries are returned, selling tree-free stationery, and recycling old notebooks, Keniry continues.

Considering the size of the orders universities make for food, cleaning supplies, office equipment, stationery supplies, appliances, heating and cooling equipment, grounds maintenance supplies, and a host of other purchases, universities are capable of creating markets for earth-friendly products if they demand from their vendors goods and services that meet high environmental standards.

Einstein makes no bones about the fact that he is training environmental activists. In order to give students a positive experience and increase the chance that their work will be acted upon, he pairs each student with a member of the university staff who has an interest in improving some aspect of the campus environment.

This kind of matchup between student researcher and administration "client" worked well recently when Einstein introduced Lori Kay, director of transportation services, to twenty-three-year-old senior honors student Tabitha Graves. Graves subsequently helped devise a number of incentives for students, faculty, and staff to use mass transit instead of driving their cars when commuting to campus.

As a first step, Graves helped designed and won administrative approval for an "emergency ride home" program, which assures staff and faculty members that if they use mass transit to come to school and an emergency requires them to return home, they can take a taxi, save the receipt, and be reimbursed for 90 percent of their expenses. Without such a guarantee many members of the staff and faculty would refuse to use mass transit, she explains. Second, she was involved in negotiating a "flex parking" program that provides a discount on campus parking fees for students who reduce by at least one day a week the number of trips they make to campus in a single-occupancy vehicle. Third, she had a hand in expanding the university's telecommuting program that permits some university staff to work at home on certain days.[2]

Looking back on it, Graves says she found the work at times frustrating because progress came so slowly, but ultimately satisfying. "This work is more rewarding than the advocacy work I did in the past because I know these programs are going to be implemented," she says. Since graduating in 1995, Graves has been working full-time for the university's transportation services department, monitoring and expanding the programs she began as a student.

This sense of accomplishment and excitement that students feel when they are instrumental in solving real-world problems also appeals to David Eagan, a doctoral candidate who is writing his thesis for the Department of Educational Administration on whether college is providing students with the teaching and learning skills they will need in the workplace. A teaching assistant, Eagan is a second connecting point for University of Wisconsin students interested in making the university more environmentally friendly.

"It is curious that in a place where inquiry is so highly prized, the environmental impact of the campus has gone virtually unquestioned," Eagan writes in the book he co-authored with Oberlin College's David Orr entitled *The Campus and Environmental Responsibility*.[3] The reason for this oversight, he contends, is that college campuses are often depicted as ivory towers walled off from the real world. Eagan takes exception to this view: "At the University of Wisconsin," he writes, "the impact of 60,000 individuals and a billion-dollar annual budget is about as real world as you can get." With 42,000 students, 17,000 faculty and staff, on a campus of 900 acres, 220 buildings, 12 miles of roads, and 4 miles of lakeshore, the University of Wisconsin at Madison constitutes Wisconsin's eighth largest city. As such the university has a huge environmental impact and students should use it as a laboratory for finding solutions to environmental dilemmas, he argues. He is well placed to do this as the teaching assistant leading the university's Institute for Environmental Studies (IES) capstone certificate seminar, which gives academic credit to students who do research on the campus environment.

During the last four years that Eagan has taught the IES capstone course, his students have done "client projects" for him that chip away at some of the more wasteful practices on campus. One student convinced the librarian at the education library to use recycled paper in the copy machines by stocking one machine with virgin paper and the other with recycled paper. Contrary to predictions, it soon became apparent that the machines jammed an equal amount and since then recycled paper has been used in the copy machines. Another student, Matt Mitby, did an informal survey with a small number of students on what happens to the 200,000 bulletins and 70,000 timetables the university prints and distributes to students and found that many of them threw these publications in the trash when they had finished with them. Armed with this information he succeeded in convincing the administration to include information on how to recycle the timetables.

Jason Frost, a soft-spoken sociology student, spent the semester working with a team of other students streamlining the way the university collects and redistributes laboratory chemicals on campus. In the past, the university collected unused chemicals from laboratories around campus and then advertised through a biannual newsletter what chemicals were available for redistribution. The system was inefficient and cried out for computerization. Frost found WiscINFO, a university-run information system that could be adapted so that information on available laboratory chemicals could be posted on the computer network. With this system chemists could be constantly updated on what was available. Unfortunately, campus cost-cutting measures interfered with this project and after a year of operation the safety department discontinued the computerized service because it lacked the staff to update it. Instead, the department is publishing a paper version of it until it has the resources to develop an on-line version once again

Admittedly, these environmental improvements of university operations are limited in scope. But the incremental approach to social change has its advocates. In an article entitled "Small Wins: Redefining the Scale of Social Issues," Karl Weick writes that "Once a small win has been accomplished, forces are set in motion that favor another small win. When a solution is put in place, the next solvable problem often becomes more visible. This occurs because new allies bring new solutions with them and old opponents change their habits. Additional resources also flow toward winners, which means that slightly larger wins can be attempted."[4]

Of course it helps if the "small win" brings some resources along with it. For instance, Einstein struck gold when he supervised a graduate student, Dan Jaffee, who examined the university's contract with a wastepaper hauler. Under the contract the university was paying $3000 a month for the privilege of trucking its wastepaper to a company that resold the material to a recycling outfit. By rewriting the contract, cutting out the middleman, and selling the wastepaper directly to the recycling company, they earned $9000 on average a month, or a total of $126,000 from January through October 1995. Jaffee now works as the paper purchasing outreach person for the SWAP program.

But these kinds of incremental improvements can only be achieved if people like Einstein and Eagan are in a position to broker projects that bring together students, faculty, and administrators. One of the reasons that practical steps are being taken to reduce the impact of the university on the environment is that Einstein has ensconced himself within the bureaucracy that makes decisions about how the physical plant of the university is operated. Specifically, he developed a good working relationship with his former boss, Duane Hickling, assistant vice chancellor for facilities planning and management.

Hickling, a transplant from the University of Southern California, is part of a new breed of facility manager on campus who sees environmental responsibility as an important part of his mission. When Donna Shalala, now secretary of Health and Human Services, was chancellor of the University of Wisconsin from 1988 to 1993, she made it clear that the university was going to become a responsible member of the community and that university facilities would be run in an environmentally efficient and accountable fashion.

There is ample evidence that Hickling took his mission seriously. When he noticed that excavation at a construction site on campus was coming dangerously close to a cluster of 125-year-old American elms, he questioned his staff about what plans there were to save the trees. He was told that the trees were scheduled to be cut down because a stormwater sewer was due to be installed where they stood. "There was a full minute of silence while I collected myself after hearing that," Hickling reports. Then, without mincing words, he told his staff that they were not going to deface the university by cutting down ancient trees in order to install a sewer. Realizing that their boss was serious about the environment, plans for the sewer were redrawn and the trees were saved.

Administrators like Hickling rarely receive full credit for making environmental improvements possible on campus, writes Julian Keniry. "While faculty become acknowledged experts and students make headlines for constructing media-genic 'trash monsters' and 'quad landfills,' it is the purchasers, facilities and personnel managers, housekeepers, office services personnel, and other staff who actually manage most of the vexing logistics—accepting the blame when approaches fail, but receiving little credit when they succeed."[5]

Through Hickling, who moved to the University of Chicago in July 1995, Einstein gained access to members of the grounds crew and custodial staff and he continues to cultivate their trust. For example, when Einstein applied for and won the governor's award for recycling, one of the people he invited to the award ceremony was John Harrod, director of physical plant. In the parking lot of the governor's mansion, just before the award presentation, Harrod requested that Einstein find a student to research methods for reducing the amount of salt used in the sand that is spread on roads and sidewalks during the winter to keep ice from forming. He was concerned that when the snow melted, the saltwater runoff was killing vegetation and polluting the freshwater of Lake Mendota. In addition, salt causes concrete to spall and metal to corrode resulting in repair and replacement costs for bridges and concrete walks.

"Here is what appears to be a simple problem," says Einstein from his modest tenth-story office overlooking the campus. "You have a pile of sand and a pile of salt. The sand is going to freeze unless you mix some salt in it. So a member of the grounds crew operates a piece of heavy equipment and dumps a bucket of salt on the sand. He then runs the front-end loader back and forth across the pile to mix it. Now you have salt and sand." But the simple problem turns complex when you try to measure the amount of salt in a pile of sand, a task that is complicated by the fact that the two materials are unevenly mixed.

Convinced that the answer to the problem lay in mixing the salt more evenly with the sand, the student investigated what kind of equipment was available for this purpose. Whether the university will ever purchase this equipment is still unclear. But building on the research done by the student, an ad hoc committee formed to deal with the road salt issue is looking into the possibility of closing off some roads and sidewalks that are not traveled much in the winter.

Through his work at Environmental Management, Einstein is now experimenting with calcium magnesium acetate (CMA) as an alternative to salt. The liquid CMA is applied with a backpack sprayer prior to a storm as a de-icing agent that impedes ice from bonding with concrete, Einstein explains. Unfortunately, pound for pound, CMA is forty times more expensive than salt, so experiments are under way to see if this material's ability to protect expensive portions of the university's infrastructure will justify its cost.

Einstein has also been working on reducing the number of trees cut on campus as the university's infrastructure expands. Since Hickling left, some valuable trees have been lost, Einstein reports. In one case, three elm and oak trees, all over 100 years in age, were killed when a chill water line was installed along University Avenue. Engineers could have moved the pipe line 10 feet over and saved the trees, but it would have cost more money. A protest against the cutting of these magnificent old trees raised the profile of the issue on campus once again.

The problem is that each construction project requires the removal of just a few trees, but over the years the loss adds up until people begin to look around and say: "Didn't we use to have more trees around here?" Einstein notes. To highlight the cumulative effect of cutting down trees every time a new building is built or road widened, Einstein initiated the campus tree

Figure 16.2
Daniel Einstein with calcium magnesium acetate (CMA) backpack sprayer. CMA is an environmentally harmless chemical used as an alternative to salt. Photograph by Miriam Grunes.

inventory mapping project that uses Geographic Information System (GIS) aerial photos that pinpoint tree locations on campus. The location of each tree is then digitalized on computer and then students are dispatched to confirm the location of the tree on the ground and other data about its species, diameter, approximate age, and condition. Information on about half of the 6,000 tree locations on the inventory has now been entered, permitting the system to be used to place the loss of specific trees in context. For example, when a 150-year-old black oak was killed due to soil compaction and salt runoff when a driveway was widened, Einstein could print out a map of how many similar trees were still standing on campus to give the issue more poignancy. Similarly, when someone poached a small hemlock from campus to use as a Christmas tree, Einstein was able to tell a newspaper reporter that it was one of only eight similar trees on campus.

The tree inventory will also be useful in the development of a natural features walk that will provide interpretive signs and a brochure that will help people learn about natural, cultural, and historic features on campus such

Figure 16.3
Daniel Einstein, director of Environmental Management at the University of Wisconsin at Madison, with his campus tree inventory map created with the help of the Geographic Information System (GIS). Photograph by Dwight Pugh.

as the 300-year-old tree still standing on campus that was used for target practice during the Civil War, Einstein adds.

Solving practical environmental puzzles delights Einstein, whose background includes directing summer Youth Conservation Corps programs in California and Oregon, leading outdoor adventure and education programs in New York State, working as an energy auditor in Massachusetts, installing energy-efficient boilers in Maine, building housing in Nicaragua, and supervising a construction crew building a chimpanzee research station in Uganda. It was in Uganda, one hot afternoon, driving a jeep through the outback, that Einstein decided he could have a greater impact as an environmental organizer working among his own people than he could in Africa.

When he returned to the United States and enrolled at the University of Wisconsin to earn his masters degree, Einstein was part of a growing number of students who felt it was important to do something practical to protect the environment. A 1993 survey shows that 29 percent of college freshmen in the United States wanted to become involved in programs to clean up the environment and 84 percent wanted the government to do more on pollution, reports Paul Rogat Loeb, in his book *Generation at the Crossroads: Apathy and Action on the American Campus* (New Brunswick, N.J.: Rutgers University Press).

But despite this evidence of interest in environmental issues among undergraduates, Einstein cautions us to keep the scale of student environmental activism in perspective. At the University of Wisconsin, Einstein estimates that there are only about fifteen dedicated activists in the University of Wisconsin Greens, forty in the Wisconsin Public Interest Research Group, and another fifteen in the local chapter of the Student Environmental Action Coalition. Thus, while he recognizes that there are many environmentalists on campus who are not affiliated with organized groups, by his calculation there are no more than seventy hard-core environmental activists out of a total of 40,000 students on campus.

Einstein attributes the relatively small number of committed student environmental activists to a profound sense of despair and powerlessness prevalent among students. As a teaching assistant, he received many course evaluations from students who wrote that they were depressed to learn that environmental problems were so bad that they didn't think they could do anything to solve them. Perversely, classes intended to expose students to useful information about the environment were leaving them paralyzed by the scope of the problem. To fight this sense of hopelessness and despair, Einstein provides students with the opportunity, the contacts, and the supervision necessary to bring about positive environmental change on campus.

Eagan's view of the student despair problem is that university courses tend to educate "only from the neck up," or, as David Orr puts it, they teach facts but not applied research and action. Einstein's and Eagan's work attempts to correct this by requiring students to see how environmental issues work out in real settings. One of Eagan's students told him that his course was the first that had required him to develop and use more than just "book skills."

For his part, Eagan not only wants students to find ways to make the university more environmentally friendly, he also wants them to see the campus as a living ecosystem, not just the place where they go to class. When they walk on the grounds of this university founded in wilderness 140 years ago, Eagan wants them to know that this land was once prairie with scattered clusters of oaks. He wants them to know that student activists brought about the restoration of an important marsh on campus. He wants them to learn that the great conservationist John Muir was a student at the university and that Aldo Leopold taught here in the 1930s and 1940s and initiated projects in prairie and woodland restoration. One of Leopold's favorite test questions was to ask students to name a native Wisconsin plant they had seen on their way to class and describe its role in Wisconsin history, Eagan notes. In the process of learning about the natural history of the campus, Eagan hopes that students will become stakeholders involved in protecting it.

Another way to bind students to the place where they live is to get them involved in physical restoration work. "When you sweat and grunt and get your hands dirty you become connected with a place," says Eagan who volunteers for work bees at a folk organization outside of Madison. Students who plant a tree on campus come back year after year to see how it is doing, he notes. To encourage this kind of physical connection with the land, he leads volunteers on "campus keepers" workshops to cut out invasive plants such as buckthorn, honeysuckle, and garlic mustard that are crowding out native species in the woods on campus.

Eagan also convinced grounds department staff to grow some thirty native species—primarily sun-loving, prairie plants—in the bare spots in the landscaped beds surrounding Science Hall. He also organized members of the building's staff and students to care for the plants in an effort to involve them in stewardship of their turf. With the building grounds no longer subject to chemical weed control, staff and students handle weeds the old-fashioned way, Eagan explains. "We pull them."

Eagan took other students into Muir Woods, a forested natural area in the heart of the campus, where they drove stakes in the ground and did surveys of all the plants, shrubs, and trees within a 20-foot radius. Their intention is to turn this into a long-term survey so that future generations understand how the local flora is changing.

This hands-on learning approach spills over into Eagan's seminar where he has asked his students to sort through the trash and recycling bins at the student union. Trash-sorting teaches them firsthand what materials are going into the trash that could be recycled and what trash is contaminating the recycling bins. This information can then be used to devise strategies for

Figure 16.4
David Eagan with native plants used in landscaping around the Science Building.
Photograph by Daniel Einstein.

reducing the waste stream. One such project looked at better recycling bin placement in a library study room.

While the campus environment may seem an obvious focus for study, in fact it is often ignored, says Eagan, who peers out from behind steel-rimmed glasses. "Professors are famous for knowing everything there is to know about the tropical rain forest, but almost nothing about what is growing from the soil right outside their windows," he says. In an effort to reverse this trend, an entomology professor started a collection of the insects and lichen on the campus.

Eagan, who takes pleasure in introducing his students to members of the university staff they would otherwise not encounter, invited Douglas Thiessen, the university pest control specialist, to talk to his class. Thiessen, who describes himself as the unofficial campus wildlife manager, told students that it is his philosophy that "if insects or animals aren't bugging anyone, then leave them alone."

Eagan is also keen to have students acknowledge their own contribution to environmental degradation and the wasting of precious resources. Dorms across America have had to be rewired and energy consumption has soared as students have brought to school personal computers, printers, TVs, mini-fridges, microwaves, hair dryers and sound systems, he notes. It is always less painful to look at problems that other people create rather than those we are responsible for ourselves, he points out.

In one well-meaning initiative, a team of students in Eagan's seminar went off and did a project on their own time. They found an area on campus where the grass had been trampled by students taking a shortcut across the green. To solve the problem they dug up some wildflowers and planted them along with commercially available packets of "wildflower seeds," Eagan reports. The idea was to create a patch of wildflowers to deter students from cutting across the grass. Unfortunately, none of the wildflowers in their packets of seeds were native to Wisconsin and the students had neglected to take into account shade from a maple tree. As a result, their wildflowers failed to sprout and within weeks, after the students had left for summer vacation, the site was a washed-out mess of mud. The grounds department employees eventually cleaned up the mess and replanted grass. Eagan now uses this sorry story as a cautionary tale to teach the need for carefully thought-out plans and follow-through in campus environmental projects.

To make sure that the data students collect are preserved for future use, Eagan put together a library of 200 research papers on various aspects of the campus environment that are available as resources for students and others working on campus environmental issues. The papers have been catalogued and indexed so that one can now see the breadth of subjects students have already looked into. There are papers on pesticide use, biohazardous waste disposal at the veterinarian building, disposal of wood pallets, the use of individual paper packets of sugar in the student union, the pros and cons of paper towels vs. hot-air hand dryers, the handling of toxic materials in the art department, student light-switching behavior in bathrooms, and a host of other issues.

While many student groups around the country have done environmental audits of their campus environment, Eagan prefers to look at the compilation of student research as constituting an environmental profile of the university "The word 'audit' is very confrontational. If you audit someone you are looking for something that is wrong. The IRS does audits. We do assessments or profiles," he says. "I don't want to sneak up on anybody and embarrass them. We are not taking an adversarial approach and trying to find dirty linen on campus. My position is that we are all in this together; it is not just us against them." So, rather than marching into the chancellor's office with a master plan for how students think the university should be reformed overnight, Eagan and Einstein have mapped out a long-term campaign where gains can be made one step at a time.

"Although it requires a stretch of the imagination," Eagan writes, "campus environmental stewardship offers students the opportunity to contribute to the construction of a new sort of monument—a sustainable campus."

Around the nation there are many other examples of students working to see that their campuses are designed, constructed, and maintained in an ecologically responsible way. In fact, since 1990 there has been a boom in environmental activism on campus.

Evidence of the extent of environmental organizing on campus surfaced at the first national student environmental conference held in 1989 at the University of North Carolina at Chapel Hill, which attracted 2000 activists, writes Keniry, senior coordinator of the campus outreach program at the National Wildlife Federation, in an article entitled "Environmental

Movement Booming on Campus."[6] A year later that number more than tripled when 7000 activists gathered at the University of Illinois at Urbana-Champaign. Today, the Student Environmental Action Coalition (SEAC) is represented on 700 campuses and a National Wildlife Federation's campus ecology program for students has registered 578 college environmental programs in the last five years. More recently, students gathered at Yale University in 1994 and held a "Campus Earth Summit" that resulted in a report entitled *Blueprint for a Green Campus* published by the Heinz Family Foundation.

This activity flies in the face of current theories that students today are more apathetic than those of the past. What may lead some to think that student activism is on the wane is that highly visible protests and sit-ins are no longer the order of the day and activism is no longer equated with conflict. Today, as we have seen in the case of the University at Wisconsin, some students work cooperatively with members of their faculty and administration in institutionalizing environmental change on campus. This solves the problem students faced previously when many environmental campaigns died after key activists graduated.

One of the tools that has shaped this movement is the campus environmental audit described in *Campus Ecology: A Guide to Assessing Environmental Quality and Creating Strategies for Change* by April Smith and the Student Environmental Action Coalition (SEAC).[7] Many student groups have used these audits to identify areas of campus activity that could be made more sustainable. Sharing the results of the audits (or profiles) has also led to the replication of successful programs.

But the proof of the vitality of this movement is not in paper audits but rather in what student environmental activists have accomplished. A number of books on student environmental activism on campus document the extent and variety of the campus ecology movement. Here are a few examples of initiatives that have taken place to date:

• An energy management plan devised by a graduate student at the University of Rochester, Morris A. Pierce, was adopted by the board of trustees. With a capital investment of $33 million, it will save the university $60 million over 20 years.

• At the State University of New York at Buffalo, energy officer Walter Simpson says that some 300 energy conservation projects saved the university $3 million between 1982 and 1993.

• At Brown University ("Brown is Green"), students successfully lobbied for the inclusion of energy-efficient lighting and machinery in the renovation of four dormitories and a number of laboratories on campus. They also succeeded in replacing the bulbs in exit signs with a more efficient variety and had low-flow showerheads installed in dorms.

• An ecolympics program at Harvard University (now known as the Green Cup) sponsored a competition among dorms to save energy and conserve water. In 1990–1991 the program is reported to have reduced energy consumption by 25 percent in residence halls, saving the university $500,000. The program has since spread to Yale, George Washington, Tufts, Texas A&M, the University of Wisconsin, and Western Washington State University.

• At the University of Colorado at Boulder a sophisticated recycling system with twenty-four grades of recyclables diverted 96 tons of materials, or 35 percent to 40 percent of the waste stream, in 1992–1993.

• At the University of Minnesota Twin Cities Campus, custodians no longer service individual wastebaskets. Instead, it is now the responsibility of the individual to presort waste into four categories—three recyclables and one trash. They then dispose of their presorted waste in a conveniently located four-bin ("quad") recycling system.

• At the University of Illinois at Urbana-Champaign students voted to increase their transportation fee to $18 a semester so they would be issued a pass to ride on the city mass transit system. Student use of mass transit rose from 3000 rides daily to 11,000. A new campus-run bus system provides an additional 26,000 rides daily.

• When students at Hendrix College in Conway, Arkansas, investigated where their food came from, they found that 90 percent of it was purchased from out-of-state sources. The Hendrix Local Food Project subsequently bought 30 percent of its food from local farmers.

• Students at Dartmouth College recently established an organic farm on campus to provide the food service with fresh herbs and vegetables. And Bates College in Lewiston, Maine, buys produce from a cooperative of local organic farmers.

• St. Charles Community College in St. Peters, Missouri, has instituted a reusable-mug campaign to reduce the use of Styrofoam and paper cups on and off campus. Local restaurants and chains such as McDonalds and Hardees offer substantial price discounts to customers who supply their own mugs.

Clearly, the network of students involved in the practical work of building ecologically sustainable campuses is extensive. From this movement we can anticipate two positive outcomes. First, the students who have been trained to do careful research focused on solving specific environmental problems will eventually graduate and spread out across the country carrying these skills with them. And second, some university communities will adopt environmentally friendly policies and practices that will be replicated by other communities.

Will communities across America become models of environmentally responsible behavior? Only if people like Einstein and Eagan are given the resources and the chance to prove that it is possible. Undergraduates, for reasons ranging from work overload to limited skills, are unlikely to make significant inroads into solving campus environmental problems without the guidance of such mentors.

Western Cattle Rancher Experiments with
Sustainable Techniques

It was a dog named Shep that started the whole thing. Back in 1983 Shep dumped a dead, weasel-like animal on the doorstep of Lucille Hogg's ranch. Lucille, who runs Lucille's Cafe, had never seen anything quite like it. So she called the game warden who passed the carcass on to some expert. Eventually it was identified as a black-footed ferret—an animal thought to be extinct.

After that, all hell broke loose in Meeteetse, Wyoming (pop. 368). Hordes of experts, environmentalists, and bureaucrats rolled into town and started to tell local cattle ranchers how to manage their operations to save what was left of this endangered species.

Jack Turnell, forty-nine-year-old manager of the Pitchfork Ranch, one of the oldest cattle producing outfits in Wyoming, was alarmed by the discovery of the black-footed ferret on his land and that of his neighbor Lucille Hogg. He knew it would mean operating under additional regulations. "When the ferret was found here I had to deal with a lot of people I didn't care to deal with," he admits. "I felt like I didn't need any input. I didn't want anybody messing around here. I just wanted them to stay the hell out."

But, like it or not, Turnell found himself on the Black-Footed Ferret Advisory Team, consulting with officials from the U.S. Forest Service, the Bureau of Land Management, the Wyoming Fish and Wildlife Service, and an assortment of environmentalists. "At first they got a little pushy," he recalls, "but I told them I must be doing something right because my ranch was one of the last remaining places that still had the black-footed ferret."

Eventually the team worked out a compromise whereby local ranchers could continue to graze cattle on lands occupied by the ferrets if they agreed not to develop oil or gas wells in the same area until further studies had been completed. The studies revealed that the sylvatic plague was killing the prairie dogs the ferrets depend upon for food. The few surviving ferrets, some two dozen, were taken into captivity where they were bred to a population of 1400 animals. Of these 228 were released into the wild but only 7 survived. Despite these low survival rates, efforts continue to save the black-footed ferret, described by some as the most endangered mammal in North America.[1]

The discovery of the black-footed ferret on Turnell's land sensitized him to environmental issues and taught him that he could negotiate a fair agreement with the authorities that would protect the ferret and permit him to run a profitable cattle operation. Since then he has made a serious study of environmentally sustainable ranching by experimenting on his own and consulting with experts in range management at the University of Wyoming. "When I took over managing the Pitchfork Ranch at age twenty-six I just did things the way they had always been done," he remembers. "I don't think we ever managed the land all that badly, but I have learned in the last few years that we can change and do things better."

A range war rages in the western states over the extent to which cattle should be allowed to graze on federal lands. Environmentalists argue that vast herds are stripping the arid western plant cover faster than the vegetation can regenerate. Cattle are damaging fragile riparian ecosystems and causing a variety of species to vanish. Outraged at this desecration of the landscape, some anti-cow activists are demanding that public lands be "cattle-free".

The question is being asked whether the public benefits from having public lands used as a feedlot for privately owned domestic animals.[2] A number of books and documentaries support the elimination of cattle from the western range. In his book *Beyond Beef: The Rise and Fall of the Cattle Culture*, Jeremy Rifkin argues that if Americans gave up eating beef and ate lower on the food chain, they would be healthier and the ecosystems of the West would flourish once again.

From the other side of the fence, ranchers are firing back by organizing their own congressional lobbying efforts and plastering their pickups with Cows Galore in '94 bumper stickers. Grazing actually improves the land, they maintain, because cattle fertilize the soil with their manure, work the earth with their hoofs, and keep the grass short and green. If ranchers are forced off the land, they warn, their water distribution systems will disintegrate and much of the land could revert to desert.

Caught in the crossfire of this war of words between ranchers and environmentalists is Jack Turnell. Turnell bridges the gulf between rancher and environmentalist at a time when the beliefs of each are becoming increasingly polarized. One of a small but growing number of ranchers experimenting with range management practices designed to minimize environmental damage, he is convinced that profitable and ecologically sustainable ranching is possible if ranchers heed the dictates of nature.

"My goal now is to prove that productive ranching and environmental concerns can be compatible," says Turnell, a short, stocky cattleman who sports a carefully cropped mustache and a pink Oxford button-down shirt. Squinting into the morning sun, he looks out across the Pitchfork Ranch that covers 120,000 acres, including a 40,000 acre U.S. Forest Service allotment. The grazing land runs up the Greybull River valley to the Absaroka Mountains and into the Shoshone National Forest. "I don't want my kids and grandchildren growing up in a polluted, dirty, broken world," he says. He is determined to pass on to his children a ranch where the natural resources are still intact.

With Washington calling for higher grazing fees and better range management practices, ranchers must begin to reform the livestock industry from within, Turnell argues. If they don't they will find their activities dictated by constantly fluctuating regulations that swing like a pendulum from strict to loose and back again. But changing ranching methods is not likely to come easily to cattlemen steeped in old ways of managing the land. Turnell's own metamorphosis from traditional to sustainable rancher did not take place overnight.

Turnell's education in eco-ranching took a giant step forward when he began to look at the impact of his cattle on the fourteen creeks that crisscross the Pitchfork Ranch. He credits environmentalists with having alerted him to the importance of protecting "riparian zones," a term he was unfamiliar with that describes the delicate ecosystem immediately surrounding streams and rivers. "They did something useful there," he concedes. While riparian zones account for less than 1 percent of the landscape, 75 to 85 percent of wildlife depend on these narrow, watered areas for their survival.

As a result, if riparian zones are degraded, it has a devastating effect on wildlife and the health of the local ecosystem.[3]

Actually, Turnell was already aware of the problem. Two decades of experience as a rancher had taught him that cattle tend to congregate in and around water sources where, as he puts it, "they beat hell out of the land." The way that livestock degrade riparian zones when improperly managed is well known. Drawn to the water to drink, they strip the surrounding area of vegetation, leaving few plants to hold the soil in place when the spring runoff rushes down from the mountains. Cattle also defecate in or near streams and trample down the banks of the streams, making the channels wider and shallower. In their wake, sediment and coliform bacteria pollute the water. Cattle also eat tree seedlings and the leaves of trees and bushes that provide shade and shelter for small birds and rodents. Eventually the water table drops and some streams dry up.

To mitigate this problem, Turnell employed a "constant herding" technique to keep his cattle from congregating too long in streams and spread them out more evenly over the land. To further reduce grazing pressure on his streams, he developed alternative sources of water for his cattle in a number of locations around the ranch by drilling wells and building reservoirs to catch runoff from the snowmelt. To trap rainfall for the cattle to drink, he installed water harvesting devices called "guzzlers" and he placed salt licks at a distance from the water to draw the cattle away from the streams.

Turnell also went looking for a new breed of cattle that would leave the stream beds and climb the mountainous Forest Service allotment where his cattle spend the summer months. After an assiduous search he found a breed of active, hardy cattle called Salers in southern France that looked promising.

"They were climbing hills like no cow I had ever seen. They looked like a bunch of goats," Turnell recalls. When he bred the Salers with his own Angus and Herefords he found that instead of lingering near the creek the new breed climbed parts of the mountain that the Forest Service officials claimed could not be grazed. To prove it he has photographs of his cattle grazing at 11,000 feet on alpine slopes.

Breeding cattle that spend a minimal amount of their time in the creek beds has substantial environmental payoffs. Cruising the dirt roads of the ranch in his Lincoln Towncar, the sun glinting off a flashy pinkie ring, Tur-

nell proudly points out the health of his fourteen creeks. Using an old photograph as a point of reference, he documents how many more willows and
cottonwoods shade a section of creek than there were years ago. "That's
a damn healthy riparian zone. Nobody can complain about that," he
declares.

Turnell's interest in the health of his creeks led him to join a group of
ranchers, environmentalists, and scientists in founding the Wyoming Riparian Association. "One of the things I respect about Jack is that he is willing to sit down and talk with people who are traditionally considered
opponents of the ranching community," says Connie Wilbert who chairs
the Wyoming Chapter of the Sierra Club and sits with Turnell on the Riparian Association board. "He is open to working with environmentalists
instead of meeting them in court." She considers him one of the most progressive ranchers in the state.

Figure 17.1
*One of the creeks on the Pitchfork Ranch in Meeteetse, Wyoming, that manager
Jack Turnell has protected from overgrazing. Photograph by Steve Lerner.*

Turnell has taken some heat for his views on sustainable ranching, says Wilbert. "When he first began to talk about environmental issues, other ranchers would just roll their eyes and ask, 'Who is this nut?,'" she recalls. "But there are a growing number of ranchers who are interested in making changes because they recognize that if they don't the federal government will eventually mandate change."

Since he began studying the ecology of riparian zones, Turnell has changed some of his management techniques. He used to trap beavers, but now permits them to build dams because the dams hold back the water allowing the creek to flow year round. As the water spreads out behind the beaver dam a wide variety of vegetation grows that attracts and supports a diversity of animal life. Turnell also discontinued the practice of permitting local residents to cut firewood from the deadwood that lines the creek beds. Leaving the debris "slows down the river and gets it to meander a bit so that when floods come our soil doesn't all end up in Louisiana," he says.

Protecting the creeks from the impact of his cattle is not the only challenge. Overgrazing and soil compaction are also damaging. When cattle crop the grass too closely, new growth is inhibited and topsoil is exposed to erosion. The compaction of soil by cattle also means that less water soaks into the ground and instead runs along the surface, causing more erosion. Overgrazing is responsible for about 30 percent of soil loss in the United States and 35 percent globally.[4]

Fragile mountain slopes are especially vulnerable to damage when large concentrations of cattle are mismanaged. As they follow one another up steep slopes, their hoofs cut into the grass cover leaving denuded paths that erode into gullies when the snow melts. This problem is exacerbated by a Forest Service practice whereby in many areas all a rancher's cattle are herded up the mountain at the same time so that forest rangers can keep an accurate count of cows on federal land. This practice results in placing too many cattle on ecologically fragile slopes.

To minimize damage of this nature, Turnell and local forest rangers experimented with "drifting" or "pulsing" small groups of cattle up the mountains to their summer grazing area to reduce the concentration of cattle. Officials at the Forest Service are so pleased with the results of this experiment on Turnell's 40,000-acre allotment in the Shoshone National Forest that they have used the Pitchfork Ranch for tours and workshops to show how ranching and protecting the environment can be compatible.

Turnell's range management plan has been further refined to protect the Forest Service lands he is permitted to graze. In the summer his herd is divided into small groups of cattle that are "pulsed" up different drainage basins and kept separate from each other by mounted cowboys. "That way the cattle drift and pick their way up the mountain just like a band of elk," says Quentin Skinner, professor of range management at the University of Wyoming, who has monitored the impact of Turnell's cattle on the land for the last five years. "The vegetation has made an amazing comeback" as a result of this practice, he observes.

To have an objective measure of what impact the cattle are having on the range, Skinner and Turnell have established fixed points where cameras are placed every year to photograph vegetative cover. These photographs are then compared with previous years to see if vegetation is becoming denser or more sparse. The number of cattle let onto any area of the range can then be manipulated to meet the needs of the land and avoid overgrazing. "I feel pretty comfortable that what Turnell has done is sustainable and that he has done a super job," Skinner says. The University of Wyoming Range Department says that the condition of riparian areas and the range on the Pitchfork Ranch have improved from a "poor to fair" condition to a "good to excellent" condition, Turnell reports.

Because he comes from the ranching community and has practical experience learning to minimize damage done to the land by cattle, Turnell is much in demand as a speaker. As he prepares for a lecture and slide show at the Rotary Club in the neighboring town of Powell, Turnell says that one of the lessons he teaches fellow cattlemen is that if you take care of the land you can profit by it. His own environmental activities have paid off handsomely: "I am making more money now than I have ever made," he says. Since his cattle now graze mountainous areas that before were inaccessible, they eat more grass and gain more weight. The Pitchfork Ranch now produces 542,000 pounds more beef than it did in 1980 despite a 65 percent cut in the federal grazing permit and several years during which he kept his cattle off the entire federal allotment.

For his efforts to make ranching environmentally sustainable, Turnell has been awarded the first Environmental Stewardship Award by the National Cattlemen's Association (NCA). The fact that the NCA decided to create such an award suggests that negative publicity about the impact of cattle on the western environment has reached a critical stage. In making the award to Turnell, NCA president Jim Wilson told 5000 cattle producers that "the environment may be the most challenging issue facing the cattle industry in the nineties." He went on to congratulate Turnell for demonstrating to the public that cattlemen are "good stewards of the resources for which we are responsible."

Turnell is modest about his award: "I am not unique among cattle producers in my concern about the land," he insists. "Ninety percent of my neighbors and friends in the industry think just like I do. They just don't like outsiders telling them what to do."

Turnell's environmental award does not mean that there is a love feast between him and all environmentalists. Nor is he someone you would ordinarily peg for an environmentalist. For one thing, the photographs of Marlboro men galloping across a western landscape on their mustangs on the wall of his ranch office would certainly not be displayed on the walls of the Sierra Club or the Audubon Society. Yet he remains proud of the fact that many of the cigarette ads were shot on the Pitchfork Ranch.

In fact, while Turnell has adopted a number of ecologically sensible ranching practices, many of his views would elicit an argument from environmentalists. For example, like many other ranchers, he is impatient with environmentalists who say that cattle production should be eliminated on public lands in the West. "Let's back up 100 years when there were 60 million buffalo on the prairie. Then they were eliminated. Now we have 25 to 30 million cows on the same lands. You don't have to be a rocket scientist to figure out that we have a whole lot less bovines tearing around the land than we did 100 years ago."

Critics of this logic, however, point out that when the buffalo roamed over the prairies there were no fences that kept them in any one area and that their nomadic movement caused less damage than current ranching practices where cattle are often left to graze on arid lands for extended periods, causing severe vegetation depletion and erosion.

But Turnell does not buy this argument. Far from doing harm, he believes that properly managed grazing leads to healthier grass on the range. "I believe we are going to improve conditions over what they were like when the buffalo were stampeding around," he says. If today's ranchers permitted their cattle to do the kind of damage that buffalo did in the past, "the public would be down our throats like a biting sow," he adds.

Television documentaries that imply that cattle ranching is ruining western lands through herds of cows overgrazing the land and compacting the soil are particularly disturbing to Turnell. "Every time there is a television show about ranchers, it shows a cow standing on some dune in a desert. That's not fair. For crying out loud, most of Wyoming is not a desert. It is a beautiful, wonderful grassland and it is well taken care of." While some might think that Wyoming's dry brown tufts of grass compare poorly with the lush green grass that covers the fields in the spring in the eastern portion of the country, Turnell sees Wyoming grass through different eyes. Pushing his black cowboy hat back on his forehead and squatting down to grab a handful of grass, he says, "This here is good grass that packs a real punch. It has as many total digestible nutrients as any grass you will ever see. That tall green grass you see in the East is some sorry grass compared to this."

Turnell admits that some of the rangeland in Wyoming has been overgrazed. "Of course there are some areas that need to be fixed—no question. Even here on the Pitchfork Ranch, as hard as we work, there are areas that we have to fix. But most ranchers are taking a bad rap." Some ranchers do a bad job and the land suffers, Turnell concedes. There are those who want to do everything the way their ancestors did and are closeminded about new ideas. They just put the cattle out on the land without thinking about what impact they have. These are the same cattlemen, Turnell observes, who later complain that their cows are not gaining enough weight and that the price of beef is too low. But while there are a few irresponsible ranchers, Turnell believes that the majority care deeply about the land. After all they live out on it every day and they depend upon it for their livelihood.

Not everyone agrees with this assessment. "Many ranchers still practice what I call the Columbus method: they turn their cattle out to graze in the spring and discover them in the fall," says Ed Chaney, director of the Northwest Resource Information Center. "Most cattlemen are still in the denial stage about the impact of their activities on the environment."

Nor does everyone agree that reform of livestock operations will be enough to restore the health of the western rangelands. "If you move cattle out of the stream bottoms and into the uplands you will still be pounding to death the springs, steeps, and creeks, not to mention contributing to soil compaction over the entire uplands," notes former park ranger George Wuerthner.[5]

In a self-published book entitled *Waste in the West: Public Lands Ranching,* Lynn Jacobs also argues that ranching reform is a contradiction in terms and that the only way to heal the damage done by livestock operations on public lands is to evict both the cows and the cowboys. "Except for a few expensive 'showcase' demonstration sites, when you get on the back forty all you see is the same old cowblasted uplands, cow-nuked riparian zones, trampled, stamped, and torn streambanks."[6]

If one looks at the big picture, the amount of pressure that domestic animals place on the world's ecosystems is clearly escalating. While human population has doubled to 5.4 billion since the 1950s, four-legged livestock has increased from 2.3 to 4.0 billion head and the number of fowl has increased from 3 to 11 billion. In this country, western herds have shrunk compared with the record numbers on the range in 1885, but that does not mean that they do not continue to do substantial damage to the land.

"The western United States is left with a sad legacy: The great cattle boom of the last century annihilated native mixed-grass ecosystems. And unsustainable practices—including overstocking and grazing cattle for too long in the same place—continue on much of the 110-million hectare area of public land the federal government leases to ranchers," write Alan B. Durning and Holly B. Brough in a Worldwatch Paper entitled *Taking Stock: Animal Farming and the Environment.* Cattle have played a prominent role in reducing the ecological productivity of drylands and caused desertification, they write.[7]

The condition of federal lands that are leased for grazing has been the subject of numerous studies. A 1990 Bureau of Land Management (BLM) study reported that only one third of its vast western holdings were in good or excellent condition and that half of U.S. rangeland is severely degraded.[8] A 1988 Government Accounting Office (GAO) report stated that "poorly managed livestock grazing is the major cause of degraded riparian habitat on federal rangelands." The GAO report went on to say that 60 percent of public range is in unsatisfactory condition and very little is being done about it.[9] A study by the Natural Resources Defense Council (NRDC) and the National Wildlife Federation (NWF) found 100 million acres of public rangeland in "unsatisfactory condition."[10]

Riparian degradation due to cattle grazing is now so widespread that most range managers have never seen a healthy stream channel, notes Wuerthner, a former Forest Service ranger who is trained as a biologist and botanist. "The rangers working now were trained to produce cows, not healthy ecosystems," observes Tom France, director of the Northern Rockies National Wildlife Federation.[11]

Not only have the 20,000 ranchers that run cattle on 270 million acres of land in eleven western states degraded the western range, they have also stuck the taxpayer with a large bill in the process. In 1989, for example, the Forest Service spent $34.5 million on range management while taking in only $10.9 million in grazing fees.

Not surprisingly, BLM officials reject this description of deteriorating conditions on the western range. In a BLM paper entitled "State of the Public Rangelands, 1990," officials noted that grazing pressure on the range had declined from a high of 22 million animal units per month (AUMs) in 1941 to 10.5 million AUMs in 1988–1989.[12] BLM officials classified 3 percent of their land as in excellent condition, 30 percent good, 36 percent fair, 16 percent poor, and 14 percent unclassified. Another BLM white paper, entitled "Range of our Vision," boasts that since 1936 areas in the highest ecological condition have doubled while the acreage in the lowest condition has been halved. Furthermore, big-game animal populations of moose, elk, deer, bighorn sheep, and antelope had expanded substantially from 1960 to 1988.

But naturalists and sportsmen are not convinced that cattle ranching is good for wildlife, pointing out that cows not only damage riparian areas on which most livestock depend, they also compete with elk, antelope, and deer for the same forage. For example, domestic sheep and cattle eat many of the same plants on which bears depend. Cattle also trample the eggs of nesting birds and crop grasses so short that many small animals are exposed to increased levels of predation.

"The potential extinction of species is merely one cost of the West's love affair with the cow and the cowboy," observes George Wuerthner in an article entitled "How the West Was Eaten." The grizzly bear, black-footed ferret, Texas tortoise, dunes tiger beetle, Bruneau Hot Springs snail, Colorado squawfish, and red wolf are among a long list of species that are either on or are candidates for the endangered species list because the western rangeland has been degraded by cattle and sheep ranching, he says.[13]

Federal funds are also spent on predator control to protect domestic livestock through the Animal Damage Control agency which kills tens of thousands of coyotes and hundreds of thousands of prairie dogs, Wuerthner points out. Some ranchers avoid the practice of having coyotes hunted by helicopter by delaying their calving season from January to mid-March when coyotes are feasting on plentiful populations of sage rats instead of on calves, he adds.

Turnell is among the group of ranchers who actively protect wildlife on their lands and he disagrees with those who hold that livestock operations and wildlife protection are incompatible. "The cattleman is wildlife's best friend," he contends, as he points out a herd of antelope browsing on the sparse brown grass in the middle distance on the Pitchfork Ranch. If the land were not used for cattle production it would likely be subdivided for summer homes, ranchettes, and resorts, which would destroy critical wildlife habitat.

Concern about wildlife conservation has a long history on the Pitchfork Ranch. The oldest ranch in the area, it was established in 1878 by Count Otto Franc von Lichtenstein, a member of the German royal family. Turnell has photographs of his wife's family bottle-feeding antelope in the 1920s and 1930s. One family photo shows bottle-fed antelope being loaded onto the Hindenburg for shipment to a German zoo. Another photo shows cowmen untangling barbed wire from the antlers of an elk: "And these guys aren't supposed to care a damn about wildlife," he remarks. To protect wildlife on the Pitchfork Ranch, Turnell has periodically made himself unpopular with local residents by closing down the hunting on his land for years at a time until the antelope herds recover from overhunting or a killoff during a particularly bad winter.

Is it possible for other ranchers to practice the same sustainable ranching techniques as Turnell or are there special circumstances that permit him this luxury? Some argue that the gas wells on Pitchfork Ranch and his outside investments in real estate and construction make it possible for him to practice a sustainable method of cattle production. "Oil and gas has certainly helped Turnell financially," admits Wyoming University professor Skinner, "but I can assure you that his model of ranching can be used by a one-family, small operation as well as a large ranch." In fact, it takes fewer cowmen to manage cattle in small groups, the way Turnell does, than to move large herds, Skinner argues.

"I think the momentum to change ranching so that it is more environmentally sustainable has just begun in our industry," Turnell says. As evidence of this he notes that there are now over 100 ranches in Wyoming alone that are practicing coordinated resource management to improve riparian and range conditions, protect wildlife, and improve cattle production. As further evidence of this momentum, Turnell points to the fact that the Wyoming Stock Growers Association added an environmental committee and that the NCA created an Environmental Stewardship Club.

While anti-cow activists and traditional livestock managers continue to snipe at each other in the media and through their lobbyists, a middle ground is emerging in the debate over how to manage western rangelands. Some environmentalists are beginning to say that some types of livestock management can work on arid western lands if it is carefully controlled.

Nancy Green, director of BLM programs for the Wilderness Society, thinks that ranchers should be given a chance to prove that they can manage their livestock responsibly: "The livestock industry, properly regulated, can have a place in the web of uses on the public lands in the West. But where, when, and how much grazing should be allowed depends entirely on what the land, habitat, and wildlife resources themselves can absorb in good health. Too often and in too many places, grazers have been allowed not merely to use the land, but to abuse it. That has got to stop." [14]

Ed Marston, publisher of *The High Country News,* a publication that has covered the debate over livestock's impact on western lands, agrees that ranchers should be given a chance to prove they can run cattle sustainably. "Were the movement for an alternative economy in the rural West an unalloyed good, I would worry less about the hasty expulsion of ranching. But, in some ways, the new, green, recreational economy—on view at Telluride, Moab, Jackson, and Vail—seems as fraught with problems as grazing. . . . Before we choose to throw away ranching, let us be certain that some of it, in certain places, in the hands of certain types of ranchers, cannot be reformed." [15]

A way to reform the management of the western rangelands is suggested in a book entitled *How Not to Be Cowed: An Owner's Manual,* published by the Natural Resources Defense Council and the Southern Utah Wilderness Alliance. This monograph describes a process by which hikers, fishermen, hunters, tourists, and concerned citizens can have more of a voice in the decision-making process about how public lands are managed—a process that has long been dominated by grazing interests.[16]

While empowering a wide variety of people to pressure federal officials into minimizing the damage done to the range by mismanagement of livestock will no doubt help, it is also clear that, if reform is to be effective, livestock managers must find their own way to become better stewards of the land. Jack Turnell's experiments with sustainable ranching techniques on the Pitchfork Ranch demonstrate that this is possible.

The Mothers of East Los Angeles Conserve Water, Protect the Neighborhood, and Create Jobs

A thunderous sound rolls out of a dumpster parked on the corner of South Mott Street in the Boyle Heights section of East Los Angeles. Inside, a man wearing goggles smashes old porcelain toilet bowls to bits with a sledge-hammer. With each stroke a deafening noise booms off the metal walls of the dumpster as another toilet splinters into a white gravel that crunches under the workman's boots. Each week two dumpster loads of crushed toilets leave this residential neighborhood and are delivered to a recycling facility where the porcelain is ground up into gravel used as an underlay-ment for the streets of Los Angeles.

The crash of toilets being pulverized outside her modest, two-story, five-bedroom, Spanish-style home is music to the ears of Juana Beatríz Gu-tierrez, the sixty-three-year-old co-founder and president of Las Madres del Este de Los Angeles–Santa Isabel. Las Madres is a nonprofit group of women activists who have organized and agitated to improve conditions in their neighborhood since 1985. The steady thump of the sledgehammer is the sound of jobs being created, water being saved, and community pro-grams being funded.

This scene on South Mott Street is the business end of the water conserva-tion programs in East Los Angeles. The old toilets being demolished, water guzzlers that required $3\frac{1}{2}$ to 7 gallons per flush, are being replaced with ultra-low-flush (ULFT) models that use only 1.6 gallons per flush. Officials at the Los Angeles Department of Water and Power (DWP) calculate that each resident who uses a ULFT toilet will save 14 gallons of water a day, potentially conserving approximately 48 million gallons of water daily for the city if all residents were similarly equipped.[1]

But water saving was not a high priority in the barrios of East Los Angeles (compared with more immediate problems such as high unem-

Figure 18.1
Water conservation worker takes apart an old toilet and recyles all the metal and porcelain. Courtesy MELA.

ployment, gang wars, drug addiction, high-school dropouts, and teenage pregnancies) until Gutierrez saw how water conservation could generate employment and become a source of funding for community programs. When Gutierrez heard that a rebate program for the installation of water-conserving toilets was underway in wealthier neighborhoods, she asked why a similar program was not available in her community. Such a program would be unlikely to work in low-income neighborhoods, she was told, and besides no one was available to take responsibility for running such a program in the barrio. "You never asked us," Gutierrez replied. "Come talk to us about it."

On 1 August 1992, Gutierrez launched the Mothers of East Los Angeles Water Conservation Program. She described the program to people in her community at meetings held in her parish. "At first people didn't believe that we were going to give away these new toilets for free," Gutierrez recalls, but gradually people began to understand how the program would work.

Los Angeles officials are willing to give away free toilets because the city is running out of water and must quickly institute widespread water conservation measures. This requires not only behavioral change (turning off the tap while you brush your teeth and sweeping the driveway instead of hosing it down) but also replacing the inefficient toilets and showerheads in residential homes with more water-efficient models. To this end, the Los Angeles DWP and the Central Basin Municipal Water District provide funding for the replacement toilets.

But who would convince local residents to make the switch to ULFT toilets and distribute them? City officials decided that this was a natural function for grassroots community groups and offered to pay a bounty of $25 for every old toilet replaced to cover the costs of running a water conservation program.

At Las Madres headquarters in Juana Gutierrez's house in East Los Angeles, a ULFT toilet is distributed free to anyone who shows up with a valid piece of identification and his or her last DWP water bill. The resident is handed instructions about how to install the new toilet and is told to return the old toilet within seven days at which time they will be given a free low-flow showerhead to further reduce water consumption and water bills.

This ingenious program has worked to everyone's benefit and since 1992 some 50,000 ULFT toilets have been installed and the old ones recovered. The money that Las Madres receives for this work is used to pay the salaries of twenty-five full-time and three part-time people to run the program. Any money left over after salaries and office expenses are paid is used to fund a variety of community programs.

But to bring change in East Los Angeles, Las Madres workers must go door to door to promote the water-saving toilets. Many residents are convinced when they learn that not only will they get the new toilet free, they will also save a minimum of $35 a year on their water bill, reports Ricardo Gutierrez, Juana's husband who is now a supervisor of the program. As for its installation, many residents can do it themselves once they are given printed instructions and a wax ring needed to seal the toilet to the floor. Las Madres refers plumbing-averse individuals to local entrepreneurs willing to install the toilet for $25.

The program is such a success in Boyle Heights, that Las Madres expanded the operation into Highland Park and the San Pedro-Wilmington area. Every Saturday, Ricardo Gutierrez and a crew of young people from

Figure 18.2
Old toilets that have been returned in exchange for a new water-conserving model.
Courtesy MELA.

the community load up a trailer with ULFT toilets and drive down the Harbor Freeway to a boys club in Wilmington where they distribute new water-conserving toilets and smash up the old ones that are returned. "At first the guy at the boys club was skeptical about the project, but when I told him I'd pay him $5 for every old toilet we received back and wrote him his first check, he quickly became a believer. Now he wants to raise $6000 so he can build a playground at the boys club with swings and slides and he has already raised half of that," says Ricardo Gutierrez.

Not only does the program conserve water, it also provides employment for local people, Gutierrez points out. In addition to the water conservation program staff of twenty-eight, the private contractors who install the toilets are also making money. But the ripple effect of the water conservation program does not stop there because profits from the program are used to hire young people to perform a variety of community services. High-school students are hired at $6 an hour to go door to door urging parents to have their children fully immunized and tested for lead poisoning, and a crew of young people is also employed in the early morning hours before school to sweep the streets and paint over walls covered with graffiti. Money generated by the water conservation program is also spent on college scholarships and on clothes and books to keep kids in high school who otherwise would drop out.

Born in the north-central region of Mexico in the town of Sombrerete, Zacatecas, Juana Gutierrez grew up in Ciudad Juarez near El Paso, Texas, before moving to the United States in 1956. She married Ricardo Gutierrez, then a Marine, who later became a nightshift warehouse supervisor in Los Angeles. Gutierrez has lived in the Boyle Heights section of East Los Angeles for the last 40 years where she and her husband raised nine children.

In 1979 Gutierrez became concerned about the safety of her children when drug dealers and gang members began to congregate in the park across the street from her house. To help keep the neighborhood safe, she

Figure 18.3
Juana Beatríz Gutierrez, co-founder and president of the Mothers of East Los Angeles (MELA). Courtesy MELA.

organized a neighborhood watch program and became the president of the local PTA. "I did this because I didn't want my kids involved with drugs and gangs," she says as she sits in a room overlooking the park. The response from the community was good, she continues, and she and her neighbors kept the neighborhood relatively crime-free. To accomplish this, however, requires constant vigilance and a willingness to be actively involved with the community children. For example, Gutierrez and her husband organized regular after-school sports activities in the park. These efforts paid off now that all her children have finished high school and gone on to college at Princeton, the University of California at Berkeley, the University of California at Santa Barbara, Loyola Marymount, and East Los Angeles College.

But protecting her children and community required more than organizing baseball after school and chasing neighborhood bullies out of the park. It also entailed mobilizing residents to fight the construction of a number of large-scale, dangerous, and toxic facilities in her community, including a state prison, a commercial hazardous waste incinerator, and an oil pipeline. It would appear, in fact, as if every type of noxious facility was slated to be located in East Los Angeles, at a safe distance from the wealthier citizens in Santa Monica, Beverly Hills, and Brentwood.

In 1985 state officials announced plans to construct a $100 million state prison on Santa Fe Avenue in East Los Angeles. That prompted Gutierrez to found the Mothers of East Los Angeles to oppose the prison's construction. With five other women she collected 1500 signatures on a petition against the prison, which she gave to state assemblywoman Gloria Molina to take to Sacramento. This initiative sparked the largest community mobilization in East Los Angeles since the Vietnam War.

Gutierrez mobilized community demonstrations against the prison at her church after Sunday Mass. She and her colleagues sold tamales to raise funds and set up a table with educational materials, T-shirts, and anti-prison buttons in Exposition Park on Earth Day 1990. During this period, Las Madres members learned how to attract media attention by holding protest marches. Their agitation proved successful in 1992 when a bill prohibiting any state prison from being built in East Los Angeles passed and became law.

Gutierrez's next campaign was to improve the quality of air in her community. Epidemiologists are concerned that air pollution is causing permanent damage to the lungs of young people in the Los Angeles region.[2] The fact that the community in which Gutierrez has lived for the last 40 years is located near the junction of five freeways doesn't help. A few blocks from her home, the Santa Monica Freeway turns into the San Bernadino Freeway and intersects with the Golden State Freeway, the Santa Ana Freeway, and the Pomona Freeway. The level of pollution generated by the passing traffic is substantial. Food processing plants, factories, and body shops add further texture, color, and odor to the air.

Threatening to further degrade the already overburdened air of East Los Angeles is a plan to site a $29 million, 22,500-ton-per-year commercial hazardous waste incinerator that would burn up to 125,000 pounds a day of hospital wastes, solvents, oils, pesticides, paints, and sludge. In addition to the air pollution the incinerator would generate it would also produce huge volumes of highly toxic ash that would have to be disposed of in a landfill. The facility was to be California's first large-scale, hazardous waste incinerator.

The site chosen for the incinerator, a mile from Gutierrez's home, was the town of Vernon, a community with only 90 residents that is dedicated almost exclusively to industry. "The city was built for industry: Leslie Salt, Oscar Mayer, Farmer John, Vienna Hot Dogs, and 7 Up and RC Cola all have warehouses or food packaging plants there," writes Rondolfo Acuna. Many less well known companies also have manufacturing facilities there.[3] Unfortunately, this highly industrialized zone is surrounded by one of the most densely populated areas in the county—East Los Angeles.

Using this industrialized town as a site for the commercial hazardous waste incinerator, an effort was made to obtain permits quickly for the facility without consulting residents from surrounding communities. Recognizing the dangers that such a facility would pose to the health of people in their community, in 1987 members of Las Madres led 500 protesters to a Department of Health Services meeting and demanded that an environmental impact review be conducted. Las Madres also threatened to sue the city of Vernon and California Thermal Treatment Systems, Inc., which was planning to build the incinerator. After a lengthy struggle, Los Angeles Mayor Tom Bradley joined the opposition to the incinerator and the project was defeated.

Like the mythical monster that sprouts seven new heads for every one you cut off, the proposals for new toxic facilities to be located in East Los

Angeles just kept multiplying. Low real estate values in the area were appealing to industrialists and it was assumed that there would be little organized opposition to controversial projects in this low-income community. These calculations proved to be wrong.

Four big oil companies had a problem. They needed to move their oil south down the coast, from Santa Barbara to processing facilities at Long Beach and Wilmington. Unfortunately, some wealthy and politically well-connected communities such as Santa Monica and Pacific Palisades stood in the path of a pipeline they planned to build. The solution was to detour the pipeline inland through the low-income communities of East Los Angeles. The pipeline, which was to be laid only 3 feet under the ground, was scheduled to pass next to several schools and directly under Hollenback Junior High School. Worried about gas leaks or a major explosion, Las Madres joined the Coalition Against the Pipeline and mobilized more than 1000 protesters. In 1987 the Los Angeles City Council was pressured into turning down the pipeline permit request.

"The state wants to place all of society's problems in our community—a prison, a pipeline, and an incinerator. But if we keep up the pressure they will have to solve these problems, not just dump them from one place onto another," Gutierrez observes.

In other protests, Las Madres joined a coalition opposing the construction of a Chem-Clear plant to treat 600,000 gallons of cyanide hexavalent chromium and other hazardous chemicals across the street from Huntington Park High School, and fought against the aerial spraying of the insecticide malathion in their community. A contingent of Las Madres members traveled to northern California to oppose the Casmalia dump site and marched against a toxic waste incinerator planned for the Latino farming community of Kettleman City. By that time Las Madres had grown into an organization with 400 members.

Initially inspired by parish priest Father John Moretta of Resurrection Church and by state assemblywoman Gloria Molina, Gutierrez and Las Madres has become a political force to be reckoned with in East Los Angeles. As Gutierrez says of the organization: "Not economically rich—but culturally wealthy; not politically powerful—but community conscious; not mainstream educated—but armed with knowledge, commitment, and determination that only a mother can possess; the Mothers of East Los Angeles–Santa Isabel is an organization striving to protect the health and safety of East Los Angeles neighborhoods."

When she is not fighting the location of toxic facilities and solving pressing environmental problems in her community, Gutierrez continues to look for employment opportunities for local people. Searching out and abating sources of lead poisoning in the homes of local residents is another possible function for Las Madres workers. Walls and ceilings covered with lead paint could be encapsulated in sheetrock, while contaminated doors and windows could be replaced, she notes. This is something local people could be trained to do that would protect community children from one of the most dangerous health threats they face.

But until such a program is funded, Gutierrez will continue to oversee the water conservation program that has made so many positive changes possible in her community and look for other practical projects that will help her people. In the meanwhile she occasionally organizes a clothing collection drive for Latino farmworkers. "You don't have to go to Central or South America to find poverty. We have it here in our own backyard," she observes.

Sustainable Agriculture Takes Root among
Family Farmers in Iowa

An old school bell hangs outside Ron and Maria Rosmann's family farm in the rolling fields of southwestern Iowa. These days Maria rings it to call her children in for supper. But years ago it called Ron's father to a one-room schoolhouse half a mile from the farm where a 50-foot linden tree now grows.

There is a deep sense of connectedness with the land at the Rosmann farm outside of Harlan, Iowa. The 480-acre farm is part of a section of land that has been in the family for four generations since Rosmann's great-grandfather founded it in 1882. Ron grew up in this rambling, two-story house where he now watches his own three sons (ages ten, twelve, and fourteen) gather to eat breakfast around the kitchen table.

Medium-sized family farms such as this are now an endangered species across America as they are squeezed out by larger agribusinesses that take advantage of economies of scale. In some ways big farms are like factories. Workers punch a time clock and drive huge machines that strip the soil of vegetation; spread nitrogen-rich, chemical fertilizers; plant monocrops of corn, wheat, or soybeans; and spray some 1.2 billion pounds of pesticides every year.[1] This system has also replaced large numbers of family-sized farms with chemical and mechanical inputs. In fact, as farms are growing larger, farming communities are becoming smaller, some of them turning into ghost towns. The percentage of farming families in the population has shrunk from 35 percent in 1910 to 2 percent today.[2] Resisting this trend are a growing number of family farmers who practice some form of what is coming to be called sustainable agriculture. Some of them are purists who grow organic food using no chemical fertilizers or pesticides; others have taken the less rigorous approach of reducing their use of agricultural chemicals to a minimum.

Ron Rosmann moved through a number of stages: he started out as a conventional farmer who used both pesticides and chemical fertilizers; subsequently he took to farming with a minimum of agrochemcials; and more recently he adopted strictly organic methods. But Rosmann's meta-morphosis has been gradual. Although his father and grandfather were farmers, Rosmann, now forty-five years old, did not always feel destined to farm. He left home to attend college at Iowa State where he graduated with a degree in biology in 1973. He also studied psychology and sociology, which led him to a job at a center for disturbed children. Briefly he consid-ered graduate school. But when his father announced that he was ready to retire, Rosmann returned to the farm and found that he liked it.

"I always liked working outdoors and the freedom and open space of farming," he says. After a season in the fields he realized that "I didn't like being a student as much as I thought I did." Since then his life has centered around the farm where what he learns is applied to the crops and livestock he raises.

"My dad, Raymond, farmed very sustainably in terms of crop rotation, use of manure, and keeping livestock," Rosmann says. "He used some chemical fertilizers when they came on the scene in the 1950s, but we never used the amounts that other people did." In fact, Rosmann's father kept a team of Belgian workhorses up to 1968 and was reluctant to give up some of the old methods such as using a threshing machine.

When Rosmann took over the farm in 1973, he farmed conventionally, using pesticides and chemical fertilizers liberally for ten years. "That was the way it was done then and I didn't know any better," he says. But even during this period he still practiced crop rotation and resisted devoting all his acreage to monocrops of corn and soybeans the way many farmers did.

Rosmann's real transformation from conventional to sustainable farm-ing began in 1979 when the fuel crisis hit. It was during the hard winter of 1980 that he became interested in alternative energy. He and his wife, Ma-ria, put together an alternative energy fair for 200 people in Shelby County. This experience prompted him to experiment with solar heating for some of his outbuildings. He painted the cinderblocks of one structure black and pulled the solar-heated air through it with fans, saving heating fuel in the process.

A newsletter published by the Center for Rural Affairs taught him about composting manure, windpower generators, and farming with fewer pesti-

Figure 19.1
Ron Rosmann at his organic family farm in Harlan, Iowa. Photograph by Steve Lerner.

cides. An article in another magazine, describing the work of two local organic farmers, caught his eye and Rosmann and a carload of farmers went to see their farms. "We were awestruck by how few weeds there were," he remembers.

While Rosmann was impressed with organic farming, he was initially convinced he couldn't make a living at it. "Ideally we should all do organic farming but it is unrealistic," Rosmann said in 1992. "Our best hope is a form of sustainable agriculture that uses as little pesticide and commercial fertilizer as possible. We want to protect the environment but at the same time keep our yields high and our net return high. Organic farming is appropriate for some, but we are trying to change mainstream farming," he added.

So instead of going into organic farming, Rosmann helped found the Practical Farmers of Iowa (PFI), a group of 500 farmers who see definite advantages in reducing their dependence on agricultural chemicals, but who want the flexibility of being able to use pesticides when necessary. "PFI farmers don't take the purist approach: most of them will use some herbicides if they think they have to," he says.

Sustainable farming, as practiced by PFI farmers, involves more than just growing food. It also requires building up the fertility of the soil so that it produces healthy, relatively pest-resistant plants. It requires crop rotation to keep infestations from becoming established. When a crop is under attack, the next line of defense is integrated pest management (IPM), a system that uses beneficial insects to fight crop-destroying insects. It is only when all these methods fail and there is a danger of substantial losses from an infestation that PFI farmers will consider applying a pesticide. This frugal approach to spraying pesticides reduces groundwater pollution and limits the pesticide residues on crops.

"We are practical as opposed to dogmatic. Whatever we do has to work and maintain productivity, profitability, and the environment," he explains. To find out what methods really work, PFI sponsors on-farm research trials to figure out how to use fewer pesticides and chemical fertilizers, improve the soil, and keep family farming alive in communities, says Rosmann, whose neighbors are also members of PFI. Their three adjacent farms have a total of 1200 contiguous acres of herbicide-free Iowa farmland.

While Rosmann was initially skeptical about whether or not he could make a living as an organic farmer, over the last three years he has taken the plunge and now 320 of his 480 acres are certified as organic by the Organic Crop Improvement Association. There are two main reasons he went organic, he says. First, he found that by using ridge tillage, crop rotation, and integrated pest management, he simply didn't need herbicides or pesticides. Only once over a 14-year period in 1992 did he use an insecticide—to battle a particularly destructive alfalfa weevil. And his own research suggested that crops were growing just as well without herbicides as with them.

The second reason for going organic was purely economic. Times are hard for small family farms in the Midwest, Rosmann notes, and economies of scale are forcing farmers to either become bigger or specialize. Rosmann found that by going organic he could access a niche market and command a significantly higher price for his produce. Along with his cousin, a neighbor, he joined the Heartland Organic Marketing Cooperative, a group of twenty-five Iowa and Missouri organic farmers who market their produce in common. Having gone through the three-year certification process and joined the co-op, Rosmann is now growing organic soybeans under contract that will be sold to the Japanese for tofu. He is also growing some organic milling-quality corn, which sells for more than traditionally grown corn and which will be used in the United States for making corn chips.

Rosmann has the advantage of having grown up on this land, having worked it since 1973, and having seen it under all kinds of conditions. As a result, he knows what the land can and cannot do. He has seen the fertility of its soil rise and fall. He has seen the cycles of boom and bust: periods when corn and soybeans were bringing in good money, and then times when neighbors went bankrupt. Through all these ups and downs, he continues to experiment with what he grows and how he grows it in an effort to evolve a type of farming that can be sustained over the long haul.

He learned, for example, that he could avoid the use of herbicides by practicing a raised-bed method of ploughing his fields known as ridge tillage. "Look at this field," he exclaims. "I'm doggone proud of it. There is hardly a weed in it and all we have done is cultivate it. See how we threw up this ridge and then skimmed off the top and planted it. When I tell people I have only used herbicides once since 1979, their mouths drop when they see how few weeds I have."

He also enhances soil fertility and discourages insect damage by practicing a complicated, long-term crop rotation method. Under his system, a field that is in soybeans this year will be planted in corn next year, then in oats, and alfalfa for two years, and then back to corn. Because the crops keep changing season after season, insects don't have a chance to become established as easily as in a field that carries the same crop year after year.

"We try to operate kind of a closed system," explains Rosmann, who composts manure from his cows and pigs. "By using decent tillage techniques, rotation methods, and compost, we have built up the soil so we don't need to rely on chemical inputs. This saves us money." His spending on chemical fertilizers decreased from $5000 in 1988, to $3212 in 1991, to zero in 1995.

Instead of relying on one or two crops such as corn and soybeans, Rosmann deliberately diversified his farm. In addition to corn and soybeans he sells rye, oats, seed oats, and some tree seedlings. He has 11 different species of evergreens growing—800 in the ground and 300 in containers. And he is experimenting with chestnuts and hazelnuts as crops. While the tree nursery is a sideline at the moment, Rosmann sees it as a potential entry point for his children should they turn out to be interested in farming.

Rosmann also expanded his livestock operation. When the price of meat went up he increased the number of pigs, cows, chickens, and broilers he raises. "We need the livestock as part of our ecosystem here in the

Midwest," Rosmann says. He feeds his legumes, almost all of his grains, and his grasses to the livestock and then uses their manure to enrich his fields. He also grows turnips for extra feed for his cows as a way to get more forage through double-cropping.

In 1980 Rosmann built centralized pigpens so that he could raise hogs instead of buying piglets and fattening them as many of his neighbors do. The pens are cleaned out every other day and the straw and manure are composted. "We have the farrowing and nurseries here. Then, when they are weaned, they go up to the old hog barn to get used to being outdoors. They come down to the grower unit and are finished in a lot when they hit 230 to 240 pounds," he says.

Still not totally satisfied with his system for raising hogs, Rosmann thinks that he relies too heavily on antibiotics in their feed to keep diseases from spreading. He is lowering the number of hogs he raises from 1200 to 900 so he can periodically leave the farrowing pens empty and interrupt the disease cycle. Rosmann is also crossbreeding his hogs with a rare British breed called Tamworth, a pig that retains some wild traits and that is well suited to outdoor hog production.

While still unsure how he will change his hog operation, Rosmann is investigating a Swedish method of raising pigs that involves some confinement but is more humane than the conventional American system. Under the Swedish method no more than ten hogs are confined in the same finishing pen. Furthermore, sows are not locked in birthing stalls for long periods: the birthing stall is removed after a week to ten days permitting the sow to mingle with other sows. The Swedish method also involves a "deep-bed" technique where straw is added to the pigpens regularly and allowed to build up. This permits pigs to practice their natural rooting and nesting behaviors during which the hogs actually compost the straw and manure by churning it. Changing to the Swedish method, however, will require modification of his facilities, so Rosmann has not yet decided whether the investment is worthwhile.

While Rosmann is not yet in a position to raise organic pork, he does plan to raise organic beef. Constantly tinkering with the way he manages the farm, he decreased the number of hogs he raises and is moving more and more into cow and calf rotational grazing. He currently has seventy-five cows and is considering feeding them out as natural beef (dosed with neither hormones nor antibiotics) or as organic beef. His remaining concerns about raising organic beef have to do with dealing with flies and worms. To suppress flies, many cattle have an insecticide implant stapled into their ear. To avoid this dose of insecticide, organic beef operations are experimenting with the release of parasitic wasps that eat the fly larvae. To prevent worms they feed cattle diatomaceous earth, a fine, siliceous earth with abrasive characteristics that irritates the stomach of the cow causing the animal to pass the worms. "These are things we will have to experiment with," Rosmann says.

To feed his cattle, Rosmann planted native prairie grass that does well in Iowa's hot dry summers and grows to 6 feet. "Our ancestors brought all sorts of plants with them such as broome and orchard grass because they didn't realize the plant wealth that was here," he says.

Raising pigs, cows, and a variety of different crops requires a lot of work—more in fact than one man can handle. The three Rosmann boys pitch in with the farm work when they can: the oldest started driving a tractor at age ten. Maria Rosmann also helps out but the amount of farm work she can do has been limited since she took a job as development director for the Shelby County parochial school system. As a result, for the heavy farm work, Rosmann must hire help. For a number of years he paid a full-time farmhand, but when he was forced to cut his hours to forty a week the man quit. He now hires a college student who works full-time during the summer and is looking for a farmer to work half-time the rest of the year. Rosmann calculates that hiring someone to work with him allows him to save on chemical inputs, but not enough to cover the wages.

Many neighboring farmers avoid hiring by increasing their use of herbicides to keep down the weeds. A farmer just a few miles away replaced a laborer with a bigger tractor and herbicide applications, Rosmann notes. In the old days there was plenty of labor because there were big farm families and "neighboring" was practiced to bring in the hay. Now it is hard to find teenagers to pay to help bring in the hay. As a result, with little affordable labor available, farmers are forced to roll the hay with a tractor and cover it with plastic—a practice that causes more spoilage than if the hay were properly stored.

Rosmann is not alone in experimenting with the best method of sustainable agriculture in his region. In concert with other members of the PFI, he is involved in a variety of field tests to determine, as scientifically as possible, what practices work best. PFI holds field tests every year that are different from those conducted on 40-by-40-foot university research plots. Using their own equipment, PFI members plant long narrow strips of crops that run the entire length of their field and take into account the variability of the soil. Once the plants have matured, PFI sponsors a picnic at the farm where the field test is taking place so that farmers can compare eight rows of plants grown with a herbicide and eight rows grown without it, count the number of weeds around the plants in the different rows, and see for themselves what works and what does not work. It is strictly the "show me" school of agricultural experimentation, Rosmann says. PFI, which won a Renew America Award in 1991, holds field days on thirty farms and meetings in the winter. Members also spread the word about sustainable farming techniques by working with the U.S. Department of Agriculture extension service and vocational agriculture high-school teachers.

More recently, PFI has begun to compare various systems of agriculture in addition to the field tests. "We have to figure out what kind of farm will make the best system of food production, not just look at isolated practices such as ridge tillage. The question is how do all these practices fit together with livestock in the operation, crop rotation, integrated pest management, and all the other techniques," Rosmann says. One of his mentors, Dick Thompson of Boone, Iowa, the first president of PFI, has tracts of land devoted to three different types of farming so that they can be compared, Rosmann adds.

Already there are organizations similar to the PFI in Illinois, Indiana, Maine, Michigan, Ohio, and Oregon that are experimenting with non-organic yet sustainable agricultural practices. In many cases these farmers are unwilling to adopt organic methods but nevertheless recognize that they can significantly reduce the amount of pesticides and chemical fertilizers used in the production of grains, potatoes, and other crops.

Evolving a sustainable method of agriculture not only provides healthier food and a reduction in the dispersal of environmentally destructive chemicals, it will also help build and preserve the vitality of rural communities, Rosmann argues. "The very definition of sustainable farming involves the preservation of the community itself," he reflects. If the farm is sustainable, it will sustain people to work it and, by extension, the local community.

Healthy small and medium-sized family farms stabilize rural communities. "The difference between one family owning 3000 acres and ten families owning 300 acres is enormous in a small community in terms of the number of businesses those extra families will support," he observes. "We want to keep farm communities vibrant in order to keep as many people as possible in rural areas, and make farming attractive to our children." Rosmann hopes that by the time his sons grow up his farm will be healthy enough to provide them with a good livelihood if they choose to farm.

Giant farms have the opposite effect. Because of economies of scale, small and medium-sized farmers find it hard to make a living and are often forced to look for work elsewhere, frequently in already overpopulated cities. Even among those Iowa farmers who do hang on, 40 percent of them now supplement their income by working off the farm, Rosmann notes. Few young people are going into farming and as a result the average Iowa farmer is now fifty-five years old and the state has one of the highest percentages of elderly people in the country, he continues. Rosmann worries that Iowa will become like Kansas and North Dakota where much of the farm work is done by hired, roving sprayers and harvesters, resulting in a proliferation of ghost towns. "We have already lost neighbors and businesses in our little community," Rosmann says. In fact, so many businesses have shut down that he must now drive 25 miles to find spare parts for his machinery.

The big farms appear to be producing food cheaply, but a closer look reveals many uncounted societal costs, Rosmann observes. Part of the hidden cost is that with the increased use of pesticides and herbicides, pests and weeds are becoming more resistant, requiring larger applications of pesticides and herbicides every year. Costly new formulations of insecticides, designed to overcome resistance, are continually marketed, making farmers increasingly chemically dependent. "The earth is forgiving but not forever," Rosmann notes. Test results of groundwater at the home of Maria Rosmann's brother, a priest, were so bad that the well had to be condemned, he says.

"I'm not saying we should all go back to sixty-acre farms, but we should find a balance," Rosmann concludes. "We are not anti-technology. Technology is very important for sustainable agriculture and it freed us from the drudgery of a lot of hand labor. We don't want to be out here hoeing our crops. Technology gave us ridge tillage and better fencing for rotational grazing." But the kind of technology Rosmann wants is the kind that does not put people out of work and does not contaminate the land.

Part of the problem in promoting sustainable agriculture is pricing, Rosmann says. There is no connection between what the farmer is paid and the price the produce sells for. "I think it is sickening. People who grow the food get only 5 percent of the price consumers pay for it. I feel used all the time: I am working hard to grow food for people and I have no financial security. This is one reason we are seeking contracts for organic food production."

Until farmers receive fair prices for the food they produce, farm subsidies will continue to be needed, he argues. But the subsidies should be a tiered system that rewards those who practice sustainable agriculture more than those who do not. The current system of subsidies penalizes farmers who practice resource conservation through crop rotation and provides incentives for them to plant their land "fence post to fence post" with a small number of monocrops. Furthermore, farm policy has been largely based on studies that do not take into account environmental degradation caused by conventional farming practices.[3]

Subsidies for conventional agricultural techniques have been a mixed blessing. On the one hand, they have kept food grown in the United States both plentiful and relatively cheap. On the other hand, subsidies for conventional agriculture have had devastating environmental effects. For example, the United States loses 3 billion tons of topsoil a year despite an active soil conservation program; and the use of pesticides has tripled since 1964 while the use of chemical fertilizers has risen by two-thirds. Conventional farming subsidized by the government has also resulted in the loss of wildlife habitat, significant groundwater pollution, and damage caused by sedimentation from agricultural runoff.[4]

Rosmann has taken this message to Washington. "A farmer like myself can really have an impact on some of these guys in Washington because they are not used to individual farmers coming in there and talking sense." Rosmann has twice had an opportunity to present his views on farming to President Clinton, and other sustainable farmers have become politicized as well. Members of the nationwide Campaign for Sustainable Agriculture are lobbying their congressional representatives to promote marketing alternatives, redirect agricultural research along sustainable lines, and ensure that farmers who rotate their crops do not lose their eligibility for price supports and other benefits. But theirs remains an uphill fight and legislation continues to favor farmers who produce monocrops using large quantities of agricultural chemicals.

Despite the challenges that face those who are trying to practice sustainable farming, Rosmann loves his work. "We have something worth fighting for here: the values of family farming, small communities, and living in harmony with the earth. On the farm I feel such a closeness to the earth and ultimately to God our Creator. What we value is our relationship to the land and our family."

Zunis Launch a Sustainable Action Plan to Manage Tribal Resources

It is early morning on Zuni tribal lands just south of Gallup, New Mexico, on the extreme western edge of the state. In the distance a pair of hawks circle some invisible breakfast in the direction of Dowa Yalanne Mesa that looms up out of the desert floor. In the cool morning air and cloudless sky the first touch of the sun's rays are welcome, but behind the pleasant warmth one can detect the approach of another relentlessly hot day.

Just outside of town a dozen men have parked their mud-splattered pickup trucks outside a pair of metal outbuildings that house the Zuni Conservation Project (ZCP). As they stand around on the hard-packed earth drinking coffee and talking quietly, a three-legged dog hops among them, sniffing at a few struggling cornstalks growing along the cyclone fence. Strung up on a utility pole above them is an impressive rack of elk horns.

By 7 A.M. forty-five watershed restoration workers are milling about the parking lot loading large vans with shovels, hoes, axes, picks, saws, and a variety of hand tools. These implements will be used to begin the arduous process of restoring the land and stemming the erosion that creases the mountains of this watershed with a network of gullies.

Wearing workshirts, jeans, heavy boots, and an assortment of hats to protect themselves from the sun, these workers sculpt the land by hand, placing obstacles in gullies, arroyos, and streambeds to slow runoff and minimize erosion from the torrential bursts of rain that occasionally visit these mountains during August and September. Drawing on ancient Zuni techniques, the restoration workers craft "silt-curtains" out of rock and brush that are permeable to water but hold back the soil. The runoff is channeled into fields where it irrigates, among other crops, the blue, red,

Figure 20.1
Watershed restoration workers building a check-dam to control erosion. Courtesy ZCP.

yellow, and white traditional folk varieties of corn that are sacred to the Zunis. Even large boulders are moved by hand by levering them onto rolling logs. This labor-intensive approach to erosion control permits the tribe to hire large numbers of people rather than purchase an expensive piece of earthmoving equipment such as a backhoe.

This small army of watershed restoration workers deployed in the Zuni mountains is following one of the most sophisticated action plans for sustainable management of resources in the country. Organized by the ZCP, the work is funded with money from a suit brought by the Zunis against the federal government for damage done to their lands at the turn of the century. And therein lies a tale that is likely to be told in Zuni circles for years to come.

Zuni aboriginal territory is estimated to have encompassed some 14.5 million acres under both the Spanish and Mexican administrations, but between 1860 and 1876 the tribe lost 9 million acres to European invaders. By 1935 the official Zuni reservation established by the United States was down to 340,000 acres; today it comprises about 474,000 acres. In addi-

Figure 20.2
Basketweave design used for site stabilization. Courtesy ZCP.

tion to having most of their lands confiscated, the area that did remain in their possession was severely damaged, says James Enote, leader of the ZCP, a tall, lean man with jet-black hair that snakes halfway down his back. Most of the damage was done when U.S. government officials unwisely decided to bring untested, "modern" agricultural techniques into this arid region by building earthen dams and reservoirs, says Enote.

Prior to that the Zunis practiced a form of sustainable, dry-land and runoff–irrigated agriculture for thousands of years in an area that is semi-arid grasslands, shrub, and woodland. Over the centuries the Zunis learned to divert runoff from the mountains onto their crops and how to read the weather to know when to plant. Dry-land farming is a subtle art that requires intimate knowledge of the soil, water, topography, seed stock, climate, and wildlife, says the thirty-eight-year-old Enote, who grew up in a farming family that also raised sheep.

Major changes in the land began in the 1890s when white settlers harvested large quantities of timber out of the mountains and shipped it off on a railroad that was built through Zuni territory. After the clearcuts, private livestock companies overgrazed the land causing further erosion. Coal was mined without compensating the tribe, roads were slashed through the woods, and even Zuni archeological sites were disturbed. As a result the fragile, arid, mountainous ecosystem that had supported the Zunis for millennia began to degrade as soil eroded from the mountains, clogged the streambeds, and filled the new dams with silt.

In the early 1900s, two earthen dams—the Ramah Dam and the Black-rock Dam—were breached, causing massive erosion that scooped out huge gullies which ran through the fields the Zunis once cultivated. In the 1920s and 1930s several smaller dams collapsed. Riverbeds were scoured and the water table dropped, making it impossible to irrigate many fields. The landscape was so changed by erosion that the old system for diverting water into the fields as it came off the mountains no longer worked and Zuni agriculture as it had been practiced for thousands of years was largely wiped out.

On the basis of these events, the Zuni Tribe sued the U.S. government for having mismanaged Zuni lands and caused massive erosion. After a costly, decade-long legal battle a settlement was reached. The United States agreed to create the Zuni Land Conservation Act of 1990, which committed the tribe to sustainable resource development. The act also promises to provide the tribe with $25 million, $17 million of which is to be placed in the Zuni Indian Resource Development Trust Fund that is expected to generate about $1 million a year in interest.

That money is being used by the ZCP to restore the damaged lands and put the Zunis back on a track of sustainable resource development, says Enote. "Sustainable development is nothing new for the Zunis. We wouldn't be here if our ancestors had not acted sustainably. What we are really talking about here is enhancing Zuni sustainable development," continues Enote, whose father is Zuni and mother is Tewa.

Enote and others see it as important to use the money for this purpose because, for a variety of reasons, many of the Zuni's time-tested sustainable practices have vanished. For example, traditional Zuni farming was decimated as it fell from 12,000 acres of land in crops in the 1850s to 1000 acres in 1991. Similarly, the ability of Zunis to provide themselves with food from hunting and fishing also suffered as the land eroded, streams became clogged with sediment, and wildlife habitat was destroyed. Faced with a severely degraded watershed and a rapidly vanishing way of living in harmony with nature, the moneys recovered from the lawsuit offer the Zunis an opportunity to invest in the restoration of, and return to a sustainable system of managing their lands.

"When I got out of college I knew I wanted to work for my people," says Enote, who attended both New Mexico State University and Colorado State University and received his B.S. in agriculture. But there were no jobs to be had in the Zuni pueblo and so he went to work as a soil conservationist for the Bureau of Indian Affairs. In the interim he did research into some of the ecological damage that had been done to Zuni lands to help with the suit against the U.S. government. After the suit was settled, Enote was approached at lunch one day by a member of the tribal council who asked him if he would like to lead the planning and implementation of the Zuni Resource Development Plan. "Let us know after lunch," he was told. "This was the break I had been waiting for so I told them I would do it," Enote recalls.

But where to start? Assigned a small office in the tribal building with no furniture, Enote decided to start by doing some research. At the library at the University of New Mexico, he found that most of the land management plans published by the Forest Service, the Bureau of Land Management, and the Bureau of Indian Affairs were standardized and applied to almost any region in the United States, while what the Zunis needed was a resource management plan geared to the fragile, arid, mountainous environment in which the Zuni reservation is located.

Looking farther afield he read restoration plans for arid lands published by the United Nations Development Programme, the Peace Corps, the Swiss Red Cross, and others. Here he found more that was relevant to the Zuni situation. In fact, Enote discovered that Zunis had a good deal in common with people from developing countries. It was during this same period that he was invited to attend the Earth Summit in Brazil where he met a network of people wrestling with the kinds of problems he faced on the Zuni reservation. Upon returning from Rio, he and his staff pieced together *The Zuni Resource Development Plan: A Program of Action for Sustainable Resource Development,* a 300-page report that constitutes one of the first major planning documents modeled on *Agenda 21,* the 1992 Earth Summit's blueprint for sustainable development.

"I had to be careful because I was in a position where I could replicate a lot of mistakes or get a program going that was considered so radical that

Figure 20.3
*James Enote, director of the Zuni Conservation Project (ZCP) in his office in Zuni,
New Mexico. Photograph by Steve Lerner.*

no one would want to participate," he says. Learning from what had worked and what had failed in various parts of the world, Enote decided to make the Zuni plan for sustainable resource management as participatory as possible. Further, he decided to send his staff into the field where they would have an opportunity to talk with the people who lived on and worked the land and come up with a consensus about how to manage the tribe's resources. Enote recognized that many Zunis still knew how to manage the land sustainably and he was determined to draw on this knowledge as an important resource.

As Enote drove around the reservation with members of his staff he was often asked if he was working for the Bureau of Indian Affairs or the Soil Conservation Service. Farmers were accustomed to seeing people from outside the tribe make decisions about the way the land would be managed. When he explained that the ZCP was doing this for the tribe and that they wanted to discuss management plans with the farmers there was general agreement that this approach was long overdue. Some of those Enote encountered said they remembered his father or grandfather and this made it easier to hold meaningful discussions about how to manage the land.

But the honeymoon did not last for long once Enote and his staff began to attack some of the more intractable problems that had largely destroyed Zuni farming. One of the issues they identified was that there were numerous land disputes that had never been settled and that the boundaries of many farms were under contention. Part of the confusion arose from the fact that the U.S. government had allocated irrigated lands to male heads of households, ignoring the fact that in the Zuni culture the land is traditionally passed on to women, Enote explains. The collapse of irrigation systems due to erosion also made boundaries less clear. "A lot of people had quit farming because they didn't want any hassles with other members of the tribe. They didn't want to erect a fence around their plot because they were afraid they would get into an argument with their neighbors," he continues.

"We told the farmers that we were not going to put a lot of energy into restoring the land if they were going to fight over who owned it," Enote says. So the project's first task was arbitrating the land disputes. This initiative brought to a boil many highly emotional disputes that had been simmering for years. But gradually a map of agricultural plots was pieced together and agreements were reached. To help resolve land disputes, a sophisticated piece of equipment—the global positioning system (GPS)—was used. The

GPS permits someone to walk the boundaries of a field with a receiver and have it plotted on a map back in the office. Mapping the disputed boundaries allowed for quicker resolution of conflicts and also helped farmers calculate how much land they had and how much seed they would need to plant it. Enote reports a significant increase in the number of young people in their twenties and thirties who have taken up farming as a result of this program.

To help rejuvenate Zuni farming the ZCP staff also did research on what crops would sell well and found that there was a market for the traditional Zuni folk crop varieties such as blue, red, yellow, and white corn. But further discussions discouraged the idea of marketing these varieties of corn. Since Zuni corn is sacred to the Zuni people, "it would be like selling your children," one religious leader observed.

The fact that most Zunis are unwilling to sell sacred corn varieties illustrates how deeply entwined the Zuni culture is with the management of resources. Zuni corn is traditionally grown and often traded between people on the reservation, but it is not sold, Enote explains. The sacred white corn is also sometimes given away on ceremonial occasions. So instead of selling their corn on the open market, farmers began to experiment with a variety of blue corn they obtained from Mexico State University. The project has also started a seed bank for traditional Zuni folk varieties of seed that are kept in a refrigerator in plastic bags marked with tape. Every year the seeds are distributed to farmers who grow them out and then return seed to the seed bank.

Just as Zunis have a noncommercial relationship with their corn, so their land ownership arrangements are different from those of most Americans. Zuni lands are not divided into a checkerboard of private plots like the landscape in most of the nation. Instead, except for the residential properties and farms, the vast majority of the roughly half-million-acre Zuni reservation is owned in common. People are given permits to raise sheep in a certain area or cut down trees, but the permits can be rescinded if the land is needed for something else. Enote is convinced that this tradition of managing land communally provides a strong base on which to build a system for sustainably managing Zuni resources.

While the communal management of Zuni lands has kept much of it from being exploited for commercial purposes, it has also led to some conflicts, Enote admits. Some members of the tribe become proprietary about land on which they have raised sheep for a long time, he explains. If they want to cut down some piñon trees to open the land up so there will be more forage for the sheep, they must go to the tribal council for permission. Enote has been lobbying to have the ZCP review proposals such as this so that unsustainable practices will not be repeated. But as it stands now some people still make an end run around the project and get permission from the council to do what they want with the land before Enote has had an opportunity to make an evaluation. "I don't want to give the impression that this is a utopian situation here," he says. "We still have problems and disagreements that we must work out."

Enote is also very careful to shape his resource management plans so that they are sensitive to Zuni traditions. To this end the wildlife management plan's conservation program makes it a priority to protect wildlife that is sacred to the Zunis by focusing on building up the deer population as well as bringing back the wild turkey and the juniper and piñon woodlands that provide a habitat for the red-shafted flicker. In a further concession to Zuni cultural practices, the plan does not require members of the tribe who hunt animals as part of a religious practice to apply for a hunting license. They are requested, however, to report to the ZCP office how many game animals they kill so the bag limit for other hunters can be adjusted accordingly.

To date, wildlife habitat enhancement efforts have included the planting of trees, the development of springs and wetlands, and the planting of watercress for wild turkeys. Recent surveys suggest that these efforts are paying off: populations of deer, turkeys, elk, bluebirds, and flickers have increased on tribal lands. Another survey, which compared populations of birds that are hunted with those that are not, shows no noticeable differences, Enote reports.

The ZCP is also introducing Zuni people to alternative technologies they might find useful. But rather than going out and buying a lot of high-tech equipment—such as photvoltaic cells, windmills, and minihydro generators—Enote is talking with the people about whether this technology fits in with the way they want to live. One possibility being explored is to install photovoltaic cells on houses in areas currently not serviced by electricity. Previously, members of the tribe had been reluctant to move into these remote areas to farm because there was no electricity for their jewelry-making equipment on which they depended to earn cash.

Young people are also being exposed through the school system to the concept of sustainable land management as practiced by the ZCP. Several high schools have started gardens at ZCP and an increasing number of science fair projects are geared toward natural resource work, Enote reports. One group of students intends to map the reservation's sewer system as part of a larger ZCP initiative to create an artificial wetlands sewage treatment system.

There is nothing flashy about what the ZCP is doing: there are no headline-grabbing, large-scale projects in the works. Instead, slow but steady gains are being made every year in terms of restoring the land. In addition to these tangible gains, Enote discerns an equally important psychological shift. "A lot of people from the outside could manage these lands," he comments. But rather than depending on outside expertise, a deliberate effort is being made to train Zunis to do the professional jobs required by the land resources management plan. For example, a Zuni hydrologist is now mapping and monitoring water resources.

"Zunis now see other Zunis taking control and doing this work and it feels really good to them," Enote observes. "All of this is leading up to some type of self-determination."

Saving the Seed: Rescuing Important Foods
and Medicinal Crops from Extinction

Kenny Ausubel's first encounter with the seed people took place in San Juan
Pueblo, New Mexico. There he met Gabriel Howearth, a seed collector and
master gardener, who was helping the inhabitants of San Juan Pueblo revive
their traditional form of agriculture by collecting heirloom seeds that had
been stored in gourds, clay pots, and the walls of adobe buildings. One of
the varieties of seeds he resurrected was the sacred red corn of San Juan
Pueblo that had not been grown for forty years.

   Ausubel, a journalist and documentary filmmaker, who was there to film
a movie about Howearth's work, knew he had stumbled onto something of
significance. What he learned at San Juan Pueblo was that there existed a
little-known network of seed savers who were building a botanical ark
against the rising tide of plant extinctions. He learned that the seed market
was coming to be dominated by large corporations selling limited numbers
of hybrid varieties of food crops. This emphasis on select hybrid seeds was
causing a wide variety of "open-pollinated" crops—plants that are pollinated in the field by insects and natural forces rather than laboratory
technicians—to become extinct. The more Ausubel learned, the more he
realized that his life was about to take a new direction: "I thought I was
there to film a movie, but it turned out I was actually there to found a seed
company."

   To find out more about the nascent movement to save traditional varieties of seed, Ausubel began to seek out other freelance seed savers. One of
them was a former teacher of Howearth, Alan Kapuler, a molecular biologist, who conducted research on the nutritional value of organic foods
grown from open-pollinated seeds, and who had grown and marketed
organic seeds since 1975. Ausubel also met Emigdio Ballon, a Quechuan

farmer who had brought to the United States thousands of seeds from his native Bolivia, and Richard Pecoraro, "an ardent seedsman."

In 1989 Ausubel took this loose network of seed collectors and forged Seeds of Change, an enterprise that sells the seeds of a wide variety of open-pollinated plants grown organically. "Fate has a way of plucking you from her garden when you are needed," says Ausubel, a gaunt, wind-burned man in his mid-forties who looks a bit like an itinerant monk just wandered in from the desert.

Seeds of Change has flourished and already it is in the top tier of alternative seed companies with sales of around $2 million, a staff of forty full-time workers, and contracts with some 100 organic growers in thirty states. And although Ausubel has moved on to other projects, the company he founded is thriving, its seeds available by catalogue and in over 1000 stores, including Home Depot. Sales in 1993 were up 65 percent over the previous year and prior growth was even greater.

Seeds of Change plays an important role within the organic movement because it provides gardeners with organically grown seed. Some 99.9 percent of all seeds available commercially are not grown organically and are often coated with a pesticide before sale, Ausubel says. By contrast, Seeds of Change sells some of the only seeds available on the market that are certified organic by the states in which they are grown, he notes.

Currently, Seeds of Change can only produce enough organic seeds to supply gardeners, and scaling up to supply farmers will not happen overnight, Ausubel concedes. "It is not like making widgets where you can go to the factory and just order more. Seeds grow out geometrically and it takes time for them to reproduce."

In addition to supplying organic gardeners with pesticide-free seed, Seeds of Change is also helping to protect from extinction a diversity of food crops. In recent decades a number of important food crops have been lost. In fact, according to Ausubel, only about 3 percent of the food varieties our grandparents ate in 1900 are available to us today.

Large seed company catalogues, which dominate the $400 million home garden seed market, offer fewer varieties of plants each season. Calculations published by the Seed Savers Exchange suggest that almost 50 percent of the nonhybrid vegetables available in 1984 had been dropped from mail-

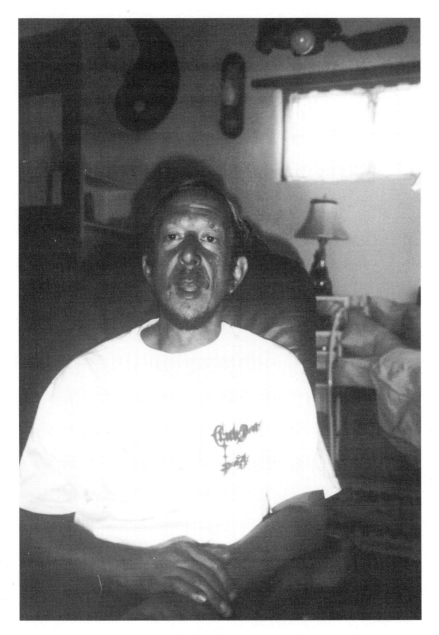

Figure 21.1
Kenny Ausubel, founder and former president of Seeds of Change, at his home in Santa Fe, New Mexico. Photograph by Steve Lerner.

order catalogues by 1991. Furthermore, the hybrids that the seed companies sell often do not reproduce true, forcing the grower to buy new seed each year.

Even more insidious, according to Ausubel, is the influence of a small group of multinational corporations. "In the last 20 years more than a thousand independent seed houses have been acquired by major chemical and pharmaceutical companies such as Monsanto, Lilly, Dow, International Chemical Industries (ICI), and Royal Dutch Shell," Ausubel writes in an article entitled "Theft of the Ark" in the Seeds of Change 1994 catalogue. By buying up small seed companies, these corporations, some of which also have major interests in the production of synthetic fertilizers and pesticides, are now positioned to sell the farmer patented hybrid seeds that require heavy doses of their agricultural chemicals. Some observers describe this as the beginnings of a vertical monopoly aimed at controlling the world food supply. Ausubel cites a study by Kevin Watkins for the Catholic Institute of International Relations, which reports that fewer than a dozen seed and pharmaceutical companies now control over 70 percent of the world seed trade.

In the past no one could corner the market on seeds because farmers grew their own. But now many hybrids are "mules" (seeds that do not reproduce true) forcing farmers to buy seed every season. These hybrids—designed to grow quickly, mature at the same time, be of uniform appearance, and transport well—are squeezing out a multiplicity of open-pollinated varieties. Grown with chemical fertilizers, pesticides, and often an abundance of water, these new hybrids temporarily boost yields, but as the pesticides kill the life in the soil and crops come to require heavy doses of synthetic fertilizers, yields and profits fall.

Bucking this trend, Seeds of Change sells a range of open-pollinated and heirloom seeds that produce a wide variety of vegetables and flowers that do not require chemical fertilizers, pesticides, or extravagant irrigation. A recent Seeds of Change catalogue offers sixty varieties of tomatoes, twenty-two chilies, twenty-two lettuces, twenty-two bush beans, twenty corns, eight varieties of amaranth (a staple grain crop originally from Mexico), and a host of other edibles.

These traditional crops, some of them rescued from the brink of extinction, are frequently higher in nutritional value and taste better than their hybrid competitors that are bred to look flawless on the supermarket shelves, transport easily, and store well. Customers for these traditional and heirloom seeds are often gardeners who remember the taste of a tomato or ear of corn they ate as a child which they thought was lost forever. Others are professional chefs looking for tasty vegetables.

Seeds of Change invites a partnership with its customers by offering them a chance to play a role in protecting biodiversity. Gardeners who buy and plant endangered or little-used varieties of food plants help create a patchwork of backyard bioshelters across the country during a period when plant extinctions have reached unprecedented levels. Growing out the seeds of endangered plant species year after year is essential because seeds cannot be stored like antiques. Stored seed eventually loses its vitality and fails to germinate, as studies of low germination rates at seed banks demonstrate. Marketing the seeds to backyard gardeners decentralizes the supply and increases the likelihood that the seeds will find a niche in climate and soil conditions where they will thrive.

Some gardeners are quite emotional about the prospect of bringing back varieties of food they knew when they were young. One offering in the Seeds of Change catalogue is for an heirloom variety of sweet corn known as Luther Hill #019. The listing reads: "This corn was the best selling sweet corn in New Jersey in the 1950s and is extra delicious. It had disappeared in the U.S. but resided in the capable hands of two Canadian members of the Seed Saver's Exchange. Alan Kapuler grew out Luther Hill, passed it on to us, and we now offer it to you. . . ." A customer writes: "I cannot wait to receive my Luther Hill Sweet Corn. I grew up and still live in Sussex County, New Jersey. As a child every picnic included Luther Hill corn, and corn has not tasted the same since. To this day you will hear someone say, 'There is no corn like that Luther Hill corn we used to get.' Bless those who have the foresight to protect our seeds."

In addition to unsolicited letters such as this, Seeds of Change employees periodically receive small quantities of heirloom seeds in the mail. "We get a lot of seeds sent to us from people who tell us that their grandmother or cousin had this in the attic," Ausubel says. For example, James Couch, a preacher from North Carolina, came across a box of heirloom bean seeds among the spare possessions of an Appalachian parishioner who died. Since no one in the family wanted the seeds, Preacher Couch sent them off to Seeds of Change. Since then he has passed along over 100 bean and corn seeds from the Appalachian region. This story underscores how fragile are the links in the chain of passing seeds from generation to generation.

Ausubel is among a growing number of people who take very seriously the loss of food and medicinal plants that have coevolved with humans over time, in some cases over thousands and even tens of thousands of years. Consider the genetic information stored in these seeds, says Ausubel. Generation after generation of farmers painstakingly selected plants that produced the most food or weathered a drought, pest, or blight better than other plants. In the process they influenced the evolution of plants that were well adapted to certain soils and climatic conditions.

In the next fifty years we will lose a quarter of the 250,000 known plants, a total of 62,500 plant extinctions, according to Ausubel. Currently, plant extinctions are running at 27,000 species a year, seventy-five a day, or three an hour—about 1000 times the natural rate, he writes. And for each plant that disappears, twenty to forty insect or animal species that depend on these plants disappear. Out of 80,000 known edible plants in the world, only about 150 are now grown in large quantity and fewer than twenty plants provide 90 percent of our food. Year after year the big seed companies are dropping more varieties from their catalogues.

To lose this many food crop species represents a significant loss because it is the diversity of plant and animal species that stabilizes nature and cushions us against the periodic convulsions that nature goes through. Furthermore, many of these plants produce food efficiently in regions of the world where other crops would fail. Protecting these traditional food crops from extinction also makes sense because we may need them some day. The vulnerability of our huge monocrops to blights, droughts, and pests argues persuasively for preserving a wide variety of potatoes, wheat, corn, rice, and fruits and vegetables so that we can return to our genetic storehouse for varieties that are resistant when widely grown varieties fall victim to pests or blights. This lesson should have been learned during the Irish potato famine (1845 to 1851), which killed a million people. At that time it proved necessary to locate an alternative variety of potato that was blight resistant. Fortunately, a wide variety of strains of potato still existed to choose from, but had these traditional strains of potato become extinct, we might have lost a major source of food.

The problem posed by the genetic uniformity of U.S. crops was underscored as recently as the 1995 growing season when a combination of bad weather and the gray leaf spot caused the corn crop to plummet by 27 percent. The potato crop was also struck, albeit less severely, by the same blight that caused the Irish potato famine 150 years ago. Despite these ominous signs, Congress refused to provide the $1 million requested by the U.S. Department of Agriculture to upgrade its seed banks. The money would have permitted the USDA's Agricultural Research Service to replant and regenerate its deteriorating seed stock—seeds that might contain the genetic material needed to breed blight-resistant food crops. Many thousands of seeds in the USDA seed banks are withering and dying and some of these have become extinct in the wild, notes Paul Raeburn in an article entitled "Seeds of Security." "When these seeds disappear any disease resistant genes they carry are gone forever," he writes.[1]

Not only are the seed banks underfunded but funds for the National Biological Service, created recently within the Department of Interior to catalogue U.S. biological resources, were cut despite a USDA report indicating that 37 of 250 species listed as threatened or endangered "carry genetic traits potentially important in crop breeding."

This highlights the importance of Ausubel's effort to collect, propagate, and market heirloom food crop seeds. As critical as reintroducing open-pollinated seeds to the market appears to be, Ausubel and company had to start from the ground up and resurrect a whole industry. "There really was no organic seed industry to speak of when we got started," Ausubel says. "What we found was that very few people know how to grow a diversity of organic seeds." It requires considerable expertise to grow seed crops side by side and keep varieties from crossing, he explains. Fortunately, there was no shortage of organic farmers willing to take on the challenge. "Many organic farmers who were previously growing baby vegetables or fancy salad greens for $16 a pound had crashed. By then the field had become very competitive and they were getting only $5 to $6 a pound. So they were very happy to get a contract to grow out our seed."

Seeds of Change began to grow out organic seed on a commercial scale on a 128-acre farm in southern New Mexico bordering the Gila National Forest. The attraction of this land, with a 250-day growing season and 60 acres they have the right to irrigate out of the Gila River, is that it is within 5 miles of ten separate eco-zones where 145 of the world's 540 plant families grow naturally.

On a midmorning in late July, Richard Pecoraro, director of seed production for Seeds of Change, walks out of a field of tomatoes as the Sonoran Desert sun beats down on his long reddish-blond hair and full beard. A five-year veteran on the farm with 15 years' experience growing out seed, Pecoraro is familiar with every crop in the field and its needs. His hand has the feel of partially cured leather and his feet are shod in thin moccasins. To watch Pecoraro squat down and uproot a deeply entrenched pigweed is to get a glimpse of the ancient tug of war between man and nature.

"Weeding is a big-time part of organic agriculture because you can't spray your weeds," declares Pecoraro as he watches a crew of twelve working their way through a field of chilies. "This is the second time through

Figure 21.2
Richard Pecoraro, former director of seed production at the Seeds of Change farm in Gila, New Mexico, shortly before it was sold. Photograph by Steve Lerner.

these fields weeding so it is getting to be more fun now because the crops are up and people can mingle with them."

The fields are laid out in long raised beds with wide furrows separating them. Down the middle of a 5-foot-wide raised bed is a shallow irrigation channel. Black plastic siphon tubes snake out of irrigation ditches whose water supply is controlled by a system of gates. Square plugs are fitted into the siphon tubes to reduce water use and keep the force of the water from washing the soil away. In a season in which water is rationed, such as this one, these water conservation methods make the most out of a precious resource.

Experience has taught Pecoraro how to raise hundreds of varieties of plants without crossing them. Chili crops, for example, are kept 300 to 500 feet apart to keep them from crossing. In addition, the crop is surrounded by a border of corn, flowers, and herbs. The fields are also separated by perennial borders made up of trees, herbs, flowers, and legumes. These borders cut the wind and soil erosion as well as provide shelter for beneficial insects. Out in the field, companion crops of leeks and cilantro and basil and tomatoes are having love affairs. Swarms of big black bees make the fields hum with life. "There is a big party going on in that field. Just the way we like it," Pecoraro says, cracking a smile from beneath his rainbow-colored visor.

Hundreds of varieties of food are being grown out at this original Seeds of Change farm. This year alone there are forty varieties of tomatoes and twenty-five of chilies. In some years trials are held in which up to 150 varieties of tomatoes are tested to determine their characteristics and see how they compare with popular hybrids. Then the best varieties are put into production, grown out, and processed in a long metal building where the plants are dried and seeds are separated.

On the farm there is a constant tension between exploring a wide variety of heirloom and traditional plants, some of which are threatened with extinction, and finding a small number of "hit" varieties that will sell well. This year the balance is about fifty-fifty between very popular varieties that have been chosen for production and less marketable varieties that have interesting characteristics worth preserving, Pecoraro says.

In 1994 Ausubel left Seeds of Change in a disagreement over the future direction of the company. The Gila farm, Ausubel reports, was subsequently sold to a Chevy dealer from Silver City, New Mexico, who turned it into a hay farm. But none of this has stopped Ausubel from pursuing his interest in preserving heirloom foodstocks. "You never know what life will bring you and sometimes when you get knocked sideways it can be quite productive," he says philosophically.

Ausubel's next project was to team up with Greg Steltenpohl, a former musician from Santa Cruz, California, who used to support his music habit by making fresh juice in his kitchen and selling it wherever he could. From these humble beginnings he founded Odwalla, a fresh juice company that posted $35 million in sales in 1995. Working as a consultant for Odwalla, Ausubel encouraged management to produce an "ultra-carrot juice" made out of three heirloom carrot varieties. Not only do these heirloom carrots have significantly better nutritional value than ordinary commercial carrots, he points out, growing them out for juice will help insure that their seed is propagated and preserved.

In a further effort to reintroduce diversity into the food system, Ausubel brought the Odwalla juice makers together with officials from a company that designed a series of "nutraceutical" (food as medicine) beverages. Among these are a multiginseng drink that combines five varieties of ginseng, a couple of which are rare. There are also plans to develop calmative and immunostimulant beverages. "Some of these are based on fairly rare or unusual plants so by marketing these new juices we hope to encourage the cultivation of these underutilized varieties," he says. Ausubel also wants to do his part to help develop an alternative food supply system and he is urging officials at Odwalla to purchase produce from family farms that practice restorative agriculture instead of from the open commodity market. Recently the company made a commitment to phase into this alternative produce supply system over the next ten years.

Through a program he founded while at Seeds of Change called the Native Scholar Program, Ausubel is also attempting to organize a network of Native American farmers involved in preserving heirloom seeds and traditional knowledge about farming. He hopes to connect them with progressive companies like Odwalla and make it possible for them to farm with heirloom seeds once again. As a first project, work is underway to grow out Native American heirloom watermelon seeds in the Southwest to produce a watermelon juice for Odwalla. This would provide a source of income for Native American farmers and an incentive to propagate heirloom seeds.

Drawing all his interests together, Ausubel hosts an annual conference called "The Bioneers: Practical Solutions for Restoring the Environment," a gathering of biological pioneers whom he describes as having both feet on the ground. Through Seeds of Change, Odwalla, the Native Scholar Program, and the Bioneers conference, Ausubel has used his success as an entrepreneur and organizer to raise the level of debate about the direction agricultural practices must take in the United States if heirloom foodstocks are to be preserved.

Cleaning Up and Reusing Abandoned and Contaminated Industrial Sites

A new wave of community organizers who call themselves eco-justice activists are changing the political landscape. Working in communities of color and impoverished areas, they are organizing local grassroots movements to protect public health and improve living conditions. Operating from the premise that heavily polluting industrial facilities, incinerators, and waste dumps are disproportionately sited in communities of color, they are working to reverse this trend.

"We see it as no accident that large numbers of burdensome facilities are placed in communities populated by African Americans and the urban poor. We say this has been done to us on purpose and that environmental racism is a justified term," says Larry Charles Sr., an environmental justice activist. Charles works in Clay Hill and North End, two predominantly African American neighborhoods in North Hartford, Connecticut, the eighth poorest city of its size in the nation.

The Clay Hill and North End communities are no strangers to the miseries of modern urban life. Residents in these neighborhoods suffer the city's highest rates of high school dropouts, teenage pregnancy, AIDS, and violent crime. The average annual income in the area is $8000. Despite these difficulties, an environmental awakening is taking place in Clay Hill and North End that defies all currently fashionable pronouncements about the apathy of the poor and working-class people.

"The environment made it to the top of the list of priorities in our neighborhoods because we are conscious of the threat that contamination represents to our health," says Charles, who is executive director of Organized North Easterners/Clay Hill and North End (ONE/CHANE).

In addition to presenting a threat to public health, the contamination in North Hartford has also stymied economic development, he continues.

Since the riots thirty years ago, there has been no significant commercial development in North Hartford and the area is lacking services often taken for granted elsewhere. There is only one supermarket, no drugstore, no butcher, and no greengrocer. Residents live amid abandoned and contaminated gas stations, garages, car washes, dry-cleaning stores, and industrial sites. The lack of money to clean up the contamination in many of these properties has complicated economic renewal efforts, he adds.

To rebuild their community, residents of North Hartford are engaged in old-fashioned, block-by-block organizing. "We go door to door looking for a person who will form the core of a block club and then we wrap all the faint-hearts around that person," Charles explains. Two years ago ONE/CHANE had seven block clubs but now its network has grown to twenty-five.

Some of these groups have been highly effective. The first night the Tower Area block association met in Cynthia R. Jennings's house, Jennings was elected block president. The second decision they made that night was to go to a public meeting on a state plan to expand a landfill by 36 acres in their community. Those seven people organized thirty people to turn out for the meeting in 24 hours and a movement to block the landfill expansion was launched.

But groups such as ONE/CHANE frequently lack the resources to deal with the complex technical and legal issues that surround a debate over siting a landfill, incinerator, or commercial hazardous waste facility. "We don't have the capacity to analyze and present our case on these issues. All we know is we don't like it. We know that real well. But if you put the burden on the community to prove the impact—that is unfair," Charles asserts.

A visit to a number of these low-budget, street-front, eco-justice offices can be inspiring. Alternatives for Community and Environment (ACE), a nonprofit organization that provides legal counsel and editorial services to environmental justice groups, is located on Warren Street in the heart of Roxbury, Massachusetts, a largely African American community in Boston. ACE is housed in a rambling wooden building bearing a weathered sign from an earlier period announcing that it is home to the "Roxbury Defenders."

Walk in the front door at ACE and there is a wall covered with butcher paper that invites local residents to write down or draw their vision of what they would like their community to become. The downstairs rooms are furnished with secondhand armchairs, trestle tables, folding chairs, recycling bins, battered files, a well-thumbed library, and old boxes of leaflets. On the wall is a bulletin board covered with flyers about upcoming meetings. In an upstairs office a staff worker holds a telephone to his ear with a hunched shoulder while he continues to tap earnestly at a blinking computer terminal. The office vibrates with the same mix of purpose, excitement, informality, commitment, realism, and energy that permeated civil rights movement offices in the early 1960s.

Outside ACE's offices, Warren Street runs downhill past a modern fortress-style courthouse into Dudley Square where a man wearing a black jacket, white shirt, and red bow tie is selling vegetarian pies to passing motorists while he hands out leaflets announcing the "Million Man March" on Washington. Around him, many handsome retail buildings stand vacant with their windows bricked up. But Dudley Square was not always an

Figure 22.1
Staff members of Alternatives for Community & Environment (ACE), outside their offices in Roxbury, Massachusetts. From left to right: Charles Lord, William Shutkin, Penn Loh, Klare Allen, and Jo-Anne Henry. Photograph by Yawu Miller.

economically depressed section of Boston. In the 18th, 19th, and the first half of the 20th century, Dudley Square was a thriving community, home to the governor and the whaling community. Waves of Irish, Italian, and then Jewish immigrants passed through Roxbury over the years. Following the white flight in the 1950s and 1960s, large numbers of businesses failed and real estate values plummeted.

Today, the three red-and-black chimneys of the old South Bay incinerator loom over the Roxbury landscape. Abandoned for the past twenty years and contaminated with dioxin, polychlorinated biphenyls (PCBs), heavy metals, and loose asbestos, the incinerator played host to homeless people who inhabited the red-brick building and frequented the 2½-acre site. "This incinerator is the gateway to Roxbury. It is a symbol of the decay, blight, and neglect that this community suffers," says attorney William Shutkin, co-founder of ACE.

ACE attorneys threatened to sue the commonwealth over the incinerator site, but instead of going to court, they negotiated an agreement between a coalition of local community groups and city, state, and federal officials. As a result of this agreement the state committed $20 million to the cleanup of the site. Now community groups are pressing for local residents to be hired in the remediation of the site and for a voice in deciding what kind of facility is to be built on it.

ACE co-founders Shutkin and Charles Lord, attorney Jo-Anne Henry, and other community activists have also been working on the 3½-acre site next to the incinerator where midnight dumpers illegally deposited a 20-foot mound of tires and scrap metal. The owner recently had the junk hauled away but the soil remains contaminated and he now intends to build an asphalt batching plant on the site. A local Coalition Against the Asphalt Plant (CAAP) brought 350 local residents to a hearing at city hall to protest the plan on the grounds that it would create a hazard to public health.

If the asphalt plant is blocked and the incinerator site is cleaned up there will be 6 acres in the center of Roxbury that could provide space for a mix of light industry, small business, retail, and residential development that would be a real asset to the community, Shutkin notes. Already some grassroots groups are canvassing the community for ideas about what residents would like built on this site and architectural drawings have been made of various alternatives.

Other eco-justice organizers are active in the area. In the adjacent community of Dorchester, Trish Settles navigates her car around huge trucks carrying garbage to be sorted at one of the three trash transfer stations located within sight of homes and neighborhood gardens. "This place is a zoning nightmare," says Settles, an environmental justice activist with the Dudley Street Neighborhood Initiative (DSNI). "We took a group of regulators on a tour to show them what they were signing off on, and many of them were amazed to find people living in areas they thought were industrial."

The Dudley Street neighborhood—an area with 24,000 residents that straddles the line between Roxbury and Dorchester—has a long commercial history that has left a legacy of contamination. The area played host to hat makers (who used mercury), lead foundries, dye manufacturers, and soap, beer, and carriage works. Subsequently, disinvestment, redlining, white flight, and arson cleared out much of the area. By 1984, when DSNI was founded, 1300 lots in the Dudley Street neighborhood stood vacant.

DSNI activists reasoned that if people could afford to own a house in the community instead of renting, they would have a greater stake in improving conditions in the neighborhood. To this end, they invoked the power of eminent domain over 30 acres of vacant lots in the Dudley Triangle area where they planned to build affordable housing. No other nonprofit had previously used the eminent domain tactic successfully, but the gambit worked, and now Settles can point to eighty new units of affordable housing—handsome gray wooden buildings with white trim—constructed on what were once vacant lots filled with lead-contaminated soil, which had to be hauled away.

In addition, DSNI activists were involved in closing down three illegal trash transfer sites in an area where there are a total of fifty-four identified uncontrolled hazardous waste sites, lobbied for the installation of white picket fences to keep illegal dumpers off the thousand vacant lots that remain scattered around the community, and insured that uncontaminated topsoil was hauled in for community gardens.

The eco-justice organizing activities described above in North Hartford, Roxbury, and Dorchester are being played out around the nation. While the struggle for environmental justice in this country can be traced back several hundred years, the movement gained visibility in 1982 when local officials established a toxic waste dump for 29,000 metric tons of dirt contaminated with PCBs in the predominantly African American town of Afton in Warren County, North Carolina.[1] Recognizing that an impoverished town was being unfairly burdened with these wastes, activists attempted to block dump trucks carrying the PCB-laden soils. Five hundred protesters were arrested.

"That was a pivotal event," notes Charles Lee of the United Church of Christ Commission on Racial Justice (UCC/CRJ). "For the first time a link was made between civil rights and environmental issues." Following the protest, in 1983, the congressional Black Caucus requested that the General Accounting Office (GAO) carry out a regional study in the southeastern states to find out where hazardous waste landfills were located. "It revealed that three out of four of these toxic dumps were located in poor, black communities," Lee says.

While these data were shocking to those who gave the matter any thought, the report of a broader UCC/CRJ survey entitled *Toxic Waste and Race in the United States: A National Report on the Racial and Socio-Economic Characteristics of Communities with Hazardous Waste Sites* confirmed that this was a nationwide pattern. The survey revealed that communities with the greatest number of commercial hazardous waste facilities had the highest concentration of minority residents. The report concluded that there exists a "striking relationship" between the location of commercial hazardous waste facilities and race, that race has been a factor in the location of these facilities, and that there is "an inordinate concentration of such sites in Black and Hispanic communities."[2]

This siting of hazardous waste dumps in communities of color "constitutes an insidious form of racism," observes Benjamin Chavis Jr., onetime executive director of UCC/CRJ, in a preface to the report. Lee, who directed the survey that resulted in the report, echoes this sentiment: "Racism is alive and well when it comes to the issue of the environment." It shows up in excessive levels of lead poisoning in inner-city children, in pesticide poisoning among minority field workers, and in the siting of a disproportionate number of highly polluting facilities in communities where people of color live, he adds.

"By the end of the decade [1980s], environmental justice struggles were going on all over the South," observes Professor Robert D. Bullard, director of the Environmental Justice Resource Center at Clark Atlanta University, who wrote *Dumping in Dixie: Race, Class, and Environmental Quality* (Boulder, Colorado: Westview Press, 1990), a book that examines five environmental justice campaigns in states ranging from Texas to West Virginia.

"The grassroots leaders in these communities conclusively dispelled the myth that African Americans are not concerned about or involved in environmental issues," Bullard writes. Among their successes, environmental justice advocates shut out hazardous waste incinerators on Chicago's South Side, East Los Angeles, and Kettleman City, California; defeated a proposed landfill on the Rosebud reservation in South Dakota and another hazardous waste landfill in Noxubee County, Mississippi; and blocked construction of a Formosa Plastics plant in Wallace, Louisiana.[3] By 1991 the First National People of Color Environmental Leadership Summit, held in Washington, D.C., attracted 750 eco-justice advocates.

People of color bring their own experiences, traditions, and understandings of what constitutes a healthy and sustainable community, Lee continues. For decades they struggled over so-called quality of life issues that plagued their neighborhoods: air pollution from factories operating adjacent to residential areas, discolored and foul-tasting water coming out of the tap, infrequent trash removal, poor enforcement of environmental regulations, midnight dumping, hazardous working conditions, underfunded mass transit, freeway construction through the heart of their communities, and a host of other problems. These issues had been largely ignored by traditional environmental groups that were focused more on rural conservation issues than on the problems of the inner city, Lee says.[4]

Environmental justice activists face a daunting challenge. Across the country, the destruction inflicted on communities by abandoned and contaminated properties continues to spread like cancer. The abandoned incinerator in Roxbury is just one symptom of this disease. A GAO report estimates that there are 130,000 to 450,000 contaminated commercial and industrial sites around the country. Cleaning up this industrial wasteland will cost an estimated $650 billion—six times the cost to the federal government of the savings-and-loan debacle.[5]

Contaminated sites come in all shapes and sizes and levels of contamination, ranging from huge abandoned steel mills to a closed gas station at a rural crossroads with a leaking underground storage tank. Some sites are so contaminated that they are listed as "superfund sites" that come under special regulations and usually require large amounts of money to remediate.

But what about all the other sites that have been abandoned or that developers shun because they are slightly to moderately contaminated or merely suspected of being contaminated? Many of these sites could be cleared of suspicion of being contaminated or, if they are somewhat contaminated, remediated at a reasonable cost. Distinguishing the heavily contaminated from the only slightly contaminated sites could free up the latter for redevelopment and revitalize many communities.

There are now so many of these sites that are either suspected of being contaminated or are slightly to moderately contaminated that bureaucrats and community activists now describe them as "brownfields" to distinguish them from pristine "greenfields" that are being developed in the suburbs and rural areas. Nowhere do these brownfields wreak greater havoc than in the communities where the poor and people of color predominate. Children and homeless people live and play on those sites not knowing that they are exposing themselves to toxins. More insidiously, dust from these sites blows through the neighborhood and unseen plumes of pollution from past dumping spread underground through the soil and in the groundwater.

Brownfields pose an economic as well as a public health hazard. Potential investors are afraid to renovate contaminated sites because of liability concerns. This in turn causes unemployment to soar. Abandoned and contaminated industrial sites also bring down real estate values and discourage buyers from investing in adjacent properties. As a result, revenues from property taxes diminish and public services dwindle.

The problem posed by brownfields is on such a large scale that a network of environmental justice advocates, no matter how dedicated and effective, will only be able to make marginal gains. Their only hope is to pioneer the way in showing what should be done to clean up and redevelop their communities, and work in a partnership with federal, state, and local governments, as well as with the private sector, to redevelop these sites. A big step in this direction was taken when President Clinton issued the Executive Order on Environmental Justice that specifically requires all departments of the federal government to help solve the problem of disproportionate environmental burdens borne by some communities.

To begin to implement this commitment, a federal government initiative, closely monitored by environmental justice advocates, is attempting to find innovative ways to jump-start the cleanup and redevelopment of brownfield sites. In January 1995, Carol Browner, Administrator of the Environmental Protection Agency (EPA), launched the Brownfields Initiative. This program attempts to encourage developers and lenders to invest in brownfield sites by streamlining administrative procedures and adjusting the enforcement of Superfund liability regulations. (The initiative is not focused on the cleanup of the 1200 most contaminated sites on the superfund's national priority list, but rather on less contaminated sites.) The EPA is also funding fifty pilot projects aimed at helping local governments eliminate barriers to the cleanup and redevelopment of these properties.

"We became aware that our enforcement of some superfund regulations had the unintended consequence of creating a barrier to brownfields redevelopment," observes Jonathan D. Weiss, formerly senior brownfields counsel at the EPA's Office of Solid Waste and Emergency Response and now with Vice President Gore's office. While superfund regulations require that polluters pay the costs of cleaning up the contamination they cause, these same regulations also hold liable innocent buyers who purchase contaminated properties and lenders who foreclose on contaminated properties.

In an effort to encourage developers and lenders to reinvest in brownfields redevelopment, EPA officials have issued new "guidance" documents clarifying liability enforcement practices and are writing "comfort letters" assuring developers that they will not be held liable if they comply with state voluntary cleanup programs. Agency officials also lobbied for changes in the Community Reinvestment Act (CRA) regulations, which require that banks reinvest a certain percentage of their loans in their host community. The revised CRA regulations are designed to boost loans to inner-city projects by giving banks CRA credit for investing in brownfield redevelopment.[6]

The EPA also learned that bankers routinely turned down loans for brownfield properties when they discovered that the property was listed on the superfund tracking list. In reviewing the properties on that list, EPA officials found that many of them had been investigated and designated as requiring no further remedial action. In a rare instance of radical regulatory surgery, EPA officials cut 25,000 of 38,000 sites from the superfund tracking list, freeing them for development.

Finally, to help bring brownfields redevelopment down to the local government level, the EPA funded fifty brownfield pilot projects at $200,000 each. These demonstration projects are aimed at helping cities inventory their brownfield sites, experiment with incentives to clean and redevelop them, and increase citizen participation in the process.

The Brownfield Action Agenda is responsible for a "tremendous mood change" among developers and lenders," asserts Timothy Fields, deputy assistant administrator at the EPA's Office of Solid Waste and Emergency Response, who has been closely monitoring the brownfields program. These EPA actions send a "powerful signal" to lenders and developers that the agency is serious about clarifying liability issues and cooperating with cities and states to facilitate brownfield redevelopment, he continues. "There is a lot of enthusiasm for the brownfields program. In my twenty-four years at the EPA I have never seen an initiative that has been so fully supported by a vast number of people from different walks of life," Fields says.

Environmental justice leaders also see real possibilities for a brownfields effort that is responsible and well designed. "What challenge faces us which is more central to our survival than revitalizing our cities?" asks Charles Lee, research director at UCC/CRJ. He sees tremendous hope in the fact that a wide spectrum of interests have flocked to the issue—a critical mass that could prove powerful enough to make the venture work.

But Lee and other environmental justice advocates also see a key problem with the program as initially structured. As Lee puts it, "The EPA's brownfield initiative was a locomotive that left the station without a major group of passengers aboard, and those are the folks struggling for environmental justice and the residents who live in the impacted communities." The program was targeted at lenders and developers, with no place at the table provided for activists and residents who actually live near the site, Lee says. Suddenly the EPA was working on brownfields, and no one in the communities of color had any idea about what the agency was talking about. Some people even found the term "brownfields" culturally insensitive: "Do they mean fields where brown people live?" Lee was asked.

Environmental justice advocates argue that the brownfields program cannot bring genuine environmental and economic improvements unless it is community driven: developer-driven projects will simply bring in unhealthy development all over again, like the asphalt batching plant proposed for Roxbury. Moreover, they point out that by starting off without consulting those who are most affected, EPA ignores a nationwide network of activists who have been deeply involved in problems of cleanup and redevelopment of contaminated sites since the early 1980s and before.

Among those who are disappointed in the way the EPA's Brownfield Action Agenda was launched is Vernice Miller, co-founder of West Harlem Environmental Action and now director of the Natural Resource Defense Council's environmental justice work. "You have to understand that in the environmental justice arena process is everything. If you screw up the process, no one will want to hear about your project," she says.

Richard Moore, coordinator of the Southwest Network for Environmental and Economic Justice, echoes this sentiment: "I have to say that a lot of our people, frankly, were not happy. A better way to say it is that they were plain pissed-off because there were initial meetings at the EPA around brownfields that involved industry and developers, while grassroots and people of color organizations were not at the table."

Fields at the EPA acknowledges these problems: "We did not have adequate community involvement in the development of the Brownfields Action Agenda." Starting in June 1995 a community involvement checkup became part of the application process for a pilot project grant, which confirms the participation of local community organizations.

Fields credits Lee and Bullard with having "sensitized" the agency to the issue and inspired a series of changes in the Brownfields Action Agenda. EPA is planning a summit of federal agencies that deal with jobs, housing, and health, in order to integrate all those considerations into the brownfields program. Moreover, Lee convinced EPA officials to co-sponsor with the National Environmental Justice Advisory Council (an advisory committee to the EPA on which he serves) a series of one-day dialogues around the nation on brownfields redevelopment. The dialogues took place in the summer of 1995 in Boston, Philadelphia, Detroit, Oakland, and Atlanta—not only to gather the recommendations of people from the neighborhoods afflicted by brownfields but also simply to hear their hopes and ideas for their communities.

Although many applaud the EPA for wading into a problem as contentious and intractable as brownfield redevelopment, several obstacles remain. Richard Moore is concerned that the program's urban focus could overlook rural problems such as pesticide dumps and uranium mining wastes on Native American lands. Administration officials note that the EPA recently established a cooperative agreement with Americans for

Indian Opportunity to establish a tribal organization to provide advice to EPA on waste issues and outreach to tribal governments on brownfield issues.

As for the impact of the program in urban areas, Vernice Miller of West Harlem Environmental Action points to a "fundamental flaw" in the structure of EPA grants for municipal pilot projects: EPA recommends that the cities receiving the grants consult with the residents who will actually be affected, but the agency has no mechanism to enforce this recommendation. "If the EPA is going to rubber-stamp decisions on brownfield redevelopment by municipalities, that is not going to help us," she contends.

At the most basic level, activists do not trust the way local land-use decisions are made. Zoning boards, planning commissions, and local development corporations have always been top-heavy with developers and bankers, observes Robert Bullard. "We have seen, time after time, zoning boards rezone residential areas as industrial areas if they happen to be communities of color," he says; these areas effectively become "sacrifice zones." What community activists want is genuine influence in shaping development decisions, he contends. If EPA's brownfields program can accomplish that, it will amount to a revolution in the way local land-use decisions are made.

From the developer's perspective, brownfield redevelopment offers unusual challenges—but also the potential for profit, especially for firms that have some environmental expertise.

Some companies already deep into brownfields development see a real possibility of economic success. Gordon M. Davidson, president of Capital Environmental, is a former EPA official who is now working on brownfields development deals in thirty states. Davidson welcomes EPA's program as a sign that the agency now recognizes that not all hazardous waste sites are "screaming emergencies," and thinks the shift is giving comfort to quite a few people in state regulators' offices and in corporate boardrooms.

More than that, Davidson points out that the funds companies must hold in reserve to cover possible environmental liabilities constitute a significant financial incentive to shed their contaminated properties. This is where Capital Environmental and other environmentally savvy developers find a niche. Davidson's firm brokers deals with these companies by bringing together all the players—investors, regulators, environmental lawyers, and community groups—necessary to craft a workable game plan for getting the site cleaned up and profitable once again.

Hemisphere Corporation takes this process one step further. This Cleveland-based company buys brownfields properties at reduced rates, taking over all the potential environmental liability, and then cleans them up and sells them at a profit. In some cases, in what president Todd Davis calls "an upside down deal," the original owners actually pay Hemisphere to take the site off their hands.

In fact, brownfields offer some serious advantages for developers, as Michael E. Porter recently noted in an article entitled "The Competitive Advantage of the Inner City."[7] First, brownfield sites can be bought cheaply. Second, they tend to be near downtown business districts and therefore make sense for a host of business support services, such as caterers, accounting firms, messenger and maintenance services, and computer repair shops. A further competitive advantage of the inner city is that there is a large pool of local people willing to work for moderate wages.

But the risks involved include significant site assessment and cleanup costs, more complicated and time-consuming loan negotiations than for greenfields development, and of course, those complex and overlapping state and federal rules on environmental liability.

Charles Bartsch at the Northeast-Midwest Institute, who has studied brownfields development for years, foresees another problem: if, on top of all these disincentives, developers believe they will face protracted negotiations with multiple community groups, they may throw up their hands and say, Why do I need these headaches? Community groups that want to compete for investment dollars, he suggests, should first develop a vision of what they want and then appoint a negotiator, so that investors can deal with a single representative.[8]

Communities "must not allow the perfect to be the enemy of the good," warns ACE attorney Shutkin. Ideally, community groups, environmental justice resource centers, and community colleges and universities should be able to assist residents in negotiating deals that will bring in jobs and protect local public health, without scaring off enterprises that could in fact do some good. Vernice Miller gives the example of a community "visioning process" for the redevelopment of the South Bronx in New York City, which brought together 250 groups of grassroots people as well as technical experts, and produced a number of proposals for development that the community would welcome and work to bring about.

But what about the environmental bottom line? Already, there are signs that the EPA may not be able to find a middle path between the demands of developers and the requirements of health. Developer Todd Davis, for instance, believes that the Brownfields Initiative does not go far enough to encourage investment. He calls for tax incentives, broad liability protection, and cleanup standards tied to future land-use plans, so that a site destined for industrial use does not have to be cleaned up to the same level as one designed for a daycare facility.

On the other side, a number of environmental advocates think that the rush to streamline rules for brownfields cleanup and redevelopment at the state level may be weakening environmental safety standards. "In some state statutes, this streamlining of regulations has meant that developers get to choose the level of cleanup standards they meet," says Rena Steinzor, director of the University of Maryland's environmental law clinic. "That makes me very uncomfortable, because state officials should be firmly in control of the level of cleanup."

Others criticize the decision to delist the 25,000 sites and tinker with liability guarantees. "The agency has not taken a studied approach to how one should go about releasing liability," says Deeohn Ferris at the Washington Office on Environmental Justice. She also believes that every community should have a say in assessing whether or not to delist each of the 25,000 sites. Ferris is now preparing a Freedom of Information Act request to find out what criteria were used in deciding which sites to drop.

Timothy Fields protests that the EPA has spent fifteen years and $500 million studying the sites on the list, and that citizens should be "very confident" about the results. But Rick Hind, director of the Greenpeace toxics campaign, does not see how the EPA could have done a thorough environmental audit of 38,000 sites for that sum—an average of $13,500 per site. "That's the best bargain I have ever seen the EPA get on these site investigations," he says. "Clearly they just eyeballed or took someone's word that a lot of these sites were not a problem," he asserts. "One wonders whether the EPA had its act together when it decided to delist these 25,000 sites," agrees Peter Montague, director of the Environmental Research Foundation and editor of *Rachel's Environment & Health Weekly*. He claims that a series of Office of Technology Assessment (OTA) studies "showed in excruciating detail" that the EPA has been inconsistent in the criteria it used in deciding which sites need to be cleaned up.

Despite these critiques, administration officials argue that the EPA will be able to walk the fine line between protecting public health and streamlining regulations to make it easier for developers to clean up sites that are only moderately contaminated. The brownfields program has been unusually successful at creating partnerships between government officials, developers, and community activists to work on returning these sites to productive use, they contend. Furthermore, the Brownfields Initiative is part of a grander administration effort, articulated through the President's Council on Sustainable Development, to bring different groups together to work to improve the economy, create jobs, and clean up the environment at the same time.

All praise, controversy, and criticism aside, EPA's Brownfields Initiative is moving ahead, with varying degrees of success in enlisting community support.

EPA's Fields says the initiative is already paying off. "In Cleveland, the brownfield project resulted in $3.2 million in new private investment, a $1 million increase in the local tax base, and more than 170 new jobs," notes EPA administrator Carol Browner.[9] President Clinton used the project as a symbol of his administration's success in marrying economic and environmental goals. In an Earth Day speech, he told the story of Bill Robinson, a forty-year-old man, who was unemployed for eight months after the company he worked for relocated. Robinson subsequently struggled to support three children while looking for work in Cleveland's depressed economy. When the Brownfields Initiative helped Dedicated Transport, Inc., open a warehouse in an old steel refinishing plant in Cleveland's industrial district, Robinson was one of the local people hired.

"Fine," says Sandra Crawford, director of the Center for Environmental Education and Training at Cuyahoga County Community College, but the EPA's pilot project in Cleveland "was a done deal by the time we heard about it." Crawford is training area residents to qualify for performing remediation work on brownfield sites and what she sees is an opportunity lost for employing her students.

Virginia Aveni of the Cuyahohga County Planning Commission (CCPC), which initiated the project, says that they held a multi-stakeholder meeting prior to the EPA pilot grant, but that the determining factor in the development of the site was simply that "an owner came along who wanted to do something with the site." Aveni says that county officials would be delighted to work with any community group that had a deal ready to go, with an end user in sight prepared to develop the property, but none of them do, Aveni says.[10]

Atlanta may stand a better chance of finding intersections between the agendas of community activists and those of developers and community planning associations. The city's first proposal for a pilot project grant had only limited community involvement, says Connie Tucker, of the Southern Organizing Committee for Economic and Social Justice. But after the EPA and Lee held one of their brownfields dialogues in Atlanta, the city agreed to revise its proposal. The new proposal includes much more input from locals—including, for example, the agreement that sites with strong community support for redevelopment should be assigned a higher priority than others.

"I am optimistic that the Brownfields Initiative can be useful in cleaning up and rebuilding abandoned and contaminated sites in Atlanta. We have a diverse group of people now willing to work together. And we have a very strong sentiment in the community that we have to work our butts off to get community people back in the process so we can build sustainable and healthy communities," Tucker says.

New Orleans, which also received a pilot project grant, is preparing a still more elaborate process for including community input. According to Dr. Beverly Wright, director of the Deep South Center for Environmental Justice at Xavier University, not only is the Mayor's Office for Environmental Affairs heavily involved in the project but the city is also setting up a consortium of stakeholders to steer the process. The consortium will include lenders, developers, city planners, agency officials, and representatives of nine or ten communities with brownfields who will be chosen in special community elections. Moreover, the choice of properties to be developed will start with the communities themselves, each of which will propose priority sites.

Another promising pilot project is taking place in Boston, which received one of the EPA grants. The city's grant proposal was carefully crafted and revised on the basis of community input, says ACE attorney Shutkin. As a result, the grant is being used to inventory sites that are priority concerns for local neighborhoods. About half the money is being spent to hire a local coordinator from one of the neighborhoods afflicted with brownfields, someone who knows the community well and can serve as a bridge for residents to communicate with the agencies and the developers. Shutkin drafted a "memorandum of understanding" (MOU) with the city of Boston, committing it to a maximum of community involvement. "This MOU became a formal part of the proposal to the EPA and calls for the EPA to mediate any disputes between the city and community partners, should they arise," Shutkin notes.

By contrast, the pilot project in Bridgeport, Connecticut, includes only a limited amount of community involvement. To date the focus has been on developing a database about which properties are contaminated, explains Kevin Gremse, project manager for brownfields economic redevelopment for Bridgeport. The city did put together a Community Linkages for Environmental Action Now (CLEAN) task force, which selected an environmental consulting firm, but the task force has not met as often as was anticipated.

"When you bring the brownfields issue to the community level there are some very sophisticated and complex issues involved, which I, as project manager, am still trying to get a handle on," Gremse says. Most people are interested in the idea of having community involvement in the brownfields issue, but when you get down to the technical specifics they often lose interest, he continues. "The redevelopment of brownfields should be neither a community-driven nor a city-driven process; all players have to join in this effort together," he concludes.

Barbara Wasko, an activist who organized community support for the cleanup of a hazardous waste site in her own backyard, was elated to be appointed to the CLEAN task force, but she has since been disappointed. "I hoped for some real involvement beyond attending a few meetings," she says.

How much good these projects can do will remain an open question for years to come. A two-year grant of $200,000 annually, awarded to each of fifty communities, cannot make much of a dent in a large nationwide problem. "Whenever there is funding for community programs the money is always dinky," comments Ferris of the Washington Office on Environmental

Justice. "When they fund brownfields redevelopment like they fund the B-2 bomber, then I'll know they are serious."

President Clinton apparently agrees and in his proposed budget for 1997 will allocate an additional $25 million to the EPA's budget to expand the Brownfields Initiative—money that will increase pilot grant awards. He also proposed a $2 billion tax incentive plan, spread out over seven years, to boost brownfields redevelopment.[11] The tax relief is projected to spur an additional $10 billion in private investment and return as many as 30,000 brownfield properties to productive use, writes EPA administrator Browner. It would be targeted "to those areas with the greatest need for redevelopment—those with a poverty rate of 20 percent or more, Empowerment Zone/Enterprise Communities and communities where EPA's brownfield pilot projects are in progress," she continues.[12]

While acknowledging that the pilot money doesn't yet amount to much, Clark Atlanta University professor Bullard suggests that if the brownfields funding is used in tandem with Empowerment Zone money, it could add up to millions of dollars and provide real leverage with city officials and developers for requiring community participation in deciding what gets built. Some of this has already come to pass: twenty-four of the forty brownfield pilot programs that have been awarded to date are in Empowerment Zone communities.

The need to coordinate government efforts in this area has not been lost on Vice President Gore who chairs a community empowerment board that includes seventeen federal agencies that oversee policies toward distressed communities. Gore is championing a community empowerment agenda that would take a more integrated approach to relief for distressed communities and improve coordination between such federal programs as the Empowerment Zone Initiative, the Brownfields Initiative, the Homeownership Initiative, and the Community Policing Initiative.

If the Brownfields Initiative is nothing else, it is at least a start. Discussing his own decision to become involved with EPA's program, Charles Lee points out that demographic projections suggest that sometime in the middle of the next century, the so-called minority citizens will outnumber the so-called majority. Given the state of race relations and the quality of urban life today, there are grave questions about what the United States will look like by then: if the harmful trends continue, will we move increasingly toward an apartheid-like society?

What brownfields offer, Lee argues, is the possibility that governments, developers, and communities will find they have shared interests in redevelopment. Out of that common interest could come the energy and resources for reversing some of these negative trends. Ultimately, American society as a whole will benefit. The stakes could not be higher, says Lee: "This is an issue of civilizational dimension."

Helping Families Minimize their Environmental Impact One Household at a Time

Americans are pigs when it comes to hogging more than their fair share of the planet's resources. While we make up only 5 percent of the world's population, we consume 25 percent of its energy and, by some estimates, 33 percent of its resources.[1] Clearly, Americans need to go on a consumption diet, but many of us are so addicted to our voracious lifestyle that we find it hard to give up our wasteful ways.

Fortunately, there is help for people who want to stop living high on the hog. David Gershon is in the business of helping American families, one household at a time, learn to use energy and resources more efficiently and develop a more ecologically sustainable lifestyle.

Founder and president of the Global Action Plan for the Earth (GAP), an immodestly named nonprofit organization headquartered in Woodstock, New York, Gershon developed a four-month behavior modification program that, to date, has enabled 3000 households in the United States (8000 households around the world) to reduce the stress their consumption patterns impose on the environment.

On average, American families participating in the GAP program reduce their volume of garbage by 42 percent, cut water consumption by 26 percent, decrease carbon dioxide output by 16 percent, use 15 percent less fuel, and save an average of $401 a year. Were these sustainable practices replicated in households across the nation, it would boost conservation efforts enormously. What is most promising about this initiative is that the emphasis is on direct action, not just talk.

By challenging us to develop sustainable lifestyle practices, forty-nine-year-old Gershon asks that we stop looking outside ourselves for the source of environmental ills and focus instead on reforming the wasteful aspects of our own lifestyle. This brings environmental damage control home and enables the individual to do something practical about it.

"We discovered that business, the perennial villain, was not the primary problem, and that government, the perennial panacea, was not the primary solution," says Gershon. "The real problem is ordinary individuals in high-consumption countries living lifestyles that are environmentally unsustainable given the finite resources of the planet. If how we live our lives is the problem, the good news is that how we live our lives can also be the solution."

Gershon, a slender, immaculately dressed activist with a diplomatic manner, makes the distinction between direct and indirect environmental action: for example, direct action entails installing a low-flow showerhead in your bathroom; indirect action might involve lobbying your local government to conserve water by fixing leaks in the city's underground system of water pipes. While both forms of activism are important, Gershon ar-

Figure 23.1
David Gershon, founder and president of the Global Action Plan (GAP) for the Earth in Woodstock, New York. Courtesy GAP.

gues that it makes sense for each of us to clean up our personal environmental act before prescribing broader political initiatives.

But how can one teach Americans to be less wasteful? To date, efforts to convince Americans to live more frugally have gone largely unheeded by the mainstream because an environmentally sensible lifestyle sounds as if it involves great sacrifice. Acknowledging this, Gershon avoids the language of sacrifice. "The sacrifice way of dealing with consumption is a loser both politically and in terms of convincing people to take the issue seriously," he observes. Sacrifice really isn't necessary, since we waste about 75 percent of the resources we consume, he argues. If we learn to use resources more intelligently, we won't need to make sacrifices or diminish our quality of life, he insists.

Using criticism and guilt to prod Americans into change, Gershon realizes, would not make eco-householding a popular movement. Nothing would squelch an interest in improving domestic environmental practices faster than having some self-righteous environmentalist show up at your house and point the finger at all your spendthrift habits. Instead, the program employs a positive spin on the subject by placing the emphasis on developing sustainable lifestyles.

"I have my own opinion about how much is enough, but it is just my own opinion," Gershon explains. "Everybody has to find their own truth. They also have to change at their own pace," he adds. GAP emphasizes the basic conservation goals and then leaves everyone to make their own choice about how to achieve them. "Moralizing polarizes and guilt paralyzes," Gershon is fond of saying. People change if they are presented with clear and reasonable alternatives to their current lifestyle and if they are supported in making a change, he adds.

Instead of trying to scare Americans into changing their lifestyle by bombarding them with grim evidence of environmental decline, Gershon took another tack. He asked himself what a sustainable American society would look like. To answer his own question, he put together a research team that emerged with fifteen measurable goals to achieve by the year 2000. Meeting these goals would not mean that we had achieved sustainability, Gershon explains, but rather that we had entered the path to sustainability.

The goals include decreasing carbon dioxide emissions by 20 percent, scaling back deforestation by 50 percent, reducing pesticide use by 75 percent, diminishing solid waste generation by 75 percent, cutting production of hazardous waste by 80 percent, whittling down water consumption by a third, and lowering the rate of global population growth by half. These targets for changes in American consumption patterns were subsequently adopted as the goals for Earth Day 1990, an event that Gershon describes as a turning point in the way the public viewed the cause of our environmental crisis.

It was in 1990 that Americans began to see that they could make changes in their own lives that would ease the environmental crisis. It was during this period that Americans bought numerous how-to books about making their lifestyles more environmentally friendly. For example, *50 Simple Things You Can Do to Save the Earth* sold 3.5 million copies in the United States While these books underscored the strong public interest in taking personal action to reverse environmental decline, Gershon found that most people remained paralyzed and overwhelmed with lists of things to do.

His own research suggests that people ask four basic questions about how to change their behavior so that it will achieve better ecological balance: (1) Where do I start?, (2) Which are the most important actions?, (3) How do I do them?, and (4) Will it make a difference? To answer these questions Gershon wrote *EcoTeam: A Program Empowering Americans to Create Earth-Friendly Lifestyles*, a guide to a four-month group process that provides people with the structure, information, and support necessary to change their lifestyle.[2]

At first glance the workbook looks a bit simplistic. Suggested activities include setting up a recycling center, cutting off junk mail, starting a compost, using cloth napkins, repairing household items instead of buying new ones, using paper on both sides, fixing water leaks, installing low-flow showerheads and toilet dams made of weighted plastic jugs, taking shorter showers, watering the lawn in the morning or evening to minimize water-loss due to evaporation, turning off lights, covering windows with shades and curtains, screwing in fluorescent bulbs, car-pooling, inflating tires properly, and limiting purchases to durable, minimally packaged items you really need that are, where possible, of local origin and ecologically efficient design.

To those of us who are already embarked on the journey of making eco-friendly changes in our lives—like screwing in a few energy efficient light bulbs and participating in curbside recycling—none of this sounds particu-

larly new. But while there is nothing novel here, few of us have system-
atically monitored our environmental impact or undertaken the broad
spectrum of actions to use resources more efficiently that the GAP program
suggests. "The question is not whether you have done these actions, but
rather how consistently you do them and at what level," Gershon says.

"Our most interesting challenge is with environmental true believers
who think they had done all of this," Gershon says. "Most of these people
have a high level of intention to make these changes and a high level of guilt
because they have not made as many of them as they thought they should."

GAP's premise is that even people who understand intellectually the
importance of making eco-friendly changes in their life need support and
feedback from a group to actually make these profound changes. Otherwise they already would have made the changes necessary to put their
household on a sustainable footing. To make the task less overwhelming,
Gershon breaks the program into bite-sized portions that can be digested
over a four-month period.

Anyone can initiate an EcoTeam, Gershon explains. For a fee of $35 per
person, GAP will provide a volunteer EcoTeam coach, coaching materials,
a newsletter, and copies of the EcoTeam workbook. The new team, usually
composed of four to eight households, is expected to meet twice a month
for four months to work on a variety of goals including reducing household
waste, conserving water and energy, practicing greater transportation efficiency, and becoming eco-wise consumers. The workbook suggests one
ecologically sustainable lifestyle practice per page with an illustration for
each action.

"We now have an updated four-month program that works that can help
people change their lifestyle," Gershon asserts. What inspires people most
to go through the program is a sense that they are making a difference
in their own community, he says. To capitalize on this motivation, GAP
adopted a community-based lifestyle campaign that operates at the neighborhood level where residents can take advantage of their proximity to
car-pool and share tools and other resources. Once enough neighborhood
groups are established, a more ambitious community-wide campaign can
take root, Gershon explains.

The goal is to initiate five community campaigns in each of ten regions in
the country over the next two years, he continues. "Our target audience is
5 to 20 percent of the American population," Gershon says. "We are not

targeting the 80 percent of the population that is least receptive." The rationale behind enlisting those who are most predisposed to this kind of lifestyle change is based on a social diffusion theory, which states that to
spread an innovation throughout society requires the participation of 5 to
20 percent of the population.

Currently GAP is working with communities in Bend and Portland, Oregon; Santa Cruz, San Jose, and San Francisco, California; Freeport, Maine;
Chattanooga, Tennessee; and Minneapolis–St. Paul, Minnesota. Already
there are signs of cooperation: the mayor of Minneapolis has promised to
spread the word about the GAP program using the network of recycling
block captains and the neighborhood revitalization program as a means of
communicating with residents about the need for household conservation
efforts. Municipal governments such as these are motivated to work with
the GAP program because GAP households reduce their demand for landfill space and municipally treated water, Gershon points out.

GAP is also spreading through the school system because there is nothing
quite as convincing as a child who wants to cut a household's waste stream
to get a family involved in resource conservation. With a grant from the Pew
Charitable Trust, GAP designed *Journey for the Planet*, a five-week behavioral change program for children that was tested with twenty-one educa-
tors and 442 children.[3] The pilot program was such a runaway success that
there is a corporate interest in providing funds to replicate the program
in 1000 schools, Gershon reports. The United Nations Environment Programme partnered with GAP on this project and created a "global hero"
patch and certificate for children who take a certain number of environmentally sustainable actions.

A number of governments are also providing funds to promote GAP programs because officials understand that empowering people to develop sustainable lifestyles complements their command-and-control regulatory
approach. The British, Irish, Dutch, Danish, Norwegian, Finnish, Swiss,
and Canadian governments are all investing in promoting GAP programs,
while in this country the Environmental Protection Agency and the President's Council on Sustainable Development have partnered with GAP
on promoting sustainable lifestyles. Internationally, GAP has been cited
as a model program by the United Nations Commission on Sustainable
Development and has worked with the United Nations Environment
Programme.

This kind of government involvement in GAP programs is important because there is only so much the individual can do without the infrastructure necessary to support sustainable activities. And only governments can put the big systems in place that will permit us to be more resource efficient. An individual can drive his or her car less, but it is only the government that can set fuel efficiency standards for vehicles and build mass transit projects. Similarly, individuals can learn to separate trash, but without a curbside-pickup recycling program, the number of people who recycle will remain small.

To promote sustainable lifestyles, Gershon is looking for opinion leaders around the country willing to urge people to make the effort to change. One of his candidates for the job is President Clinton who announced that he plans a "greening of the White House" to get the nation's house in ecological order. Gershon wants Clinton to extend this effort by challenging Americans to live a more ecologically sustainable lifestyle and tell them how to find out how to do it.

The biggest obstacle to this kind of lifestyle change is that many people initially say they don't have the time to go through the program. People have to be put in touch with their values and come to realize that they care deeply about nature and the future of their children and grandchildren, Gershon says. Once they realize that they place a high priority on developing sustainable lifestyle practices, then the idea of attending a total of eight two-hour meetings over a four month period does not seem too demanding, he adds. A number of GAP graduates have said that they are grateful that the program slowed them down enough to realize they had been living at a "convenience level" without having taken the time to learn to live more sustainably, Gershon says.

Another obstacle to lifestyle change is that a lot of people just don't want to spend their free time joining yet another group—especially one where the discussion centers around weighing garbage. Some of this anti-group feeling may come from a sense of alienation or lack of community that many people experience, Gershon says. But those who go through the GAP program find it builds a sense of community. As people talk together about their difficulty making changes and share tips about how to use resources more efficiently, a bond develops between them and it can be a very positive interaction, he continues. In GAP groups people overcome their sense of isolation in tackling environmental problems, they hear new ideas for sustainable practices, and they get to know their neighbors better.

When you actually monitor the volume of resources you waste it is something you cannot forget, Gershon notes. Spreading your garbage out on a plastic garbage bag to see exactly what you throw away can be very revealing and lead to ideas about how to reduce your waste stream. Similarly, weighing your garbage before taking it out to the trash can provides feedback about whether you are making progress. Practices such as these promote an awareness that keeps people improving their sustainable behavior—not just for four months but for a lifetime. Once your systems are in place to reduce waste, it is hard to revert to your previously profligate practices, he adds.

A two-year longitudinal study conducted at the University of Leiden in the Netherlands confirms that GAP's behavior change program has lasting power. The study, carried out by the university's faculty of social and behavioral science in 1994, revealed the following. First, nine months after graduating from a GAP program, participants were, on average, still taking twenty-one of an original twenty-two sustainable actions and had added four new actions on their own. Second, resource savings were not only sustained but actually improved by 10 to 20 percent in the post-program period. And third, about half of those who graduated from the program transferred what they learned either to their workplace or to their community or were helping to build the program elsewhere.

Among those who have participated in the GAP program are Seth Weaver Kahan and Christine Cranston, both of whom work as consultants at the World Bank in Washington, D.C. Kahan and Cranston put together a group of eleven World Bank employees who, at the end of the program, had reduced the amount of garbage they sent to the landfill by 3276 pounds a year, water consumption by 99,280 gallons a year, carbon dioxide output by 17,914 pounds, and gasoline for transport by 558 gallons. They also saved $2867 a year.

Kahan, a computer consultant for the World Bank who is also involved in experimental theater, says learning what was in his garbage by following GAP's garbage monitoring system was particularly revealing. "I never knew how much paper went through my house. This was a real wake up call for me. Before I started the GAP program I thought I was recycling by putting out newspaper, glass, cans, and plastic," he says. "But now I am at a whole new level of recycling and most of it is paper."

Kahan now stores cardboard cartons, catalogues, phone books, and wastepaper on his screened porch until the day once a month when he makes the one hour roundtrip to the Fairfax County recycling center. "It's a treat to find a facility that is appropriate to your needs," Kahan says. "You can whip out your wax-coated catalogue and there is a bin for it. You have a sense that this paper will be recycled and become tomorrow's toilet paper or something. And that it is going to help a little bit."

In addition to increasing the amount of resources he recycles, Kahan stopped letting the water run while he brushed his teeth and stopped flushing the toilet as often. He also started monitoring his water and electric bills for the first time. He then lowered his thermostat 3 degrees to 68 degrees and reduced the thermostat setting on his hot-water heater by 20 degrees. The water was still plenty hot and he saved both money and electricity. "I realize that turning my thermostat down by 3 degrees isn't going to stop global warming," Kahan comments, "but it's a start."

"I liked the little concrete things you can do to save energy and use resources efficiently," says Christine Cranston, a human resources and staff development consultant at the World Bank, who co-led the World Bank GAP group with Kahan. "Every little step you take you can say: I did this instead of just thinking about it or reading an article about it." Cranston, age thiry-six, reduced her water consumption by switching to a water-saving showerhead. "I don't mind it. I don't feel deprived. I still get a good shower," she reports.

When she turned down the thermostat Cranston found that she had to wear a couple more layers of clothes around the house in the middle of the winter to be comfortable. "But what is the big deal about that? Some Americans are so indulgent about the way they want to live," she observes. Cranston turns the furnace off at night to save fuel. "It's kind of cold in the morning before the heat comes back up, but a hot shower helps."

A lot of the GAP lifestyle changes go against the whole culture and the way we have been brought up to behave, Cranston notes. "We have been trained to get in the car and drive around all the time, to go to the mall and buy a lot of heavily wrapped stuff we don't need, keep the house really warm, buy a lot of clothes, and have a lot of appliances for every little thing we want to do in the kitchen." Yet in the midst of this consumer culture Cranston has a yearning for a simpler lifestyle. She is making an effort to eat dinner out less often and develop what she calls "an honoring relationship with the earth."

Not only are there gains in terms of the amount of resources and energy conserved by people who go through the GAP program, there will be other benefits as well. For example, the people at the World Bank who have been sensitized to the amount of energy and resources they consume will consider these issues in a new light as they shape World Bank projects around the world.

Gershon's efforts to help Americans reduce their unsustainable rate of consumption comes none too soon.

Studies peg U.S. municipal wastes at 250 million tons per year,[4] and a further 180 million tons of hazardous waste plus 11 billion tons of industrial wastes are generated annually by American business.[5] Despite efforts to promote recycling, generation of municipal waste has actually expanded by 10 percent since 1988 from 3.5 to 4.3 pounds per person per day.[6] Meanwhile, landfill space is shrinking by 8 percent a year and it is expected that by the year 2009 about 80 percent of our landfills will be closed.[7] Clearly, Americans place an enormous burden on the environment not only in the United States, but also in foreign lands where metals are mined, trees felled, and factories belch smoke to feed the insatiable American appetite for goods.

This high-consumption American lifestyle is exhausting resources that either should be left undisturbed, are needed elsewhere around the world, or will be needed by future generations. In his book *How Much Is Enough? The Consumer Society and the Future of the Earth,* Alan Durning observes that our current consumer society is a passing phase that the earth can ill afford. We must move toward a society of permanence, a sustainable society, which lives within its means and does not borrow from the ability of future generations to meet their legitimate needs, he adds. "Unless we climb down the consumption ladder a few rungs our grandchildren will inherit a planetary home impoverished by our affluence," he concludes.[8]

With only 8000 households (some 20,000 people) having graduated from the program thus far, Gershon has a long way to go before his behavioral change program becomes statistically significant. Looking at these relatively small numbers, it would be easy to dismiss Gershon as a dreamer who suffers from grandiose visions. To those who suggest to him that he is dreaming if he thinks he can get the bulk of the American population to live in an ecologically sustainable fashion he, asks: "What is the alternative?"

While the GAP program is a modest start on promoting sustainable life-styles, there is the potential for a ripple effect in this country as people visit GAP households that are set up for energy efficiency, sophisticated re-cycling, and reuse of resources. Our relatives, friends, and neighbors learn more from us by what we do than from what we say. Visiting an eco-efficient house can teach people more in five minutes than they would learn from long, impassioned diatribes about the importance of "saving the planet."

Beware: eco-friendly lifestyles are contagious and may soon change the way you and your family lead your lives.

Two Approaches to Restoring Trashed Urban Rivers

A sudden summer squall sweeps across a steaming parking lot at a shopping mall on the outskirts of Washington, D.C. Sheets of rain wash across the asphalt carrying motor oil, candy wrappers, and plastic bags down a storm drain. Shoppers hopscotch across the pavement as they try to keep their feet dry.

To most people this is nothing more than a typical suburban scene. But to Thomas R. Schueler, president of the Center for Watershed Protection in Silver Spring, Maryland, the sight of water collecting in a parking lot and flooding down a storm drain is as menacing as a scene from the movie *Jaws*.

To Schueler a parking lot is a river killer. Because its impervious surface does not permit rainwater to soak into the ground, a large volume of water collects on the asphalt, swirls down a storm drain, and funnels through an underground pipe, which empties into a stream. At this "outfall" the torrent of fast-moving, toxic, trash-laden stormwater rips at the stream banks, causing massive erosion that silts up the river downstream. As a result, many urban streams that used to harbor abundant wildlife have become gullies, stripped of vegetation and devoid of life.

The injury to streams and rivers done by this urban runoff is vividly on display along the Anacostia River, which runs from its headwaters in Maryland through the heart of Washington, D.C. Owing to intense development, sections of the 170-square mile Anacostia watershed (where 600,000 people live) are now more than 30 percent to 50 percent impermeable and flash floods of stormwater runoff are taking their toll. Large deposits of sediment have caused the river to become dramatically shallower and a wide variety of fish to disappear. During the hot Washington summers, conditions deteriorate as runoff passing over impervious surfaces heats up and becomes oxygen-poor, causing fish in the river to drown. The river is further burdened with toxic chemical pollution from a variety of industrial sources. As a result, the Anacostia has been listed by the conservation group American Rivers as the fourth most endangered river in the country.[1]

Recently, however, restoration efforts have begun to clean up the water and bring fish back to the Anacostia. For the last ten years an impressive coalition of individuals, nonprofit groups, and government agencies has been working to make the Anacostia River swimmable and fishable by the year 2000. This effort has taken an innovative watershed approach to solving the river's ills. Rather than concentrating on picking the trash out of the river, the strategy is to reconstruct those ecological services in the watershed that have been lost to incremental development. To this end, millions of dollars have been spent on building stormwater detention ponds, shaping and planting new wetlands, restoring urban streams, constructing wildlife habitat, and removing barriers to fish migration.

The history of how the Anacostia was trashed is a tale that could be told, with minor variations, about many urban rivers. Once a river that twisted and turned on its way to joining the Potomac, the Anacostia harbored thousands of acres of wetlands and beds of wild rice that supported a dense population of waterfowl. On the eastern bank of the river is a site that was occupied by the Macotchtanks for 3000 years. These sedentary, semiagricultural people, also known as Nascotines and Anacostines, were part of the Algonquin linguistic stock who inhabited the Potomac tidewater region.[2] When early European settlers arrived on the scene the ecosystem was flourishing. The settlers wrote home about huge flocks of birds darkening the skies along the river and the sounds of ducks, the honking of geese, and the croaking of heron. Large schools of herring came up the river to spawn.

The river's problems began in earnest during colonial times when tobacco farmers plowed up to its banks permitting soil to wash into the river and silt it up. Gradually, the Anacostia's depth dropped from 40 feet to 8 feet impeding the passage of oceangoing vessels that previously traveled up the river to the tobacco port of Bladensburg, Maryland. Subsequently, intense flooding in the area resulted in a public demand that the river be fixed.

Enter the U.S. Army Corps of Engineers, which dredged the riverbed to keep it navigable and channeled it for flood control purposes. "When you compare maps of the Anacostia 100 years ago and today you can see that the Army Corps of Engineers completely reconfigured the channel," says Scott Faber, director of the flood plain program at American Rivers. The Corps straightened the kinks out of the river, and drained and filled thousands of acres of wetlands along the Anacostia in an effort to eradicate malarial swamps. In 1931 alone, 41,000 tons of stone were used to harden the banks of the river and keep them from eroding.[3] Some tributaries were paved and of the original area covered by marsh, 90 percent of the tidal wetlands and 70 percent of the nontidal wetlands were destroyed.

But now the river is being given a second chance. The restoration effort along the Anacostia is one of the most ambitious river restoration efforts in the nation and dates back to 1979 when the Metropolitan Washington Council of Governments (MWCOG) made the river's restoration a priority. In 1984 the District of Columbia and Maryland signed an Anacostia Watershed Agreement. By 1987 the two counties in Maryland that contain the upstream portion of the watershed—Montgomery and Prince George's counties—joined the coalition and established the Anacostia Watershed Restoration Committee. Subsequently the U.S. Army Corps of Engineers and the Enviormental Protection Agency became involved in this urban river restoration effort. Out of this coalition emerged a six-point plan to reduce pollutant loads, restore the ecological integrity of the river, increase its natural filtering capacity by expanding wetlands, increase forest cover in the watershed, restore the spawning range of anadromous fish to its historical limits, and provide greater public awareness about watershed restoration activities.

But something was needed to kick-start the practical work of river restoration. Upon becoming chief of staff at MWCOG for the Anacostia restoration team, Thomas Schueler launched an inventory of where river restoration projects could be carried out within the watershed. Members of the restoration team spent two years poring over maps and walking much of the watershed on foot as they put together a list of over 450 recommended retrofit projects. To date, about a quarter of these restoration projects are either in progress or completed.

While 75 percent of the proposed Anacostia restoration projects still remain to be acted on, Schueler argues that the 109 watershed restoration projects either completed or underway constitute a major accomplishment. The millions of dollars committed to restoration work and the number of agencies involved make the Anacostia restoration effort one of the longest running and most comprehensive in the country, he says.[4] "Restoring the Anacostia is the work of a generation," observes Schueler, who quit his job with MWCOG and the Anacostia restoration team to found the Center for Watershed Protection. "It took generations to destroy this watershed and it will take generations to restore it."

One of the most impressive of the stormwater retrofit projects already functioning within the Anacostia watershed is located near the Wheaton regional shopping center, a complex in Maryland comprised of 40 acres of parking lots and shopping malls.[5] The Wheaton restoration project was designed to allow stormwater to soak into the ground and to trap pollutants in the runoff. To accomplish this, engineers excavated a three-cell wetpond covering 5.9 acres. These interconnected detention ponds drain 800 acres and reduce runoff damage by gradually releasing water into the soil and downstream. The planting of seven species of native wetland plants along

Figure 24.1
Wheaton Branch stormwater retrofit pond #2. Courtesy Metropolitan Washington Council of Governments (MWCOG).

the fringes of the pond also filters out pollutants and improves water quality by 60 percent. The stormwater ponds and their vegetation provide an inviting habitat for great blue herons, migratory ducks, and a variety of waders, Schueler notes. The Wheaton Branch stormwater retrofit project is one of fifty similar projects now in progress or completed. The total of 159 stormwater retrofit sites listed in the inventory are estimated to cost $27.6 million.

Downstream from the Wheaton Branch stormwater settling pond a number of innovative techniques have been used to restore a "blown-out" stream. Before the restoration work, the stream had become wider and shallower as fast-moving runoff eroded its banks. As the banks crumbled, shade trees tumbled into the stream exposing it to the sun and causing the river to nearly dry up in the summer. This left little habitat for fish and birds.

To restore the stream and re-create some of the lost wildlife habitat, the restoration work was designed to narrow and deepen the river as well as bring back trees that would shade it, Schueler says. As a first step the stream was sculpted with heavy machinery to deepen it and direct its flow. Then, 200- to 300-pound rocks were used to stabilize banks that were vulnerable to erosion. Rootwads (the segments of trees that contain the roots) were anchored to the banks to prevent erosion and stone deflectors were constructed to narrow the channel. Notched log drops were placed across the stream to create scour holes and provide deep pools for fish. Boulders were placed in the stream to create eddies that attract fish. Riffle areas were devised so that insect life could thrive and form the base of the aquatic food chain. Along the edge of the stream, vernal ponds were constructed and stocked with spring peepers. And trees were planted to provide shade and help hold the soil in place.

As a result of these measures, and the trapping of pollutants by the wetland plants around the stormwater detention pond upstream, vegetation is beginning to grow back along the streambanks and a variety of fish are returning to Wheaton Branch. Before the restoration work, monitoring of the fish population revealed only two varieties of minnows known to be resistant to pollution along with a couple of escaped goldfish. Since the restoration work eleven species of fish have returned to the stream, Schueler reports. This is one of eight stream restoration projects either in progress or already completed in the Anacostia watershed. The sixty stream restoration projects listed in the inventory are estimated to cost $8 million.

Figure 24.2
Stormwater runoff and stream restoration project on the Anacostia River, Wheaton Branch work area #2, in the fall of 1991 after it had been restored. Note the log drop placed across the river to slow the water. Courtesy MWCOG.

While these restoration projects vary in size and cost, the megaproject on the Anacostia was carried out by the United States Army Corps of Engineers. Conscious of its less-than-sterling environmental record, the corps is conducting one of its first "green experiments" on the Anacostia. Instead of filling in wetlands as it has in the past, the corps is now using 155,000 cubic yards of dredge material to reconstruct 32 acres of emergent tidal wetlands in an area known as the Kenilworth Marsh. One of the selling points of this restoration project is that building the marsh with dredge material costs less than it would to truck the material to an upland site and dispose of it. In addition to using dredge materials to shape an area at the proper elevation to support a tidal wetland, the corps subcontracted the planting of some 350,000 seedlings of seventeen species of native plants, including pickerelweed, swamp rose-mallow, buttonbush, lizard's-tail, and salt marsh bulrush. The corps sent a $19 million plan to Congress including 123 projects to restore 80 acres of wetlands, 5 miles of streams, and 33 acres of bottomland habitat.[6]

"There is no cookbook for how to restore wetlands. We are doing this by the seat of our pants," says Steve Garbarino, the corps' project manager. But the three years of demonstration projects he undertook before the restoration work gave him confidence it could work. "Emergent wetlands are vital to the restoration of the Anacostia River," he says. "Wetlands serve as filters. They tend to absorb a number of pollutants and contaminants. As the tide rises it covers the wetlands and then as the water recedes the wetlands act like a sieve." Bringing back these wetlands also re-creates an important habitat for birds and fish. This wetland restoration project is one of ten in progress or already completed within the Anacostia watershed. The thirty-four wetland recreation projects listed in the inventory are estimated to cost $7.3 million.

Elsewhere along the Anacostia, barriers to fish migration are being eliminated. One of the problems faced in reintroducing fish into the river is that since they have been blocked from the upper reaches of the river for so long, they no longer have the "memory" of their home portion of the river that impelled them to swim upstream. To reacquaint them with the farther reaches of the river, fish are caught, transported upstream, and then released. This strategy seems to be paying off. There are six fish passage projects either in progress or already completed along the Anacostia. The thirty-one similar projects listed in the inventory are estimated to cost $1.1 million.

Another critical restoration project along the Anacostia is the effort to stem the flow of raw sewage into the river from the combined sewer overflow (CSO) system that serves the District of Columbia. Like many other cities along the East Coast, Washington has a sewer system that is connected to its storm drain system. As a result, when it rains a combination of stormwater runoff and raw sewage empties into the Anacostia. A 1981 study carried out by the D.C. Water Resources Management Division of the Department of Public Works estimates that 12 million gallons of CSO discharge are entering the Anacostia every year containing about 4 million gallons of raw sewage. Fixing this problem will not be cheap. The district has already invested $17 million into constructing a huge device called a "swirl concentrator" that separates sewage from the stormwater. A second phase of the project, which may cost the city up to $100 million, remains under consideration.

Schueler calculates that in a watershed as degraded as the Anacostia's, restoration costs a million dollars a square mile. He also figures that within the watershed, a minimum of 2 percent of the land area must be devoted to sites where restoration projects can be carried out. The high price of restoration provides a powerful incentive for communities to protect environmental services within their watershed, he argues.

"The federal government couldn't print money fast enough to restore all the watersheds in the country," Schueler says. Funding watershed restoration work will require that each locality create a "stormwater utility" similar in structure to our water and sewer utilities, he argues. Already there are some 120 stormwater utilities around the country that charge customers based on the area of imperviousness on their property. Schueler also advocates that local "watershed stewards" be elected or appointed to look out for the interests of the watershed and see that ecological services provided by marshes, ponds, streams, rivers, springs, seeps, and forests are not incrementally destroyed by development, or in cases where they are sacrificed that mitigative measures be required.

Despite these heroic and expensive measures taken to restore the Anacostia, there are some who remain skeptical about the long-term prognosis for the river's recovery. "What is ironic about the restoration efforts being led by MWCOG is that we are spending all this money and the restoration effort is being held up as a model, but it won't work," says Scott Faber at American Rivers. It won't work because not enough is being done to prevent future development from undermining restoration efforts. "You can win a lot of battles along the Anacostia and still lose the war to incremental development and incremental changes in the watershed," Faber warns, adding that the real problem the Anacostia faces is poor land management. County governments along the river are under pressure to raise revenues by attracting more development and in the process environmental restrictions are eased, he notes. Schueler admits that the Anacostia restoration team's main thrust was on restoration as opposed to watershed protection and reforming land use. "I, myself, am not certain whether the great strides we made in restoration will be, to some degree, overwhelmed as a result of new impacts from development over the next two decades," he allows.

In fact, there are a number of critics of the Anacostia restoration team's efforts, but Schueler takes them philosophically: "I look at it as an extended family where there are always one or two relatives who are angry with you," he says. The consummate insider, Schueler has been busy making sure that new restoration projects are in the pipeline, that permits have been obtained for the construction of stormwater settling ponds, and that papers are moving from one desk to another across the dozens of agencies involved in these projects. He is not surprised that this multiagency approach to river restoration appears from the outside to proceed at a glacial pace. "Typically it takes two years from the initiation of a restoration project to its completion," he notes.

The restoration work taking place along the Anacostia is part of a maturing movement around the nation to save trashed urban rivers, many of which have been transformed over the last century into little more than concrete flood-control channels and industrial drainage ditches. In the 1930s, the U.S. Army Corps of Engineers channeled and paved the Los Angeles River for flood control. This method became so popular from the 1940s to 1970s that "channel improvement projects were constructed on 35,000 miles of waterways. A report to Congress in 1973 estimated that another 200,000 miles of waterways had been modified by state and local authorities."[7]

Examples of the grassroots rebellion against the paving of rivers and streams can be found across the country. In California, for example, rivers and streams that have been buried in underground culverts are being dug up ("daylighted") and restored. Elsewhere, citizens are demanding that a more natural approach be taken to flood control. For example, along the stretch of the Platte River that runs by Denver, rather than channel the river with levees or hardened banks, flood control has been achieved by creating a 625-acre park that is used as a recreation area and nature preserve when the river is low and as a floodplain when the river rises. Similar activities are taking place along other rivers, including the Charles in Boston, Mingo Creek in Tulsa, and the Chattahoochee in Atlanta.[8]

While the total number of citizen groups and government agencies involved in this kind of work is unknown, the Rivers, Trails and Conservation Assistance Program run by the National Park Service publishes a *Rivers Conservation Directory* that lists over 1640 organizations and agencies. This list is less than comprehensive: in California alone there are some 350 projects designed to protect and restore urban streams. Also testifying to the surge of interest in urban rivers is the recent formation of the Coalition to Restore Urban Waters (CRUW) which hosted a Friends of Trashed Rivers conference in San Francisco at which 300 activists traded war stories and innovative restoration techniques.

In fact, the urban river restoration movement is nothing new, points out Ann L. Riley, regional director of CRUW on the West Coast and the author of a upcoming book entitled *Restore with Nature: Urban Stream Restoration, Alternatives to Conventional Engineering* (Washington, D.C. and Covelo, California: Island Press, 1997). The movement can be traced back to the 1920s when fishermen on the upper Mississippi began to organize river restoration efforts, she reports. It was also during the 1920s that a group of women stopped the San Antonio River in Texas from being covered by a highway, she points out. Subsequently the Works Progress Administration (WPA) and the Civilian Conservation Corps (CCC) took up stream restoration during the depression. A new wave of activism on urban rivers began in 1969 when the Izaak Walton League of America started the Save Our Streams (SOS) program. It was then that a man named David Whitney drove an SOS Winnebego—the "water wagon"—stuffed with water monitoring equipment all around the country. In the process, he and a number of other activists trained a small army of Americans in the esoteric art of biological and water quality monitoring. Armed with new information about the degraded state of their urban waters, many local groups then formed chapters dedicated to protecting and restoring their streams.

Subsequently, in California, where the destruction of urban rivers peaked with the paving of the Los Angeles River, a breakthrough occurred in 1984 when the Urban Creeks Council successfully lobbied the state legislature to pass the Urban Stream Restoration and Flood Control Act. This act provided grants to community groups and nonprofits to carry out low-cost, environmentally sensitive flood control and river restoration.

One of the places these moneys were spent was at Wildcat Creek where a $26 million project, which would have controlled flooding with an 800 foot concrete culvert, was rejected by the local community as lacking in environmental sensitivity. Instead of paving the banks of the stream, the community hired consultants who redesigned the project along more natural lines. Instead of a concrete box, the cross-section of the river was reconfigured into a low-flow channel, a flood plain, and a berm to contain the river at its highest level.

The Coalition to Restore Urban Waters is now pushing for legislation that would provide funding for similar projects on a national level. One bill would redefine the mandate of the Soil Conservation Corps and the Army Corps of Engineers so that they would provide grants to community groups to do environmentally sensitive, labor-intensive restoration work on urban waterways.

"There is a real groundswell of people who are fed up with the status quo surrounding urban rivers," says Alan Turnbull at the National Park Service's Technical Assistance Program. When these people look at an urban river "they want to see more than mowed grass with a paved ditch."

Cleaning up America's urban rivers can be heavy lifting. During his nine-year effort to restore the Anacostia River, Robert Boone has wrestled with some spectacular pieces of river junk, including parts of a cement truck, a telephone booth, and countless hot water heaters.

An impatient activist, Boone, who is president of the Anacostia Watershed Society (AWS), has mobilized 7776 volunteers to do ninety cleanups since 1989 to haul 65 tons of trash. During that period 3166 tires were pulled out of the Anacostia and donated to a tire recycling company. Boone also enlisted 275 youths from the D.C. Service Corps and planted 5800 trees along the riverbanks.[9] Often, volunteers travel in canoe flotillas as they pick up the trash and in the process AWS has provided 1339 volunteers with a boating experience on the river. "There is a tremendous support for river restoration," Boone notes. "People see how bad it is and they don't want to leave this to their kids."

Some disparage these efforts to pick trash out of the river, arguing that the next rainstorm will inevitably wash more down from the streets. But psychologically it is important to keep picking trash out of the river because, more than sedimentation or engineered banks, the presence of litter repulses people and keeps them from visiting or caring about a river, Boone explains. "And the fact that volunteers are willing to spend their precious time picking trash out of the river sends an electric message of hope through the community. That is why we focused on the trash first." AWS is lobbying to get booms installed across a number of feeder streams to inhibit the

Figure 24.3
Robert Boone, president of the Anacostia Watershed Society, is a fierce advocate for cleaning up and protecting the river. Photograph by Steve Lerner.

spread of litter and make its collection easier, but so far none of the jurisdictions involved have agreed to build them.

A lot of these agencies charged with cleaning up the river are dragging their feet, Boone complains. "Bureaucrats don't really want to clean up the river because then they would be out of a job. "It's like the psychiatrist who thinks: 'Hell, I don't want to heal this patient because if I do I'll have to find a new one.'"

Furthermore, the astronomical estimates of how many millions of dollars it will cost to restore the river are vastly inflated in Boone's opinion. It does not have to cost much if you organize volunteers to do these streambank restorations. "To say that it costs a million dollars a square mile to restore a watershed is nonsense and perpetuates a defeatist, it-can't-happen-here attitude. A lot of this money goes to slick consultants and paper shufflers who don't do anything except pack their pockets with money. These estimates are a contractor's dream. When people hear that it will cost this much to restore the river it loots them of their idealism, and convinces them that they don't have a prayer of restoring the river."

The fifty-six-year-old Boone, who has a masters degree in psychology, approaches river restoration as a sacred task, not just an engineering job. A man who has lived in a teepee in North Carolina and in the mountains of New England, Boone is involved with river restoration for very personal reasons. Eleven years ago when he moved from the Appalachian Mountains ("where you could drink from the streams") to the less than pristine lowlands near Washington D.C., he was determined that his son Oakley not just grow up amid the works of man. "I wanted to put him in a situation where he could learn his place in nature," Boone says.

When Oakley reached the age of six, Boone began to hike with him along the Anacostia River and its tributaries. But as father and son explored the riverbanks, they discovered abandoned vehicles, refrigerators, and rusting shopping carts along with egrets and great blue herons. The streams that lead into the Anacostia were so littered with trash and polluted with the rainbow hues created when petroleum mixes with water that Boone began to drive his son to the more distant but cleaner banks of Chesapeake Bay and the Patuxent River. But commuting to nature and fighting the traffic to spend some peaceful time along a relatively clean river soon became too tiresome and absurd for Boone. Instead of giving up on the Anacostia, he decided to stand and fight for it. Why not clean up the Anacostia instead of driving for hours to get to a cleaner river? he asked himself.

"When I see these pollution sites along the Anacostia it is as if a fire has burned a hole through the web of life," Boone says. But the great thing about nature is its regenerative capacity. "That is why I do this work, because there is this tremendous regenerative healing force behind me."

Impatient with the pace of restoration work being done by the MWCOG, Boone refuses to wait for the bureaucrats to act before wading in and cleaning up the river. He plays the archtypical outside-agitator role to Schueler's insider game.

"Up until now the government has said, 'Don't worry about it, we will take care of the rivers.' But we are fed up with that. If ordinary people don't clean up the river it won't get done. If we don't get involved we won't own it and people want ownership of their environment because they see it going to hell," Boone asserts.

Expensive restoration work will be for naught if the overall load of pollutants entering the Anacostia is not significantly reduced, Boone argues. The Anacostia and many other urban rivers are severely degraded today because regulators turned a blind eye while they were polluted. A kind of triage was practiced as regulators gave up on urban rivers, which they assumed could never be made swimmable and fishable. But now Boone and others are determined to see regulatory action brought to bear on those who pollute the river.

After pinpointing some of the worst sources of pollution along the twelve tributaries of the river, Boone enlisted the help of the Institute for Public Representation (IPR) at Georgetown Law School. Hope Babcock, associate director of IPR and a visiting professor at the law school, took the AWS on as a client and assigned her students to make a map of sources of pollution along the river. The map was designed to help them focus regulatory action on, or bring legal action against, large-scale polluters. "As a result of our work we have finally brought the Environmental Protection Agency (EPA) into the District of Columbia," says Babcock, former counsel general for the National Audubon Society. EPA officials have promised tougher enforcement of pollution laws to reduce the amount of raw sewage flowing into the river and to track down illegal dumpers. But coming up with the money to repair the combined sewer overflow system will be difficult.[10]

One of the sites being monitored by state and federal agencies is Joseph Smith & Sons, Inc., one of the largest auto recycling operations on the East Coast, which straddles the banks of lower Beaverdam Creek, a tributary of the Anacostia.[11] Boone has photographs of cars and car parts cascading down the banks and into the river from this site. The company, he explains, employs a machine that shreds a car and reduces it to ferrous metals, nonferrous metals, and "fluff"—the shredded plastic, seat padding, and other auto parts that are nonmetallic. Where the cars are shredded, oil, brake fluid, and a variety of other substances seep into the ground and some of this toxic brew eventually makes its way to the river, Boone observes.

Tests by Maryland and EPA officials found polychlorinated biphenyls (PCBs) and heavy metals both on site and in the sediment of the river near the operation. They also discovered a biological "dead zone" just off the site, Boone says. When the scrap metal operation applied for a special exemption zoning permit, which would have permitted the facility to expand, IPR attorneys opposed the granting of the zoning variance and prevailed. Recently, after AWS brought suit against the company, a judge in federal bankruptcy court ruled in favor of AWS and required officials of the auto recycling facility to sit down with AWS and its environmental consultants to develop a floodplain management plan that will reduce the environmental impact of their business.

The scrap metal yard is not the only source of toxic chemical pollution on the river. The Anacostia is intoxicated by petrochemical products and heavy metals from a variety of sources. Tests show that its sediment is laced with lead, cadmium, and mercury, as well as polycyclic aromatic hydrocarbons, PCBs, and chlordane (possibly from chemicals used to prevent termite infestations).[12] Another study revealed that some fifty chemicals were detected in one or more fish species.[13] Signs have now been posted warning people to limit their consumption of fish caught in the river, particularly bottom-feeding fish such as carp, catfish, and eels.

"We will stop the major polluters of the Anacostia," Babcock declares, but it will not happen all at once. In the urban setting there is no one major decisive battle but rather a series of small battles within the watershed over shutting off sources of pollution, preserving a wetland, removing a leaking storage tank, or requiring development to proceed in an environmentally sensitive fashion, she continues. With each battle a different community along the river becomes involved and consciousness is raised about the need to protect the river and its watershed, she concludes.

Along the Anacostia, within the District of Columbia, a number of communities are becoming increasingly aware that the river is a valuable neighborhood resource that deserves protection. The Izaak Walton League of America's (IWLA) SOS program involves district public school students in monitoring the Anacostia and other urban rivers and streams. The AWS also works with students who fan out across the city and stencil Don't Dump: Anacostia River Drainage next to sewer grates at 223 locations. AWS is working with the University of Maryland and the district's high schools, and grade schools in conducting water quality monitoring for fecal coliforms, pH, and dissolved oxygen.

Each school involved in the program also adopts a nearby river or stream and teaches students how to monitor its health. Students use chemical test kits and nets provided by the IWLA that permit them to do a chemical analysis of the water and biological monitoring of macroinvertebrates. The results of this monitoring are sent to the National Park Service and district officials charged with protecting water quality.

"In the city we talk to students about sewersheds in addition to watersheds" because so many city streams have been buried underground, says Karen Firehock, program director of SOS. Unlike rural students, for whom an excursion to a nearby stream is a relatively simple exercise, students in

Washington, D.C., are sent out into their neighborhood to examine the landscape and determine where a river or stream would run if it were not covered with streets and buildings. Then city maps are obtained to see if the students have correctly identified where underground sewers carry urban streams.

Students who live along the Ancostia are also becoming involved with the river by rowing and sculling on it. In 1988 Robert Day, a retired foreign service officer who crewed at Cornell, founded the Organization for Anacostia Rowing and Sculling (OARS) with a number of friends. Since then OARS has taught 150 young people to row and scull on the Anacostia, reports Carl Cole, who has been involved with the program from the outset. Students involved in the program participate in river cleanups and the two annual regattas that OARS sponsors also attract a large number of families to the river and focus community awareness on the need to clean it up. The Anacostia is ideal for rowing and sculling because it is a slow-moving river without rapids, it is protected from the wind by hills, and there are long stretches along the banks where spectators can watch a race from start to finish, Cole observes. "The Anacostia could be a world class rowing river, the equivalent of the Thames," Babcock says, "but you're not going to do that through floatable trash."

Some of the trash along the river was collected by the D.C. Service Corps, which for two years fielded a team of twelve young district residents. Working in collaboration with the AWS, this team picked up 800 bags of garbage along the river over a three-week period, reports Suchita Amin, project coordinator at the D.C. Service Corps. The Department of Public Works also has a $3 million a year debris removal program that employs a skimmer on the tidal portion of the river.

While these efforts are commendable, critics of the Anacostia restoration effort suggest that young, unemployed district residents should be hired and trained to do more than just pick up the trash and plant trees. They point to a program in North Richmond, California, where young, unemployed, inner-city residents are trained and hired to restore the river that runs through their neighborhood. "This is a big part of what we are trying to do to address the social justice and environmental justice issue," says Mike Houk, one of the founders of Coalition to Restore Urban Waters. Training inner-city youth in sophisticated river and watershed restoration techniques could prepare them for a career, he continues. "Soil bioengineering is a skill that will be marketable," Houk argues, while just mobilizing young people to pick up trash along the river or plant a tree does not necessarily lead to a career.

But if the Anacostia river restoration effort has been unable to find funding for a program that hires and trains inner-city youth to do river restoration, there is nevertheless a considerable amount of inner-city community activity focused on protecting the river and the parklands along its banks.

Recently, some elements within the predominantly African American community living along the Anacostia have focused on the disproportionate number of large and environmentally high-impact facilities along their home river. "All you have to do is look at a map of D.C. to see that a disproportionate number of these facilities are located along the Anacostia in predominantly African American communities and not along Rock Creek in predominantly Caucasian and affluent communities," observes Herbert Harris Jr., past president of the Kingman Park Civic Association and an advisory neighborhood commissioner in that district.

Harris joined with Robert Boone and a coalition of other activists in opposing plans to locate three more large-tract facilities along the Anacostia: the 76,000-seat Jack Kent Cooke football stadium, the Barney Circle Freeway, and a theme park. The last of these is a proposal to build a $100 million theme park in the middle of the Anacostia River on Children's Island, a 45.5-acre parcel of National Park Service land constructed out of dredge material by the Army Corps of Engineers in 1916. Planners project that if the theme park is built it will attract an estimated 3.3 million visitors a year. One of the components of the development plan most objected to by environmentalists calls for constructing watergates across the Anacostia above and below Children's Island to control the normal 3-foot rise and fall of the tidal river.

"The theme park will be educational in nature and not a bunch of rides like roller coasters and carousels," Carroll B. Harvey, board member of National Children's Island, Inc., told the D.C. city council. Although visitors will have to pay to enter the theme park, no admission fee will be charged at the 13.5-acre playground designed by the American Society of Landscape Architects which features an Earth Place, Life Place, River Place, Time Place, and Resource Place where various products made of recycled materials will be displayed.

Jodean Marks of the D.C. Greens remains unimpressed by "Resource Places" and the like, which she describes as little more than environmental window dressing: "We need a clean river, not a River Place; we need clean soil, not an Earth Place; we need a natural environmental education center for children; we need to plant indigenous trees and other species on the island so children can come there and learn to plant and tend them; we need a swimmable, fishable river."

Harris agrees. "Not everything in the D.C. should be for sale." The phenomenon of environmental racism, in which toxic or high-impact facilities are located in minority communities, has been well documented, he observes. Less well understood is the fact that development is permitted on federal parklands near minority or disadvantaged communities—development that would not be permitted in more affluent neighborhoods.

To block the theme park, a number of environmental groups filed a lawsuit in federal court against the National Park Service in the national capitol region. The suit contested the transfer of the jurisdiction of Children's Island from the National Park Service to the D.C. government without an environmental impact statement as required by the National Environmental Policy Act.

"There seems to be emerging a pattern in this region that is an attempt to give away our precious green space for private development" despite language in the D.C. masterplan which calls for Anacostia parkland to be maintained in a natural condition and for a Great Anacostia Park on both shorelines of the river, Boone told D.C. city council members. "Our city fathers had a vision of a great Anacostia Park. Let's keep that dream alive. The river is finally getting some long deserved attention and if this effort is sustained the Anacostia will be restored like its big sister, the Potomac," he says.

As an alternative to the theme park development Boone wants to see Children's Island become a nature preserve. "This could be a great recreational resource for the kids who now have nowhere to play except the parking lots and the malls. It would be a nature preserve for hiking, biking, canoeing, bird watching, environmental education studies, wetland studies, and studies of urban impacts on stream systems." He also wants to locate a nature center on Children's Island, a nursery that grows trees for the city's streets, and an urban camping area.

Despite the various challenges posed to river restoration, Boone finds cause for optimism in the fact that the amount of sediment and bacteria entering the river is being reduced and an increase in the abundance and diversity of fish has been noted. Brown trout are reproducing in the Paint Branch tributary of the Anacostia. "In the spring we have herring coming up the river. Their return is getting stronger each year. I can imagine the herring coming back in great multitudes," Boone enthuses.

"I have a vision of the Anacostia River being a healthy river right in the middle of Washington, D.C. We can have healthy fish, a fishing contest, and kids swimming, boating, and water-skiing on the Anacostia. I don't think this is a pipe dream. I have already seen little snippets of this dream happening," Boone says.

On the 4-mile stretch of river between the Bladensburg marina and Anacostia Park, Boone sets his canoe paddle across his lap and gazes out as a

Figure 24.4
One of the least spoiled sections of the Anacostia River. Photograph by Steve Lerner.

great blue heron takes flight from the bank. "There are more blue heron than ever before. Some days I see forty or fifty of them, while nine years ago, when I first started canoeing down the river, I was thrilled to see one or two." Elsewhere along the river egrets, kingfishers, cormorants, geese, and ducks make their way between the floating debris. Turtles, beavers, foxes, muskrat, and mink have also been reported to be returning to the riverbanks. Thanks to the efforts of many individuals and groups, the Anacostia appears to be making a slow recovery.

"They tried to throw the Anacostia River away," Boone says, "but you can't throw a river away. It won't go away. It's still here and people still care about it. We want to make the Anacostia swimmable and fishable by the year 2000."

A Green Priest Preaches About the Need
to Protect God's Creation

It is another day on the road for the Reverend Jeffrey M. Golliher, an itinerant environmental organizer from New York City, who is traveling from parish to parish in New England in his blue 1985 Ford station wagon talking about what it means to be a Christian in an era of environmental limits.

This afternoon he is speaking to two dozen Episcopalians in an overheated library in Claremont, New Hampshire. Golliher must tread carefully here in New England because some locals (who make their living skidding logs out of the hills, cutting ski trails through the mountains, or spraying apple orchards with pesticides) are naturally resistant to listening to an urban intellectual lecture them on the merits of conservation.

But the environmental minister, whose Irish-American roots are in the hills of North Carolina, proves surprisingly adept at rural diplomacy as he begins to talk about landfills, recycling, and jobs vs. the environment. "I grew up poor in the hills of Appalachia," says Golliher, a slender, intense, forty-two-year-old priest wearing tan chinos and a clerical collar that peeks above a colorful sweater. He explains that his great-grandfather was a coal miner, his grandfather and father worked in a furniture factory, and he himself worked summers at the same factory "breathing the sawdust with everybody else" before becoming the first member of his family to earn a college degree and subsequently a Ph.D. in anthropology.

"It would be hypocritical of me to talk about jobs vs. the environment without acknowledging that my family worked in mining and timber. We recognize in our family the need for jobs in poor areas like Appalachia, the need of people to make a living from the land. But people who live in rural areas must come to terms with the fact that we can no longer continue to exploit natural resources the way we have in the past," he says.

As one of the first Episcopal priests to officially dedicate himself to an environmental ministry, Golliher is in the vanguard of a growing movement within the religious community that emphasizes the sacred nature of creation and the crucial role humans must play as its stewards.

In hundreds of churches and synagogues across the country, priests, ministers, and rabbis are organizing proenvironmental activities and speaking out against the incremental destruction of nature. Their sermons add a new voice with considerable moral authority to the chorus of environmentalists and scientists who warn of encroaching environmental degradation.

"Church basements will be increasingly green," predicts Paul Gorman, executive director of the National Religious Partnership for the Environment, which raised $4.5 million in its first three years of a mobilizing effort aimed at integrating environmental themes and activities into virtually all aspects of congregational life. The partnership mailed 105,000 education action kits that reached 67,300 congregations—including every Catholic parish in the country. The kits will help rabbis and ministers identify those themes that are both scriptural and ecological.

The size and strength of this movement is already impressive. Hundreds of congregations are listed in *A Directory of Environmental Activities and Resources in the North American Religious Community*, published by the National Religious Partnership for the Environment. The partnership has documented activity in 2000 model congregational projects and held leadership training sessions for 2000 clergy and laypeople on how to integrate religious and environmental issues This surge in interest in church environmental activity suggests that just as the religious community played a critical role in the civil rights and anti-war movements, so today it is poised to be an important player in the environmental movement.

Protestant theologian Robert Kinloch Massie Jr., who teaches at the Harvard Divinity School, goes a step further: "The churches have been busy for decades on behalf of civil rights, nuclear war, and economic justice . . . All these questions have been and remain vital, and they are interrelated. Together they form the prologue for the greatest challenge humans have ever faced, that of working together to avert the (ecological) tragedy to which our behavior has destined us." [1]

For his part, Golliher, who is known as the "green priest" at the Cathedral of Saint John the Divine in New York where he is based, has focused on organizing the Episcopal Church in New York and New England to develop what he calls "a bio-regional vision." In "Making Connections: Steps to a Sustainable Earth Ministry," Golliher writes that parishes located in the same or adjacent ecosystems ought to collaborate on safeguarding their local river or watershed as a specific project in stewardship. Along the Connecticut River, for instance, parishes should be "watchful of its use and its regenerative capacity." As he moves from town to town and from church to church, he urges parishioners to become familiar and protective of the slice of creation their parish encompasses. To this end he advocates reviving the old practice of walking the parish boundaries once a year to get a feel for the land. He speaks of the need to be "willing to work for the renewal of God's creation" and notes that the "stewardship of all God's creation is a fundamental part of Christian ministry."[2]

This sense of the sacredness of God's creation and all of God's creatures has a venerable history in the church as stories about Saint Francis of Assisi clearly attest. One of the most eloquent eco-theologians was priest-author Thomas Merton who wrote about humans finally taking their appropriate place in God's "huge chorus of living beings." Like Golliher, Merton saw a clear role for green priests:

> Monks are, and I am, in my own mind, the remnant of desperate conservationists. You ought to know what hundreds of pine saplings I have planted, myself and with the novices, only to see them bulldozed by some ass a year later. In a word, to my mind, the monk is one of those who not only saves the world in a theological sense but saves it literally, protecting it against the destructiveness of the rampaging city of greed, war, etc. And this loving care for natural creatures becomes, in a sense, a warrant of his theological mission and ministry as a man of contemplation.[3]

Kentucky farmer-poet Wendell Berry is equally emphatic. "The ecological teaching of the Bible is inescapable. God made the world. He thinks the world is good. He loves it. It's his world. He has never relinquished title to it. And he has never revoked the conditions that oblige us to take excellent care of it," Berry said at the first North American Conference on Christianity and Ecology in North Webster, Indiana, in 1987.[4]

Envisioning the kind of culture that could emerge from an earth ethic, eco-theologian Father Thomas Berry writes of a new era in which a "mutually enhancing human-earth relationship is being established. . . . As soon as we demand environmental impact statements on new projects affecting the environment, we are beginning to move beyond democracy to biocracy, to participation of the larger life community in our human decision-making process."[5]

Others are now joining the chorus: "We Christians have a responsibility to take a lead in trying to take care of the earth," said evangelist Billy Graham; "It is essential that Jews work with others for radical changes . . . based on the important Biblical mandate to work with God in preserving the earth," wrote Richard H. Schwartz in his book *Judaism and Global Survival;* and the Quakers are now publishing environmental manifestos, notes Pat Stone in an article entitled "Christian Ecology: A Growing Force in the Environmental Movement."[6] Stephen B. Scharper and Hillary Cunningham, the editors of *The Green Bible,* write that "Many persons and institutions rooted in the Judeo-Christian heritage have begun to make a connection between faith and the fate of the earth."[7]

Politicians, such as Vice President Gore, are coming to similar conclusions: "The more deeply I search for the roots of the global environmental crisis, the more I am convinced that it is an outer manifestation of an inner crisis that is, for lack of a better word, spiritual," he writes in his best-selling book *Earth in the Balance.*[8]

Amid this profusion of green religious perspectives, Golliher sees parishioners along the Connecticut River beginning to take practical actions to protect God's creation. "There is a clearer understanding in the churches that a watershed is a living thing and that parish life is related to the condition of the watershed. That is new and it puts the Church back into Creation the way it was a few hundred years ago before the industrial revolution," he says. Some parishioners are organizing river cleanups while others are trying to reduce the amount of wastes flowing into the river, he reports.

Golliher also works with a group of environmental leaders from various churches in New England who meet quarterly to coordinate their activities on regional issues. They have held training sessions about the northern forest and the pressure it is coming under from increased logging.

In addition to his organizing work, Golliher signed on as an environmental consultant for the White Mountain School, an Episcopal preparatory school in Littleton, New Hampshire. The environmental program he helped shape there involves students in building compost boxes for kitchen scraps, installing energy-saving lights, visiting the local landfill, and participating in an energy and waste stream audit of the school. Students hear the views of the radical environmental movement Earth First! but are also sent out into the field to watch local loggers' operations and listen to them explain their methods.[9]

Golliher is aware that the kind of organizing he is doing through his environmental ministry can pit parishioners against one another, especially when some of them are owners of factories, farms, textile mills, logging operations, landfills, housing developments, or ski resorts. As a result of his activist environmental ministry, Golliher has received phone calls warning him to stop meddling in the affairs of New Englanders and go back to New York City.

"The course that the environmental ministry will take over the next few years will involve it in some serious confrontations with the powers that be if it is going to be a genuine environmental ministry at all," Golliher says. "Part of the reason for the mess we have gotten into as a planet today is that we have fostered ways of thinking that have encouraged people to exploit each other. That is sin, that is domination, that is taking the power of pride and self-centeredness and claiming that we are sovereign rather than God. One of the most important things the church can do is preach the gospel about the misuse of power."

Justice and environmental issues come together in minority communities and this is not going unnoticed within the religious community, Golliher notes. It is no accident that some of the congregations most active on environmental issues today are in African American and Latino communities, he continues. "African American congregations know about environmental justice, they know about the increased incidence of pollution-induced asthma in minority communities, lead poisoning, and the hazards of living next to waste treatment plants. These issues are extremely personal and close to them," he observes.

As a complement to his regional organizing activities, Golliher works on global environmental issues at the United Nations where he is the associate for the environment for the Right Reverend James Ottley, the Anglican observer at the United Nations. In this capacity he has authored a series of research papers that are circulated to UN representatives and to Anglican churches around the world. One, entitled "Poison in Poverty's Wound: The International Trade in Hazardous Substances," provided grist for a lecture Golliher gave at the New York General Theological Seminary, from which he graduated several years ago.

"What does Jesus's second great commandment love thy neighbor as thyself mean in the context of the export of hazardous waste from the U.S. to Africa," he asks twenty seminarians? Whether or not you know someone, in a Christian sense they are still your neighbor, he argues. "Therefore, if I dump my waste so it goes downstream and pollutes your well, that is a sinful violation of Jesus's second great commandment," he told the class.

As he digs deeper into issues such as the international traffic in toxic waste and incidents of environmental injustice, Golliher says that he often finds himself dealing with what he has come to call "ecological grief," an intuitive understanding on the part of some parishioners that some aspects of nature as we have known it are dying. "Grieving is one phase in the growth process, and in the churches we need to understand the causes of this ecological grieving in order to better assist the process of rebirth."[10]

Golliher is now focusing his environmental ministry in New York City, where he was recently named canon for the environment at the Cathedral of Saint John the Divine, an honorary title that he says signifies the cathedral's commitment to an environmental ministry. He has been involved with organizing a group of leaders from institutions in the Morningside Heights neighborhood on the upper West Side of Manhattan to improve the efficiency in the way in which these institutions use energy and resources. A number of officials from large institutions are committed to this effort, including representatives from Columbia University, Saint Luke's–Roosevelt Hospital, Hunter College, the Museum of Natural History, and the Cathedral of Saint John the Divine, among others.

One of the first tasks these institutional leaders have taken on is to reduce the amount of water consumed in their facilities, Golliher reports. In the past, people who live in the Catskill Mountains, near the reservoirs from which New York City draws its water, were asked to make sacrifices to protect the watershed, he continues. It is high time New Yorkers did their part by reducing their rate of water consumption, he says. "There is a real effort here to understand what an urban ministry means, taking into account the watershed in which it exists," he says. That involves reaching out to people in the countryside to learn from them and form a partnership with them in taking care of Creation, he adds.

In addition, the group of institutional leaders is working with Paul Mankiewicz of the Gaia Institute to find a way to compost institutional food waste using his superefficient, rubber-inflatable, composting units. Still to be decided is whether each institution will do its own composting or if it will be centralized on the grounds of several facilities. The institutional leaders group also initiated a tree planting program by inviting members of the American Foresters Association, which was holding its annual meeting in New York, to come to the Cathedral of Saint John the Divine to celebrate the spiritual component of their work and join in planting trees in front of the cathedral.

As the coordinator of the René Dubos Consortium for Sacred Ecology at the Cathedral of Saint John the Divine, Golliher also works to improve environmental practices in the offices and maintenance department of the cathedral. One of his projects involved enlisting ten- and eleven-year-old students at the Cathedral School in lecturing staff workers about the merits of recycling. Along with their pep talk, the grade-school students presented members of the staff with brightly painted recycling boxes they had decorated. Students also designed a coffee mug that Golliher helped them sell to staff workers in an effort to cut down on the use of paper cups.

Golliher has also been instrumental in establishing a "green tour" at the cathedral for the half-million tourists who visit every year, turning their visit into an occasion for environmental education. Tourists can either sign up for a guided green tour or conduct a self-guided tour with the aid of a leaflet that directs them to some twenty environmental exhibits or activities around the cathedral.

Part of his work as an environmental priest, Golliher says, is to urge people to look within for the solutions to environmental problems. "We must get beyond seeing the environment as something outside ourselves that we could fix if only we had the right tool," he says. Through self-examination he asks parishioners to conduct an environmental impact assessment of their work, what they eat, how they dress, what they buy, and how they treat others. "The greenest thing I can do as a priest is to help people make connections between these areas of their lives that they usually think about in a fragmented way."

The way people dress, for example, is both an environmental and a spiritual issue, he maintains. People who buy too many clothes not only waste resources and place an unnecessary burden on the environment, their dependence on buying an excessive number of new clothes to enhance their sense of self-worth also suggests a spiritual imbalance. Some people who lack a core sense of themselves or their place in the cosmos cover this over by dressing up their exterior, Golliher observes. In his book *Message and Existence* Langdon Gilkey expands on this insight: "The central dynamic of infinite demand is not anxiety about security but desire, lust for more and more, and impatience and dissatisfaction with what is now possessed, a sense of yearning emptiness if more is not gained. . . ."[11] This is exactly the kind of deep-seated, environmentally destructive, behavioral problem that the environmental ministry may prove more effective at changing than other sectors of the environmental movement.

"Jeff Golliher understands the kind of humble tasks that are required of us in order to protect the environment," says the Very Reverend James Parks Morton, dean of the Cathedral of Saint John the Divine, who pioneered the integration of environmental activities into the Episcopal Church. "It is the way you wash your clothes. It is the kind of toilet paper you use. It is very fundamental stuff. It is not just what you say because that can be a lie. It is what you do. This requires radical behavioral change and people must come to see that we are talking about a complete turnaround in lifestyle," he adds.

The trick is not to scold people for living in an ecologically incorrect fashion, but rather to show them the way they are connected to the whole of God's creation. Rather than shame people into change, Golliher encourages them to expand their vision of themselves and their place in the world. Many people are looking for a way to live a sustainable lifestyle and abandon destructive patterns, he observes, but the structures of society— including the church—are not very helpful in this regard. The job of the environmental ministry is to reorient the church so that it helps people see the connection between religion and the environment and the possibility of living in a more grounded and sustainable way.

The religious community is becoming increasingly inventive at integrating environmental issues into congregational life. For example, The National Religious Partnership for the Environment has now sent out over 100,000 education and action manuals to congregations. These packets suggest that in the "prayers for the people" section of the Christian liturgy, that congregations may wish to pray for the biosphere or for a different endangered species every week; give householder tips about how congregations can conserve energy, set recycling targets, car-pool, and make environmentally sensible decisions about land-use planning; and encourage congregations to make commitments about changing purchasing and consumption patterns, forming coalitions with local environmental groups, and becoming active in public policy advocacy.

But beyond the practical lifestyle changes that individuals can make and the political pressure that the 100 million Americans who go to church, synagogue, or other religious services can bring to bear, religious organizations can also teach the message of the sacredness of creation through symbols and a reinterpretation of sacred texts.

To date, one of the most advanced efforts to use the liturgy to highlight the sacredness of all creation has come out of the Cathedral of Saint John the Divine where, on Saint Francis Day, a procession of animals are brought into the cathedral to be blessed at the alter by the bishop. In planning the first Saint Francis Day at the cathedral, Paul Gorman recalls that he and Dean Morton tried to anticipate what people's reaction would be to a procession of animals coming into the cathedral. Would people laugh and get nervous and make the animals jittery? Would flashbulbs go popping off scaring the animals? Would some people be uncomfortable and feel that sacred space was being violated?

Instead, Gorman reports, when the two ton bronze doors of the cathedral were opened and the procession of animals came in—a collection that now includes, among other species, an elephant, a camel, rats, roaches, birds, fish, algae, snakes, dogs, cats, and a blue crab—"people all over the cathedral were in tears. The feeling was almost like: How could we have left these creatures out of the Church in the first place? Who closed the doors on them?" Recognizing the sacredness of creation has an extraordinary potential for the renewal of religious life, Gorman continues. "Creation is not just here for us to take care of," he points out. "As anyone who has taken a walk by a stream or a hike up a mountain can tell you, creation is here to awaken and replenish our faith."

Not all factions of the church are enthusiastic about turning the church into a modern-day equivalent of Noah's Ark or in permitting the scripture to be reinterpreted from an eco-theological perspective. In an Earth Day sermon, John Cardinal O'Connor, Archbishop of New York, said that environmental concerns should not overshadow the centrality of human beings in

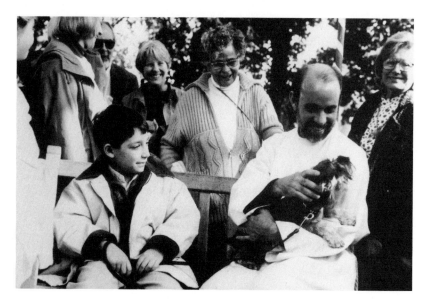

Figure 25.1
Reverend Jeffrey M. Golliher at the Saint Francis Day blessing of the animals conducted at the Cathedral of Saint John the Divine in New York City. Photograph by Asha Golliher.

God's plan: "The earth exists for human persons and not vice versa," he said from the pulpit of Saint Patrick's Cathedral in the heart of Manhattan.[12]

Other voices within the church worry that interpreting all of creation as sacred is leading to a New Age form of paganism in which trees and other aspects of nature will be worshipped as God. Golliher dismisses this argument, saying that while creation is sacred, it is not God. He adds that he has never met anyone who worships nature as God, not even when he was an anthropologist living among tribal people in the rain forests of Costa Rica.

While such eco-theological debates may sound ethereal, in fact they have a powerful hold on the role the church plays in society. Recognizing this, a number of religious scholars are busy finding biblical backing for an environmental ethic, such as Noah's instructions from God to build the ark. "And of every living thing of all flesh, two of every sort shalt thou bring unto the ark, to keep them alive with thee; they shall be male and female. Of fowls after their kind, and of cattle after their kind, of every creeping thing of the earth after his kind, two of every sort shall come unto thee, to keep them alive" (Genesis 6:19–20). "If you translated that story into modern terminology, you would frame it as saving threatened species," observes Calvin B. DeWitt, director of the Au Sable Institute for Christian Environment and Education.[13]

Deeply involved in these eco-theological debates, Golliher is particularly disturbed by the interpretation of the passage in the Bible that speaks of man having dominion over nature. "And God blessed them, and God said unto them, 'Be fruitful, and multiply, and replenish the earth, and subdue it: and have dominion over the fish of the sea, and over the fowl of the air, and over every living thing that moveth upon the earth'" (Genesis 1:28). This passage, he points out, has been used for centuries as a spiritual carte blanche for the exploitation of nature. But the idea of a self-giving Christ suggests that God is in service of creation, he continues. "Our place in Creation is defined much more by our responsibility to care for it and for each other than it is by how Creation might be useful to us," says Golliher.

Golliher also maintains that it is a mistake to believe that because human beings were made in God's image, somehow the rest of creation is devalued. All of creation is sacred, he insists. As evidence, he cites the passages in Genesis that describe how God created the world. God paused at the end of each day and judged that what he had made was good. "These solemn blessings suggest to me that all God's creation is, in some way, sacramental and holy," Golliher writes.

Albert J. LaChance echoes this sentiment: "Praising the Creator while we ruin creation is blasphemy."[14] Golliher puts it even more boldly: "You can't love God and destroy Creation. The times demand that we speak clearly about this: It is a sin to destroy God's Creation."

Looking Ahead to Green Taxes and a Green Plan

Eco-pioneers can only go so far toward transforming the United States into a sustainable culture. Without an environmentally sensible infrastructure it will be difficult for people to meet their legitimate needs without doing harm to nature.

Consider a trip to the store to buy food for dinner. If there is not a good mass transit system in your neighborhood or if you do not live in a community with a good mix of houses and stores within walking distance, then you will likely have to drive and add to the pollution and congestion that automotive transport entails. (Of course you could take a bike with a basket, but with two kids under six that can be tricky.) Policies that promote transit-oriented design and mixed-use neighborhoods can go a long way toward making it possible for people to become less dependent on their cars, but transforming existing communities into this configuration will take a significant investment and require important and controversial changes in our zoning ordinances.

Even if you can find environmentally sensible transportation, once at the store even the most eco-conscious consumer will find it hard to go home without a sizable amount of unnecessary packaging. Only regulations that require minimal packaging will solve that problem. And what about the pesticide residues in the food offered for sale? Only growing your own food or finding an outlet for organic meat and produce will permit shoppers to avoid taking home toxic chemicals for dinner. Without agricultural policies that reward farmers for adopting sustainable practices that minimize the use of pesticides, our food will continue to be raised with unnecessarily large doses of chemicals. Similar examples abound of areas of commerce in which unsustainable practices are built in. For instance, if you want to buy a house, few energy-efficient homes built with nontoxic materials are available. Only the relatively wealthy, who can afford to have a house built to green specifications, or the heroically virtuous, who are willing to build their own eco-home on a shoestring, can avoid standard housing.

The point is that there is only so much an individual can do to live lightly on the planet in a society that ignores or discounts environmental costs. Even individuals who are working on solutions to environmental problems, such as the eco-pioneers described in this book, will remain only marginally effective until an environmentally sustainable infrastructure is put in place—an event that will only be possible once a consensus is reached that a certain level of planning and regulation is necessary to insure that we maintain a healthy ecosystem while meeting our needs.

Looking ahead it appears as if two significant changes must take place if we are to establish a more sustainable culture in the United States. First, we must shift the way we price goods and services to reflect more accurately their environmental costs. And second, we must develop a "green plan" that sets up a context within which commerce can take place within the limits of nature. Changing our laws, taxes, regulations, subsidies, and economic indicators so that they promote sustainable activities now looms as the next great task of the environmental movement.

The fact that our economy is based on economic indicators such as the gross national product (GNP) and gross domestic product (GDP), which only ascribe value to a forest once it is cut down and sawed up into lumber, should give us all a serious case of the heebie-jeebies. After all, forests provide a host of priceless ecological services at no cost. Forests supply a carbon sink that slows global warming; act as the lungs of the world by converting carbon dioxide to oxygen through the magic of photosynthesis; hold huge amounts of water in the ground, thus averting floods; prevent erosion; modify temperatures; and provide a habitat for a diversity of species. All of these functions are extremely expensive or impossible to replace once a forest is cut down, yet nowhere does the value of these ecological services show up in our national accounts.

That economic indicators, such as the GNP and GDP do not take into account natural capital depreciation worries a number of economists: "The fact that earth's productivity is nearly everywhere in decline and that this decline is registered as global economic growth is one of the madder economic conventions," Paul Ekins observes.[1] Eco-economist Herman Daly agrees, noting that we are managing the earth as if it were "a business in liquidation."

Alternative measures of economic prosperity are possible, argues David Pearce: "We should measure income as the flow of goods and services that the economy could generate without reducing its productive capacity—i.e. the income it could produce indefinitely."[2] This kind of sustainable accounting system would, among other factors, take into account the atmosphere's capacity to absorb greenhouse gases without triggering global warming, the environment's ability to neutralize toxic wastes, and the minimum conservation areas needed to keep species alive, Ekins adds. Lester Brown, president of the Worldwatch Institute, fills out this sketch of what an ecologically balanced economy might look like in his introduction to *State of the World, 1996.*

> In a sustainable economy human births and human deaths are in balance, soil erosion does not exceed the natural rate of new soil formation, tree cutting does not exceed tree planting, the fish caught do not exceed the sustainable yield of fisheries, the cattle on the range do not exceed its carrying capacity, and water pumping does not exceed aquifer recharge. It is an economy where carbon emissions and carbon fixation are also in balance. The number of plant and animal species lost does not exceed the rate at which new species evolve.[3]

Bringing about such a sustainable culture will require the best efforts of eco-pioneers from all over the world who are experimenting with or reviving techniques that allow humans to provide for their legitimate needs without doing substantial harm to nature. But the transition to a sustainable economy will also likely require a change in the way we place a price on goods and services.

"Getting the price right" by partially replacing income taxes with environmental taxes has become a rallying cry for those who believe that taxing environmentally destructive practices more than we tax income makes good sense. Even the President's Council on Sustainable Development—an amalgam of corporate executives, heads of environmental organizations, and members of the cabinet—recommends that the United States "begin the long-term process of shifting to tax policies that—without increasing overall tax burdens—encourage employment and economic activity while discouraging environmentally damaging production and consumption decisions."[4]

Why this call for a shift to green taxes? It is argued that only a restructuring of prices to reflect environmental costs will signal consumers about the true cost (including the environmental costs) of goods and services available to them. This, in turn, will place pressure on manufacturers to change their manufacturing and distribution processes so that they are more naturally efficient and do less damage to nature.

To make this concept less abstract, consider what would be required to internalize the true ecological costs of producing a book so they are reflected in its price. One way to do this would be a partial shift from taxing income to taxing goods and services that cause substantial environmental degradation. Under this green tax system, book publishers would have a choice. They could either buy heavily taxed paper derived from unsustainably harvested timber, bleached with chlorine, and printed with poisonous inks; or they could purchase more environmentally friendly, unbleached, recycled paper printed with vegetable inks that is taxed at a lower rate. The choice would be a no-brainer: in order to keep the price of their books competitive, publishers would select the materials taxed at the lower rate, thus minimizing their costs and creating a demand for unbleached, recycled paper and vegetable inks. Similarly, environmental taxes might also be levied on the use of pesticides, the generation of toxic wastes, the use of virgin raw materials, the conversion of croplands to nonfarm use, and carbon emissions, suggests Lester Brown.[5]

Unfortunately, green taxes such as these are not on the books today and our current pricing system leaves eco-pioneers at a critical disadvantage because the good they do for the environment remains largely unrewarded in the marketplace. Many manufacturers lower their costs and the price at which they sell their product by externalizing environmental and health costs to society at large. In their book, *For the Common Good: Redirecting the Economy Toward Community, the Environment, and a Sustainable Future*, Herman Daly and John B. Cobb Jr. use the example of coal mining companies that externalize the cost of black lung disease among coal miners. Instead of internalizing the cost of preventing and treating this occupational disease into the price of coal, it is externalized and taxpayers foot the bill.[6]

By externalizing environmental costs, companies can undercut the market for goods produced sustainably. While it is true that some people will buy a product that is advertised as produced in an environmentally sensible fashion, many consumers buy goods on the basis of price, quality, or impulse, with little thought for the environmental damage incurred during the product's manufacture. Thus, while eco-pioneers protect nature for all of us, they remain largely unpaid for this service.

Take, for example, the work of Pliny Fisk III, the Texan architect. By building structures out of locally available materials that are sustainably harvested, Fisk limits the amount of unsustainably harvested wood that is transported over long distances by diesel-burning trucks. In a rational world there would be a stiff tax on the unsustainably harvested lumber and on diesel fuel, making Fisk's locally available building materials highly attractive.

Under the current economic system, however, not only do we fail to tax environmentally unsustainable practices, we actually subsidize them. One of the most egregious examples of this are the taxes spent to build logging roads on federally owned land so that trees can be sold at a loss to lumber companies that practice clearcutting. Taxes are also spent to subsidize ecologically destructive agricultural practices instead of using that money to provide incentives for sustainable farming.

The failure to shift taxes and subsidies to encourage sustainable enterprises is not only a tragedy in terms of our ability to keep American ecosystems healthy and stable, it also makes the job of eco-pioneers infinitely more difficult. The inventions of eco-pioneers would quickly be recognized as cost effective if the price structure were recalibrated to internalize environmental costs and favor sustainable economic activity through a change in our tax and subsidy system. If, for example, there were a significant tax levied on the use of chlorine—a chemical that can spawn trihalomethanes and other toxic compounds—then John Todd's living machines would become instantly more attractive to communities sensitive to the cost of treating sewage.

While green taxes have yet to be enacted on a significant scale, some economists point out that a less visible environmental and economic revolution is already underway. In his book *The Environmental Economic Revolution: How Business Will Thrive and the Earth Survive in the Years to Come,* Michael Silverstein points out that some 2 million Americans are already employed in the field of environmental protection and some $130 billion is spent annually on environmental cleanup in this country. This decade $1.2 to $1.5 trillion will be spent domestically on environmental cleanup and there is a $50 billion international trade in the environmental cleanup and services market.[7]

Perhaps even more significant than this growing domestic and international market in environmental cleanup is the fact that some of the biggest and best-managed companies around the world are in the midst of a transformation aimed at reducing waste and energy inefficiency. The 3M company, for example, has spent $600 million in recent years on reducing their generation of wastes and many other large corporations are following their lead. On the energy front, some $5 billion is invested annually in energy conservation, the Rocky Mountain Institute estimates. This trend is driven not by altruism but by the bottom line, Silverstein notes. Pollution and waste equal inefficiency and additional costs; cutting them out just makes good business sense.

Despite these gains, "changes in the tax code, better regulatory practices, and more finely honed litigation targets" might hasten the transformation to a more environmentally sustainable economy, Silverstein continues. "Even creating a green national product measuring system to complement rather than replace traditional gross national product and gross domestic product systems" might be useful in consciousness raising, he adds. But the central reality that must be kept in mind is that the environmental and economic revolution is already underway.

While green taxes such as the carbon tax have at least been considered in Congress, a discussion about a "green plan" for the United States is still below the political radar screen. Putting together those two words— "green" and "plan"—remains a political anathema because it smacks of a planned economy and gives free market traders a case of the jitters. This political taboo, however, has not deterred Huey D. Johnson, president of the Resource Renewal Institute in San Francisco and author of *Green Plans: Greeprint for Sustainability.*[8] "The time is ripe for the United States to adopt an environmental policy as new and sweeping as a green plan. It would be an historic proposal that would inspire the nation to solve the environmental problem," Johnson asserts.

The evidence that long-term green planning is more than just a pleasant dream can be found abroad, Johnson writes. His book, the first major effort to promote green plans in this country, makes the following argument.

Figure 26.1
Huey Johnson, president of the Resource Renewal Institute (RRI) in San Francisco, argues that the time is ripe for the United States to adopt a green plan. Courtesy RRI.

While the practice is still in its infancy, green planning has taken root in the Netherlands, New Zealand, and Canada, among other countries. These nations "commit every segment of society—and large sums of money—to a national environmental recovery program." For example, the Dutch invested $9.5 billion in activities stemming from their green plan in 1994 and the Canadians are committed to spending $2.2 billion over six years. In a sense, these countries are just making a virtue out of necessity because the limited carrying capacity of the world's ecosystems will eventually force all major industrialized nations to adopt green plans.

While there is no single prototype for green plans and each nation must mix its own unique cocktail, there are certain elements common to successful green plans wherever they are found. One goal they all share is to create a sustainable economy in which resources are managed in a way and at a rate that enables people to meet their needs without compromising the ability of future generations to meet their own needs.

Another distinguishing feature of green plans is that they are comprehensive, recognizing that environmental problems are interrelated and that solving one problem at a time won't work. In the past, the government's piecemeal approach failed because it merely shifted environmental problems from one medium to another. For example, one solution to the disposal of hazardous waste has been to burn it in incinerators fitted with scrubbers to filter out toxic emissions. But many of the toxins remain in the incinerator ash and in its filters, and this material is subsequently disposed of in landfills, where it may end up contaminating groundwater supplies. Thus, while the toxins may be temporarily kept out of the air, they eventually end up either in the ground or in the water supply. By contrast, green planning focuses on preventing the generation of hazardous waste.

Another shared characteristic of successful green plans is that they call on all segments of society to contribute to environmental solutions, instead of engaging in utopian planning sessions for environmentalists. Typically, all the stakeholders are invited to join in the process of developing, implementing, and maintaining a green plan. A new, cooperative relationship between government, the business sector, and environmentalists is a key element of successful green plans. Green plans are also goal-oriented: they practice management by objective rather than relying on standard regulatory micromanagement of the process.

That green planning blossomed first in the Netherlands should come as no surprise, given that the Dutch have intimate experience with long-term planning, having spent three centuries building an extensive system of dikes to reclaim land from the sea. Green planning really took hold in the Netherlands after Queen Beatrix (also known as the Green Queen) galvanized public opinion with her 1988 Christmas message, in which she warned that "The earth is slowly dying, and the inconceivable—the end of life itself—is actually becoming conceivable." Her speech was based on a sobering report entitled *Concern for Tomorrow*, which detailed what was wrong with the environment in the Netherlands and what needed to be done.

Faced with the fact that they had reached a point at which tough choices would have to be made, the Dutch didn't flinch. In 1989 they passed the National Environmental Policy Plan (NEPP) entitled *To Choose or to Lose*, which required 223 policy changes. NEPP made it clear for the first time that all sectors of society, not just the government, were responsible for environmental quality.

The Dutch green plan has been extraordinarily successful and has enjoyed overwhelming backing by the public. The NEPP bolstered standard pollution reduction measures, contributing to such accomplishments as a 30 percent reduction in emissions of sulfur dioxide, nitrogen oxides, ammonia, and volatile organic compounds from 1980 to 1990; a 20 percent reduction in pesticide use from 1988 levels by 1992; and the cleaning of 10 million tons of contaminated soil between 1991 and 1994. In addition, the NEPP spurred plans to plant 99,000 acres of forest and restore 800,000 acres of wetlands, and to double the life span of products and stabilize the use of energy by the year 2000. "These kinds of large-scale projects would not be possible without a comprehensive and integrated plan," Johnson notes. Above all, with its comprehensive perspective, long-term time frame, and built-in incentives for environmental improvement, the Dutch green plan encourages businesses to rethink their entire product life cycle, from raw materials to disposal.

New Zealand enacted its own green plan, known as the Resource Management Act, which redefined political districts along watershed lines to facilitate environmental management. Unlike the Netherlands, New Zealand is not densely populated or highly industrialized; its ecological threat is from unsustainable agricultural, forestry, fishing, and mining practices. Rather than try to reform the existing tangle of inadequate environmental regulations, government officials and planners in New Zealand decided to start from scratch and develop an integrated plan. After three years of meetings with all the stakeholders from around the country, and months of parliamentary debate, the Resource Management Act became law in 1991.

Passing a major green plan such as those in the Netherlands and New Zealand requires canny political maneuvering. "It is a little like war or chess," Johnson observes. "You don't necessarily want to take on your opponent's strong point. You may want to by-pass a well fortified castle and attack a weaker point," he says, taking a page from Lao Tzu's *The Art of War*. This is what happened in New Zealand, where the mining industry put up a fierce fight against green planning. Those pushing the program simply ignored them and got everybody else on board until the mining interests were isolated and were eventually forced to reach an agreement.

While at first some businesses opposed green planning, eventually key players decided that the plan was to their advantage. In particular, New Zealand's forestry and agriculture industries decided that they could create a "green market" for products raised or made using ecologically sustainable methods. To other industrial groups, one of the key ingredients that made green planning palatable was the simplification of the permit process.

Since the green plan has been adopted in New Zealand, the country's remaining original forests have been placed on permanent preserve status by channeling forestry activity into intensive tree farming. From now on, all forests under preservation status must be logged at a rate that does not exceed the growth rate. Further, by ending subsidies for agriculture, the use of chemical fertilizers and pesticides plummeted.

Green planning has been realized most fully in the Netherlands, New Zealand and, to a more limited extent, Canada. Other nations developing some version of a green plan include Norway, Sweden, Denmark, Austria, the United Kingdom, Germany, Singapore, and the European Union.

Some will doubtless scoff that while green planning may work in small, relatively homogeneous countries such as the Netherlands and New Zealand, it is just not practical in a demi-continent–sized nation such as the United States, with its varied geography and heterogeneous population. But Johnson disagrees: "I came to the conclusion that I was unlikely to come up with a better strategy for solving the environmental dilemma than the green plan approach that is being demonstrated in the Netherlands and New Zealand."

One of the key features of successful green plans is the new dynamic they establish between the business community and government. Instead of being part of the source of environmental problems, they can become part of the solution to cleaning them up and preventing them. Furthermore, industries have an incentive to become partners in the green plan process because it offers them a streamlined regulatory system and freedom from micromanagement by government. It also offers them a strong participatory role and a stable and predictable regulatory environment that allows them to do long-term planning and budgeting. In most cases green plans permit businesses to adjust their practices gradually and invest in cleaner technologies that will help them meet future environmental requirements. And although the green plan's environmental objectives are set by government, the methods used to achieve these goals are often left up to the industry involved.

Businesses in green plan nations are not alone in deciding to change the corporate attitude toward environmental management. Officials at a wide variety of American corporate giants, having concluded that being anti-environmental is both bad for business and for their image, have already evolved progressive environmental policies of their own. For example, electronic firms in California's Silicon Valley reduced emissions of pollutants by 85 percent between 1985 and 1990, Johnson notes. Furthermore, 3M, Dow, DuPont, Hewlett-Packard, Intel, and a number of other firms now operate on a "beyond compliance" basis, which means they do better than meet current regulatory requirements.

Corporate-level green plans will stimulate the invention of environmentally sophisticated technologies that will bring enormous economic benefits to those who pioneer them, Johnson observes. Companies that are energy efficient, materials efficient, create a minimum of waste, handle their pollution responsibly, and create durable, high-quality goods will win in the marketplace, he predicts. But corporate green plans alone are not enough, he continues, pointing out that the Dutch covenants could not work without the framework established by the NEPP. The essence of green planning is that it is comprehensive, he says. It is about dealing with the whole picture of environmental decline, not just certain pieces of it.

Will the United States be left behind in the move to green planning? Johnson thinks not, yet he points out the irony of the fact that much of the groundwork for green planning originated in the United States before being put into practice abroad. For example, the Carter administration's *Global 2000 Report to the President,* which looked at possible solutions to international environmental problems, inspired much of the thinking behind New Zealand's environmental reforms. The work of American environmentalists such as David Brower, Barry Commoner, Aldo Leopold, and Rachel Carson "have also had a greater impact on the rest of the world than they have apparently had on us," Johnson notes.

Among the reasons that the United States has not been in the forefront of green planning is the fact that its large land mass and abundant resources have permitted Americans the luxury of avoiding tough choices about how to efficiently manage resources. In addition, Johnson notes, the few, long-term planning efforts that have been attempted in the United States have frequently not been acted on, leaving the public cynical about these exercises.

But while the United States may be slow to come to green planning, eventually we will come around to it, he predicts. And although we currently may be a long way from enacting a national green plan, the concept is already being tested out in a number of states. New Jersey, for example, has adopted some of the strictest environmental standards in the country and initiated a state planning act that includes a comprehensive land-use planning strategy. Oregon developed a risk-based strategy called "benchmarks" that sets up five-, ten-, fifteen-, and twenty-year goals on some thirty-five environmental issues. Borrowing from this experiment the Minnesota Milestones program set up numerical goals such as a 22 percent reduction in the annual per capita energy consumption by the year 2020. Similar green plan strategies are underway in Florida and Connecticut, and Washington's Environment 2010 is a long-range planning exercise that has resulted in a number of "managed growth" laws.

Once green plans have been shown to be a success at the state level, federal officials should involve the public in compiling a national "greenprint" for the United States, Johnson says. But when will the political climate be ripe for such an initiative? Recent attempts by members of the Republican congressional delegation to roll back environmental regulations might suggest that a comprehensive national environmental recovery plan is not politically feasible at present. But strong public support for environmental protections, reflected in the proenvironmental stand taken by a significant number of moderate Republicans, argues otherwise. Green planning's potential appeal to responsible members of the business community also strengthens its prospects. Johnson, for one, remains undismayed by the political climate. Understanding that timing is a big factor in politics, he continues to polish the idea of a green plan for the United States until a national leader emerges with the foresight and political muscle to make it happen. "Eventually, green planning will be taken up by leaders around the world, and it will be immensely successful. It has been tested, and has shown its worth. The idea is gaining momentum and as it continues to move forward it will find leaders to champion it," he says.

"The power of green plans is their scope. People take hope from such a large-scale effort because they can see their government is serious in its commitment," Johnson says. When citizens see their government taking a comprehensive approach to solving environmental problems, they are more willing to make sacrifices such as paying increased taxes, because they come to recognize that the payoff is a livable future for their grandchildren. "What is happening in countries like the Netherlands and New Zealand is the politics of hope. The whole nation becomes involved. And that is what it will take here in the United States as well," he believes. While a green plan for the United States may seem like a distant hope to some, Johnson contends that it is the only answer that ultimately makes sense. "No other approach has shown such a potential for achieving long-term sustainability," he says.

Whether or not a comprehensive plan to put the U.S. economy on a sustainable footing takes shape before or after ecological disasters begin to befall us will depend upon the ingenuity and resourcefulness of Americans in all walks of life. This work can begin at the community level, in our cities and towns, but ultimately it will require the building of a national consensus to create an integrated plan to bring about an ecologically sustainable culture in the United States. A central part of this plan will necessarily be a mechanism for including environmental costs in the price of goods and services so that consumers have a financial incentive for buying ecologically sensible products. Eco-pioneers will play a crucial role in this process by developing cost-effective ways to meet human needs while causing a minimum of damage to nature.

Notes

Introduction

1 Jonathan Rowe, "Honey We Shrank the Economy," *Yes: A Journal of Positive Futures,* B1 (spring/summer 1996): 29.

2 Nancy Jack Todd and John Todd, *Bioshelters, Ocean Arks, City Farming: Ecology as the Basis of Design* (San Francisco: Sierra Club Books, 1984), p. 107.

3 Ibid., p. 1.

4 Steve Lerner, *Earth Summit: Conversations with Architects of an Ecologically Sustainable Future* (Bolinas, Calif.: Common Knowledge Press, 1991); and *Beyond the Earth Summit: Conversations with Advocates of Sustainable Development* (Bolinas, Calif.: Common Knowledge Press, 1992). Available through Commonweal, Box 316, Bolinas, CA 94924.

5 William E. Rees, "Ecological Footprints," *Yes: A Journal of Positive Futures,* v. B1(spring/summer 1996): 26–27. Rees is also author of *Our Ecological Footprint: Reducing Human Impact on the Earth* (Philadelphia: New Society Publishers, 1995).

6 Gordon Shepherd, "Campaigns and Treaties, World Wildlife Fund," *E & D File,* v. 2, no. 15 (December 1993): 2.

7 Paul Ekins, Mayer Hillman, and Robert Hutchison, *The Gaia Atlas of Green Economics* (London: Anchor Books, 1992), p. 33.

8 Thomas L. Friedman, "Gardening with Beijing," *The New York Times,* 30 April 1996, p. A23.

9 A good example of this is the way in which Baltimore Mayor Kurt L. Schmoke learned how to improve child immunization rates in his city by copying a Kenyan program. The U.S. Agency for International Development's "Lessons Without Borders" program promotes this type of cross-cultural learning process by sharing international urban improvement techniques with officials of cities in the United States. See Kelly Couturier, "Baltimore, U.S. Cities Find Community-Building Models in Developing Countries," *The Washington Post,* 10 June 1996, p. A20.

10 Lester Brown et al., *State of the World* (New York: W.W. Norton, 1996), pp. 3, 4, 7.

11 Herman E. Daly and John B. Cobb Jr., *For the Common Good: Redirecting the Economy Toward Community, the Environment and a Sustainable Future* (Boston: Beacon Press, 1989), pp. 143–4.

12 Ekins, *The Gaia Atlas,* p. 66.

13 Quoted in James Campion, "Once Upon a Universe. . .," *Maryknoll,* 87, no. 3 (March 1993): 16.

14 Lester Brown, *Building a Sustainable Society* (New York: W.W. Norton, 1981). Quoted in Nancy Jack Todd and John Todd, *Bioshelters, Ocean Arks, City Farming: Ecology as the Basis of Design* (San Francisco: Sierra Club Books, 1984), p. 165.

Chapter 1   *Pliny Fisk III*   The Search for Low Impact Building Materials and Techniques

1 Robert Bryce, "The Mad, Mad World of Pliny Fisk: From Fringe to Center," *The Austin (Texas) Chronicle,* 29 April 1994, p. 50.

2 Patricia Leigh Brown, "Mr. Fisk Builds His Green House," *The New York Times,* 15 February 1996, p. C1.

3 Eugenia Bone, "The House that Max Built," *Metropolis,* December 1996, 40–41.

4 Ibid., p. 42.

5 UNESCO's "Man in the Biosphere" program has a list of these groups. Also check with the International Union for the Conservation of Nature (IUCN).

Chapter 2   *Lorrie Otto*   Bringing Native Plants Back to the American Lawn

1 Jim Wilson, *Landscaping with Wildflowers: An Environmental Approach to Gardening* (Boston: Houghton Mifflin, 1992), pp. 78–82. Other useful books on this subject include Roger Holmes, ed., *Taylor's Guide to Natural Gardening* (New York: Houghton Mifflin, 1993); Viki Ferrreniea, *Wildflowers in Your Garden: A Gardener's Guide* (New York: Random House, 1993); and William H. W. Wilson, *Landscaping with Wildflowers and Native Plants* (San Ramon, Calif.: Ortho Books, 1984).

2 Michael Pollan, author of *Second Nature: A Gardener's Education* (New York: Dell, 1991) makes a convincing argument that plant species, like humans, migrate from one part of the world to another over time. He contends that it makes little sense to attempt to keep alien species out unless, like kudzu, they are destructive of other species. Many so-called alien species can enrich an ecosystem by making it more diverse, he points out. His argument is made most cogently in "Against Nativism: Horticultural Formalism May Be out, But the New American Garden, Free of Foreign Flora and Human Artifice, Isn't as Natural as Advocates Claim," *The New York Times Magazine,* 15 May 1994, pp. 50–5.

3 William K. Stevens, "Botanists Contrive Comebacks for Threatened Plants," *The New York Times,* 11 May 1993, p. C1.

4 Estimates of acreage in the United States taken up by lawns vary. In Charles Fen-yvesi, "Ready, Set, Mow: Seasonal Start-Up for Environmentally Friendly Lawns," *Washington Post,* (Home Section) 7 April 1994, p. 18. Ann E. McClure, executive vice president of the Professional Lawn Care Association of America is quoted as saying that there are more than 25 million acres of lawn in the United States, 100 million people operate 61 million mowers, and Americans spend approximately a billion hours on lawn care every year. Another estimate of 30 million acres of lawn in the United States is made in Anne Simon-Moffat, Marc Schiler, and the staff of Green Living, *Energy-Efficient and Environmental Landscaping: Cut Your Utility Bills by up to 30 Percent and Create a Natural, Healthy Yard* (South Newfane, Vt.: Appropriate Solutions Press, 1993), p. 103.

5 Sara Stein, *Noah's Garden: Restoring the Ecology of Our Own Back Yards* (Boston: Houghton Mifflin, 1993), p. 16.

6 Monica Michael Willis, "Save Our Countryside: Backyard Wildlife Habitats Is a Progressive 20-Year-Old Gardening Program Whose Time Has Finally Come," *Country Living,* July 1993, 16.

7 Eliot C. Roberts and Beverly Roberts, *Lawn and Sports Turf Benefits,* (Pleasant Hill, Tenn.: The Lawn Institute, n.d.), p. 1.

8 Anne Simon-Moffat, Marc Schiler, and the staff of Green Living, *Energy Efficient and Environmental Landscaping* (Newfane, Vt.: Appropriate Solutions Press, 1993), p. 103.

9 Brenda Biondo, "Greening Grass: Cultivating Lush Lawns Needn't Add Up to Polluting the World Around You," *The Washington Post,* 11 April 1995, p. C5.

10 Caroline E. Mayer, "The Shrinking American Lawn," *The Washington Post,* 29 April 1995, p 1.

11 Roberts and Roberts, *Lawn and Sports Turf Benefits,* 28–9.

12 Mayer, "The Shrinking American Lawn."

13 The best article to date on Lorrie Otto is by Paul G. Hayes, "Doing What Comes Naturally: How Lorrie Otto, the Prairie Lady, Evolved into Environmental Activism," *The Milwaukee Journal,* 28 June 1992, p. 10.

14 Mayer, "The Shrinking American Lawn."

15 Mariuanne Kyriakos, "More Companies Take Natural Care Approach," *The Wall Street Journal,* 3 September 1994, p. 1.

Chapter 3 John Todd Greenhouse Treatment of Municipal Sewage

1 Nancy Jack Todd and John Todd, *Bioshelters, Ocean Arks, City Farming: Ecology as the Basis of Design* (San Francisco: Sierra Club Books, 1984), p. 29.

2 "If You Leave it to Mother Nature, Says Biologist John Todd, Sewage Doesn't Have to go to Waste," *People,* 27 November 1989.

3 Nancy Jack Todd and John Todd , *Bioshelters,* p. 32.

4 Edie Clark, "Troubled Waters: Part Four. Voices of the Future," *Yankee,* July 1990, 70.

5 See "Harnessing Greenhouses to Purify Sewage," *National Geographic,* October 1990; "Living Machines Offer Alternative Treatment Methods," *Architecture,* December 1990, 100; "Putting Nature to Work," *Discover,* December 1991, 72; Chu, Dan, "If You Leave It To Mother Nature, Says Biologist John Todd, Sewage Doesn't Have To Go To Waste," *People,* 27 November 1989. 133.

6 Clark, "Troubled Waters," p. 70.

7 Beth Josephson and John Todd, "Ocean Arks Report: Performance Report for Frederick, Maryland Living Machine," *Annals of Earth,* 14, no. 3, (1966): 7.

8 Ibid., 11.

9 "Interim Report: Evaluation of Advanced Ecological Engineering System "Living Machine" Waste water Treatment Technology, Frederick, MD" (EPA 832-B-96-002), September 1996.

10 Ted Gregory, "Wasting Away Again: A Sewage Reuse Resource," *Chicago Tribune,* 22 July 1994.

11 Janet Marinelli, "After the Flush: The Next Generation," *Garbage,* January/February, 1990, 94.

Chapter 4   Vicki Robin and Joe Dominguez The New Frugality Movement Promotes Living Better by Consuming Less

1 Robin attributes this concept to Robert Muller, former assistant secretary of the United Nations and current advisor to the New Road Map Foundation, Seattle.

2 Joe Dominguez and Vicki Robin, *Your Money or Your Life: Transforming Your Relationship with Money and Achieving Financial Independence* (New York: Penguin, 1992).

3 Alan Thein Durning, *How Much Is Enough? The Consumer Society and the Future of the Earth* (New York: W. W. Norton, 1992), p. 156. Cited in *All-Consuming Passion: Waking Up From the American Dream* (Seattle: New Road Map Foundation, 1993).

4 Laurence Shames, *The Hunger for More* (New York: Times Books, 1989), p. 43. Cited in *All Consuming Passion.*

5 Carey Goldberg, "Choosing the Joys of the Simplified Life," *The New York Times,* 21 September 1995, p. C1.

6 Juliet B. Schor, *The Overworked American* (New York: Basic Books, 1993).

7 Sandy Bauers, "Study: Save Earth; Have Fewer Children," *Philadelphia Inquirer,* 17 January 1990. Cited in *All Consuming Passion,* p. 10.

8 Alan Thein Durning, "Asking How Much is Enough," in Lester R. Brown et al., *State of the World 1991* (New York: W. W. Norton, 1991), p. 156. Cited in *All Consuming Passion,* p. 11.

9 John Young, "The New Materialism: A Matter of Policy," *World Watch,* v. 7, no. 5 (September/October, 1994): 30.

10 Alan Thein Durning, "Redesigning the Forest Economy," in Lester R. Brown et al., *State of the World 1994* (New York: W.W. Norton, 1994), p. 36; "Harper's In-

dex," *Harper's,* March 1992, 13.; Durning, *Saving the Forests: What Will it Take?, Worldwatch Paper,* no. 117 (December 1993): 33; cited in *All Consuming Passion,* p. 3.

Chapter 5 Scott Bernstein Environmental Solutions to Inner-City Problems

1 Center for Neighborhood Technology, *Building Sustainable Communities* (April 1989): 3.

2 Ibid.

3 Center for Neighborhood Technology, "Engineering An Energy-Efficient Future, in *Annual Report 1990,* 4.

4 See Valjean McLenighan, *Sustainable Manufacturing: Saving Jobs, Saving the Environment* (Chicago: Center for Neighborhood Technology, 1990), pp. 36–41.

5 Center for Neighborhood Technology, "President Hails Location Efficient Mortgage," *Place Matters* (fall 1995): 1.

Chapter 6 S. David Freeman A Utility Company Switches from Nuclear Power to Energy Conservation, Renewable Energy, and Electric Vehicles

1 Christopher Flavin and Nicholas Lenssen, "Powering the Future: Blueprint for a Sustainable Electricity Industry," *Worldwatch Paper,* no. 119 (June 1994): 33.

2 Peter Asmus, "Saving Energy Becomes Company Policy," *The Amicus Journal* (winter 1993): 38–42.

3 Flavin and Lenssen, "Powering the Future," 15.

Chapter 7 Sally Fox Breeding Naturally Colored Organic Cotton Eliminates the Need for Toxic Dyes and Pesticides

1 Sally Fox, "In Search of Colored Cotton," *Spin-Off,* December 1987, 48.

2 Suzanne Oliver, "Seeds of Success," *Forbes,* August 1994, 98. This article is the best source on the chronology of Fox's development in the early years.

3 Seventh Generation calculates that 16 million pounds of pesticides are used on the U.S. cotton crop each year, whereas Ecosport claims 20 million pounds are used for this purpose.

4 Greenpeace claims 25 percent of all pesticides are used on cotton in the United States, whereas Joel Makower, editor of the *Green Consumer Newsletter,* claims that 50 percent of the pesticides used in United States are applied to cotton.

5 An excellent description of the environmental costs of cotton growing can be found in Paul Schneider, "The Cotton Brief," *The New York Times,* 20 June 1993, p. 1.

6 Seventh Generation's *Earth Impact* newsletter, 2, no. 6 (December/January 1993): 3.

7 Ibid., *Earth Impact* newsletter, 3.

8 An exception to this is a naturally dyed cotton fabric in indigo offered by Seventh Generation, which uses plant colors, no heavy metals, no phosphates, and only bio-degradable surfacants.

Chapter 8   Daniel Knapp Mining the Discard Supply

1 Dan Knapp and Tom Brandt, *Rattlesnake Recycling* (Oregon Appropriate Technology, 1979).

2 Daniel Knapp, *The Berkeley Burn Plant Papers*, ed. May Lou Van Deventer, ill. Nancy Bechtle and Mark Gorrell (Berkeley, Calif.: Materials World Publishing, 1981).

3 David Kirkpatrick et al., "North Carolina Business Study," North Carolina Department of the Environment and Natural Resources, July 1995.

4 David Stern and Daniel Knapp, "Reuse, Recycling, Refuse and the Local Economy: A Case Study of the Berkeley Serial MRF," Urban Ore and Center for Neighborhood Technology, October 1993.

5 Daniel Knapp, David Stern, and Mary Lou Van Deventer, "Processing Discards Sequentially Using Serial MRFs," *MSW Management,* March/April 1996, 50.

Chapter 9   **Walton Smith** Returning to Selective Forestry After the Failure of Clearcutting

1 Ted Morgan, *FDR:A Biography* (New York: Simon & Schuster, 1985), p. 379.

2 Waton R. Smith, "The Fallacy of Preferred Species," *Southern Journal of Applied Forestry,* 12, no. 2 (May 1988): 79–84.

3 Steven T. Taylor, "If a Tree Is Stolen from the Woods. . .Would the Clinton Administration Make Any Noise?," *The Washington Post,* 29 October 1995, p. C2.

4 Bill McKibben, "What Good Is a Forest? Economists Now Agree with Ecologists: Forests Are Worth More Standing Than Logged," *Audubon Magazine,* May/June 1996, 60–3. Since 1969, through selective harvesting, Smith has doubled the value of both his trees and his land according to a recent county tax assessment.

5 Alan Thein Durning, "Saving The Forest: What Will it Take?" Worldwatch Paper no. 117 (December 1993): 25.

6 Randal O'Toole, *Reforming the Forest Service* (Washington, D.C.: Island Press, 1989) p. xi.

7 Ibid., p. 199.

8 Susan Seager, "The Other Logging Dispute Rages in the Forest Primeval of New England," *The New York Times,* 21 November 1995, p. A15. See also David Dobbs and Richard Ober, *The Northern Forest* (White River Junction, Vermont: Chelsea Green Publishing Company, 1995).

9 McKibben, "What Good Is a Forest?," p. 60.

10 Gordon Robinson, *The Forest and the Trees: A Guide to Excellent Forestry* (Washington, D.C.: Island Press, 1988), p. 25.

11 Op. cit., O'Toole, *Reforming the Forest Service*, p. 74.

12 Op. cit., Robinson, *The Forest and the Trees*, p. 52.

13 Ibid., p. 29.

14 Ibid., p. 82.

Chapter 10   Christopher Nagel and William Haney III Transforming Hazardous Waste into Useful Industrial Materials

1 Robert Correa, "Gore: Technology Can Clean Up Waste," *The Providence Journal*, 19 April 1995, p. 1.

2 John H. Sheridan, "Molten Metal Technology, Inc.: Catalytic Extraction Processing Converts Hazardous and Toxic Wastes into Valuable Materials," *Industry Week*, 20 December 1993, 34.

3 See Elizabeth Corcoran, "A Waste Not, Want Not Goal: Fledgling Firm Takes a Lesson from Steelmakers to Recycle Industrial Byproducts." *The Washington Post*, 22 February 1994, p. 1.

4 For a technical analysis of the CEP technology, see *Metal Recovery, Environmental Regulation and Hazardous Waste: An Analysis of Federal RCRA Subtitle C Regulations Affecting Metal Recovery from Hazardous Wastes*. Report prepared by Paul Boerst for Congress, U.S. Environmental Protection Agency, Office of the Comptroller, Office of Solid Waste and Emergency Response, June 1994.

5 Christopher J. Nagel, Claire A. Chanenchuk, Esther W. Wong, and Robert D. Bach, "Catalytic Extraction Processing: An Elemental Recycling (TM) Technology," *Environmental Science and Technology*, 30, no. 7 (1996): 2155–2167.

6 Laura Johannes, "Molten Metal Technology Shares Plummet 49% As Agency Declines to Renew Contract," *The Wall Street Journal*, 22 October 1966.

7 "Deputy Assistant Secretary of Energy Frank Makes Statement," *Bloomberg Business News*, 22 October 1996.

8 "DOE Disputes Report that Pushed Down Molten Metal Shares," *Bloomberg Business News*, 22 October 1996.

9 A discussion on how to regulate new technologies such as MMT, which can recycle some components of a waste stream, may be found in David Reishus, "Use of an Economic Test for Distinguishing Legitimate Recycling Activities," (Cambridge, Mass: Economics Resource Group, July 1993). The paper was commissioned by the Massachusetts Department of Environmental Protection (MDEP).

Chapter 11   Paul Mankiewicz Urban Rooftop Agriculture

1 William Clairbourne, "Greens, Browns Find Common Ground in World's Cities," *The Washington Post*, 26 September 1994, p. A3.

2 The World Resources Institute, the United Nations Environment Programme, the United Nations Development Programme, and the World Bank, *The Urban Environment: World Resource 1996–97* (New York: Oxford University Press, 1996), p. 8.

3 Nancy Jack Todd and John Todd, *Bioshelters, Ocean Arks, City Farming: Ecology as the Basis of Design* (San Francisco: Sierra Club Books, 1984), p. 120.

4 Jerry Orabon, "Up on a Roof: Could Cities Ever Feed Themselves? Yes Says the Developer of a Revolutionary Rooftop Greenhouse," *New Age Journal,* November/December, 1990, 38.

5 Paul Mankiewicz, Ron Hays, William Kinsinger, and Julie Downey, *The Urban Rooftop Greenhouse Project* (New York: The Gaia Institute, October 1989).

Chapter 12   *David Crockett* Transforming Chattanooga into an Environmental City

1 For an excellent article on the cleanup of Chattanooga, see Vernon Summerlin, "Chattanooga: Becoming the Environmental City," *The Tennessee Conservationist,* September/October 1994, 14–7.

2 Peter Montague, Traci Darnell, and Brian Hunt, "Yearning to Breathe Free: A Citizen's Perspective on Environmental Problems of South Chattanooga, Tennessee," Annapolis, Maryland, November 1991, 44.

3 *A History of Air Pollution Controls in Chattanooga and Hamilton County,* (Chattanooga, Tenn.: Air Pollution Control Bureau, 1991), p. 10.

4 It should be noted, however, that Chattanooga continues to have an air pollution problem with airborne particulates of 10 micrometers or less. In a study of 239 cities, the Natural Resources Defense Council found that Chattanooga ranked twentieth among the cities with the most fine-particle pollution in the country. See Philip J. Hilts, "Fine Particles In Air Cause Many Deaths, Study Suggests," *The New York Times,* 9 May 1996, p. B10.

5 Jay Walljasper, "Chattanooga Chooses: The Revitalization of This Once Dying City Shows Urban Decline Is Not Inevitable," *Utne Reader,* March/April, 1994, 15–6.

6 Frances Moore Lappe and Paul Martin Du Bois, *The Quickening of America: Rebuilding Our Nation, Remaking Our Lives* (San Francisco: Jossey-Bass, 1994), p. 174.

7 See Nancy Bearden Henderson, "Behind the Scenes at the Tennessee Aquarium," *Chattanooga,* 3, no. 1 (spring 1994): 18.

8 Anne Funderburg, "Unused for 14 Years, Century-Old Span to Reopen. Restored Chattanooga Bridge Will Serve Pedestrians," *Historic Preservation News,* October 1992, 12.

9 Calthorpe Associates and William McDonough Architects, *A Comprehensive Revitalization Strategy: The South Central Business District Plan. Chattanooga, Tennessee,* January 1995. Prepared for Rivery Valley Partners.

10 Nancy Jack Todd and John Todd, *Bioshelter, Ocean Arks, and City Farming: Ecology as the Basis of Design* (San Francisco: Sierra Club Books, 1984), pp. 115–6.

11 John Todd, *A Proposal for Chattanooga Bioshelter, Urban Farm and Market* (Falmouth, Mass., Ocean Arks International, n.d.).

Chapter 13 William McDonough Redesigning Buildings and Building Materials for Environmentally Intelligent Architecture

1 William McDonough, "Design, Ecology, Ethics and the Making of Things." A centennial sermon delivered at the Cathedral of Saint John the Divine, New York City, 7 February 1993, 1.

2 Michael D. Lemonick, "Architecture Goes Green: An Array of New Projects Proves that Buildings Can Be Ecologically Correct, Cost Efficient and Beautiful As Well," *Time,* 5 April 1993, 40.

3 See David Goldstein, "Easing Wal-Mart Into the Environment: A Store in Kansas Blends Ecologically Correct Elements," *The New York Times,* 5 February 1995, p. 28. See also "Wal-Mart Store Built From 'Forest-Friendly' Wood Products," *Initiatives for a Sustainable Forestry Industry* 1, no. 1, (May 1993): 1.

4 Michael Wagner, "Creative Catalyst: A Proposal for a New Attitude Toward Technologies and Aesthetics Brings Man-Made Forces into Harmony with Nature," *Interiors,* March 1993, 53–67.

5 Donella Meadows, "Malling America: How to Stop a Superstore Takeover," *Amicus Journal,* 12.

6 Rose Gutfeld, "Will Poland Plant a Forest to Satisfy a U.S. Architect?", *The Wall Street Journal,* 23 October 1989, p. 1.

7 William McDonough, "A Boat for Thoreau: Architecture, Ethics, and the Making of Things," *Business Ethics,* May/June 1993, 27.

8 William McDonough, "Environmentally Intelligent Textiles: The William McDonough Collection," DesignTex Inc., 1995.

9 Gene Bylinsky, "Manufactured for Reuse," *Fortune,* 6 February 1995, 102–12.

10 Paul Hawken, and William McDonough, "Seven Steps to Doing Good Business," *INC.,* November 1993, 79–90.

Chapter 14 Pattonsburg, Missouri Moving Out of the Flood Plain and Designing an Environmentally Sustainable Community

1 Patricia Leigh Brown, "Higher and Drier, Illinois Town is Reborn," *The New York Times,* 6 May 1966, p. B7.

2 Timothy Egan, "California Storm Brings Rethinking of Development: Some Officials Begin Exploring How to Keep People from Building in Flood Zones," *The New York Times,* 15 January 1995, p. 1.

3 Ibid., p. 1.

4 Nancy Skinner and Bill Becker, *Pattonsburg, Missouri: On Higher Ground*. Case study for the President's Council on Sustainable Development, Sustainable Communities Task Force, January 1995. This is the best source on the background and design team process for New Pattonsburg.

5 These details are drawn from a draft, "Declaration of Community Responsibilities, Covenants and Restrictions," by Dan Slone of McGuire, Woods, Battle & Boothe. Richmond, Virginia.

Chapter 15   Alana Probst Promoting Ecologically Sustainable Businesses in West Coast Temperate Rain Forests

1 Spencer B. Beebe, "Conservation in Temperate and Tropical Rain Forests: The Search for an Ecosystem Approach to Sustainability." Paper presented at the fifty-sixth North American Wildlife and Natural Resources Conference, 1991, pp. 595–603.

2 Edward C. Wolf, *A Tidewater Place: Portait of the Willapa Ecosystem* (Long Beach, Washington: Willapa Alliance, 1993).

3 See Betina Von Hagen and Erin Kellog, "ShoreTrust," *Yes: A Journal of Positive Futures*, B1 (spring/summer 1996): 10.

4 For an excellent account of the Schmitz business and other similar ventures in Willapa Bay, see Janet Maughan "Beyond the Spotted Owl: Investing in 'Green Market' Enterprises Can Be Good for Both Business and the Environment," *Ford Foundation Report* (winter 1995): pp. 4–11.

5 Alana Probst and Glen Gibson, "Oregon Marketplace: Building Local Markets," *Commentary* (winter 1987).

Chapter 16   Daniel Einstein and David Eagan Students Swap Protests for Practical Work Building an Ecologically Sustainable Campus

1 Julian Keniry, *Ecodemia: Campus Environmental Stewardship at the Turn of the 21st Century* (Washington, D.C.: National Wildlife Federation, 1995), pp. 5–23.

2 Tabitha Graves, *Transportation Demand Management (TDM) Programs: Profiles of Selected Universities*, University of Wisconsin—Madison, Environmental Management, Campus Ecology Research Project, report no. 5, December 1993.

3 David J. Eagan and David W. Orr, *The Campus and Environmental Responsibility* (San Francisco: Jossey-Bass, 1992).

4 Karl Weick, "Redefining the Scale of Social Issues," *American Psychologist*, 39, No. 1 (January 1984): 36.

5 Keniry, *Ecodemia*, p. 1.

6 Julian Keniry, "Environmental Movement Booming on Campus," *Change*, September/October 1993, 44.

7 April Smith and the Student Environmental Action Coalition (SEAC), *Campus Ecology: A Guide to Assessing Environmental Quality and Creating Strategies for Change* (Los Angeles: Living Planet Press, 1993). For copies, call 919-967-4600 or 1-800-700-SEAC.

Chapter 17 Jack Turnell Western Cattle Rancher Experiments with Sustainable Techniques

1 Catherine Dold, "Breeding Plan Helps Black-Footed Ferrets in Making Comeback," *The New York Times*, 31 October 1995, p. C4.

2 George Wuerthner, "How the West Was Eaten," *Wilderness*, spring 1991, 28–37.

3 Elizabeth Royce, "Showdown in Cattle Country," *The New York Times Magazine*, 16 December 1990, p. 60.

4 *The 1993 Information Please Environmental Almanac* (Boston: Houghton Mifflin, 1993), pp. 324–7.

5 Wuerthner, "How the West Was Eaten."

6 Ibid., 37.

7 Alan B. Durning and Holly B. Brough, "Taking Stock: Animal Farming and the Environment," *Worldwatch Paper*, no. 103 (July 1991): 7, 20, 22. Durning and Brough cite Ed Chaney et al., *Livestock Grazing on Western Riparian Areas* (Eagle, Idaho: Northwest Resource Information Center, 1990), and Johanna Wald and David Alberswerth, *Our Ailing Public Rangelands: Still Ailing! Condition Report, 1989*, Washington, D.C.: National Wildlife Federation; San Francisco: Natural Resource Defense Council, October 1989).

8 Durning and Brough, "Taking Stock," p. 24.

9 Richard Conniff, "Once the Secret Domain of Miners and Ranchers, the BLM is Going Public," *Smithsonian*, September 1990, 30–46.

10 Wuerthner, "How the West Was Eaten."

11 Royce, "Showdown in Cattle Country."

12 *State of the Public Rangelands, 1990: The Range of Our Vision*, U.S. Department of Interior, Bureau of Land Management, p. 2.

13 Wuerthner, "How the West Was Eaten."

14 Ibid.

15 Ed Marston, "Three Books Take on Cows and Cowboys," *The High Country News*, 23 March 1992, p. 27.

16 Johanna Wald, Ken Rait, Rose Strickland, and Joe Feller, *How Not to Be Cowed: An Owner's Manual* (Natural Resource Defense Council and Southern Utah Wilderness Alliance, 1991).

Chapter 18 Juana Gutierrez The Mothers of East Los Angeles Conserve Water, Protect the Neighborhood, and Create Jobs

1 "A Guide to Water Conservation," Los Angeles Department of Water and Power, n.d., p. 8.

2 For an excellent history of the struggle to improve the quality of the air in Los Angeles, see Eric Mann, *El Aire Mortal de Los Angeles: Nuevas Estrategias para Formular Politicas, Organizarse y Actuar* (Los Angeles: Labor Community Strategy Center, 1993).

3 Rondolfo Acuna, "The Armageddon in Our Back Yard," *Los Angeles Herald Examiner,* 7 June 1989.

Chapter 19 Ron Rosmann Sustainable Agriculture Takes Root Among Family Farmers in Iowa

1 "Use of Pesticides in U.S. on the Rise, EPA Claims," *The Washington Post,* 29 May 1996, p. A16.

2 Verlyn Klinkenborg, "A Farming Revolution: Sustainable Agriculture," *National Geographic,* December 1995, 87.

3 Paul Faeth, Robert Repetto, Kim Kroll, Qi Dai, and Glen Helmers, *Paying the Farm Bill: U.S. Agricultural Policy and the Transition to Sustainable Agriculture* (Washington, D.C.: World Resources Institute, March 1991), p. 5.

4 Ibid., pp. vii, 2, 5.

Chapter 21 Kenny Ausubel Saving the Seed: Rescuing Important Foods and Medicinal Crops from Extinction

1 Paul Raeburn, "Seeds of Security," *The Washington Post,* 6 December 1995, p. A25.

Chapter 22 Eco-Justice Activists Cleaning Up and Reusing Abandoned and Contaminated Industrial Sites

1 Aaron Sachs, "Eco-Justice: Linking Human Rights and the Environment," *Worldwatch Paper,* no. 127 (December 1995): 26.

2 Benjamin F. Chavis Jr. and Charles Lee, *Toxic Waste and Race in The United States: A National Report on the Racial and Socio-Economic Characteristics of Communities with Hazardous Waste Sites* (New York: Commission for Racial Justice, United Church of Christ, 1987). See also Benjamin A. Goldman and Laura Fitton, *Toxic Waste and Race Revisited: An Update of the 1987 Report on the Racial and Socioeconomic Characteristics of Communities with Hazardous Waste Sites* (Washington, D.C.: Center for Policy Alternatives, 1994).

3 Robert D. Bullard, "Grassroots Flowering: The Environmental Justice Movement Comes of Age," *Amicus Journal,* spring 1994, 33. Also, see Bullard's excellent "Environmental Equity: Examining the Evidence of Environmental Racism," *Land Use Forum,* winter 1993, 6–10.

4 See Richard Moore, "Confronting Environmental Racism," *Forward Motion/ Crossroads,* April 1992, 6.

5 According to Robert Bartsch, senior policy analyst at the Northeast-Midwest Institute in Washington, D.C.

6 Jaret Seiberg, "Financing of Inner City Cleanups Can Now Lead to CRA Credits," *The American Banker,* 8 May 1995.

7 Michael E. Porter, "The Competitive Advantage of the Inner City." *Harvard Business Review,* May/June 1995, 55.

8 See Charles Bartsch and Elizabeth Collaton, *Coping with the Challenges of Brownfields,* (Washington, D.C.: Northeast Midwest Institute, April 1995); and Charles Bartsch and Richard Munson, "Restoring Contaminated Industrial Sites," *Issues in Science and Technology,* spring 1994, 74–8.

9 Carol M. Browner, "Brownfields Agenda: Solving an Urban Pollution Problem," *The New York Times,* 22 April 1996, p. A14.

10 See "Cuyahoga County Brownfields Reuse Strategies: Working Group Report," (Cleveland: Cuyahoga County Planning Commission, 13 July 1993).

11 Gary Lee, "Breathing New Life into Brownfields: Incentives Lure Firms to Contaminated Sites," *The Washington Post,* 11 March 1996, p. A4; and Mitchell, Alison, "Clinton Asks Tax Breaks for Toxic Waste Cleanup," *The New York Times,* 12 March 1996, p. 4.

12 Browner, "Brownfields Agenda," p. A14.

Chapter 23   David Gershon Helping Families Minimize Their Environmental Impact One Household at a Time

1 Cited in Michael Silverstein, *Environmental Economics* (New York: St. Martin's Press, 1993), p. 110; and GAP information sheet.

2 David Gershon and Robert Gilman, *EcoTeam: A Program Empowering Americans to Create Earth-Friendly Lifestyles* (Woodstock, New York: Global Action Plan, 1995). For copies, call (914) 679-4830.

3 David Gershon and Steven Connolly, *Journey for the Planet: A Kids Five Week Adventure to Create an Earth-Friendly Life* (Woodstock, New York: Global Action Plan, 1994).

4 World Resources Institute, *The 1992 Information Please Environmental Almanac,* (Boston: Houghton Mifflin, 1992), p. 107.

5 Lester Brown et al., *State of the World 1995* (New York: W.W. Norton, 1995), p. 81.

6 Allen Hershkowitz, "Is Garbage Really A Problem," *Garbage,* spring 1994, 9.

7 *The 1992 Information Please Environmental Almanac,* p. 110; and *The Communicator,* Dade County (Florida), winter 1994.

8 Alan Thein Durning, *How Much Is Enough? The Consumer Society and the Future of the Earth* (New York, W.W. Norton, 1992).

Chapter 24   Thomas Schueler and Robert Boone Two Approaches to Restoring Trashed Urban Rivers

1 Kevin J. Coyle, Dale Pontius, and Randy Showstack, *Endangered Rivers of America: A Report on the Nations Ten Most Endangered Rivers and Fifteen Most Threatened Rivers of 1993* (Washington, D.C.: American Rivers, spring 1993), p. 19.

2 *In the Anacostia Watershed,* 2, no. 2 (Rockville, Md.: Interstate Commission on the Potomac River Basin, summer 1994): 4.

3 Op. cit. *Endangered Rivers of America,* p. 19.

4 Anacostia Restoration Team, "A Blueprint for the Restoration of the Anacostia Watershed," draft, 10 July 1992.

5 "Wheaton Branch Stream Restoration Project," *Audubon Naturalist News,* 19, No. 7 (September 1993): 7.

6 "Corps Study Recommends Anacostia Projects," *Potomac Basin Reporter,* 50, no. 7, Interstate Commission on the Potomac River Basin, September/October, 1994.

7 Sarah Pollock, "Consciousness of Streams," *California,* May 1990, 78.

8 *A Casebook in Managing Rivers for Multiple Uses,* Association of State Wetlands Managers, Association of State Floodplain Managers, and National Park Service, October 1991.

9 Barbara Ruben, "Restoring Washington's Watersheds," *Environmental Action,* spring 1993, 32. The statistics cited above were updated by Robert Boone, president of the Anacostia Watershed Society.

10 D'Vera Cohn, "EPA Pledges Campaign and Money for Anacostia River Cleanup," *The Washington Post,* 1 October 1994, p. 10.

11 Anacostia Watershed Society press release, 19 February 1993.

12 *In the Anacostia Watershed,* 7.

13 David J. Velinsky and James D. Cummins, *Distribution of Chemical Contaminants in Wild Fish Species in Washington, D.C.,* Interstate Commission on the Potomac River Basin, June 1994, Rockville, Maryland, Report no. 94–1.

Chapter 25   The Reverend Jeffrey Golliher A Green Priest Preaches About the Need to Protect God's Creation

1 James Campion, "Once Upon a Universe: Church Leaders Urge Change in Our Approach to Creation—How We Tell the Story and Our Role in It," *Maryknoll,* March 1993, 14 , 16.

2 Jeffrey M. Golliher, "Poison in Poverty's Wound: The International Trade in Hazardous Substances," *Anglican Observer at the UN,* 1993, 28.

3 Thomas Merton, *The Hidden Ground of Love: The Letters of Thomas Merton on Religious Experience and Social Concerns,* ed. William H. Shannon (New York: Farrar, Straus Giroux, 1985), p. 503. Cited in Robert E. Daggy, "Choirs of Millions: A Reflection on Thomas Merton and God's Creatures," *Cistercian Studies Quarterly* 28, no. 1 (1993): 95.

4 Pat Stone, "Christian Ecology: A Growing Force in the Environmental Movement," *Mother Earth News,* January/February 1989. Quoted in *Utne Reader,* November/December, 1989, 78.

5 Thomas Berry, *The Dream of the Earth* (San Francisco: Sierra Club Books, 1988), p. xiii.

6 Stone, "Christian Ecology," p. 79.

7 Stephen B. Scharper and Hillary Cunningham, eds., *The Green Bible* (Maryknoll, New York: Orbis, 1993).

8 Albert Gore Jr., *Earth in the Balance* (Boston: Houghton Mifflin, 1992), p. 12.

9 Jeffrey Golliher has written a background paper on the thinking behind the environmental program at the White Mountain School entitled "Learning How to Live: Steps Toward a Program of Integrative Environmental Education at the White Mountain School," Northeast Environmental Education Project and the René Dubos Consortium for Sacred Ecology at the Cathedral of Saint John the Divine, New York, fall 1992.

10 Jeffrey M. Golliher, "Training for Stewards: The Beginning of an Ecological Education," in *Green Cathedral,* (New York: The René Dubos Consortium for Sacred Ecology at the Cathedral of Saint John the Divine, 1992): 4.

11 Langdon Gilkey, *Message and Existence* (Seabury Press, 1977), p. 152.

12 Ari L. Goldman, "Churches Joining Green Movement," *The New York Times,* 21 May 1992.

13 Alan AtKisson, "Thou Shalt Care for the Earth," *Utne Reader,* March/April 1995, 15.

14 Albert J. Chance, *Embracing Earth: Catholic Approaches to Ecology* (Maryknoll, New York: Orbis, 1994).

26 Conclusion Looking Ahead to Green Taxes and a Green Plan

1 Paul Ekins, Mayer Hillman, and Robert Hutchison, *The Gaia Atlas of Green Economics* (New York: Anchor Books, 1992), p. 66.

2 David Pearce et al., *Blueprint for a Green Economy* (London, Earthscan, 1989), quoted in Ekins et al., *The Gaia Atlas,* p. 67.

3 Lester Brown et al., *State of the World 1996: A Worldwatch Institute Report on Progress Toward a Sustainable Society* (New York: W.W. Norton, 1996), p. 11.

4 President's Council on Sustainable Development, *Sustainable America: A New Consensus for Prosperity, Opportunity, and the Future,* (Washington, D.C.: U.S. Government Printing Office, February 1966), p. 47.

5 Lester Brown et al., *State of the World 1996,* p. 20.

6 Herman Daly and John B. Cobb, Jr., *For the Common Good: Redirecting the Economy Toward Community, the Environment, and a Sustainable Future* (Boston: Beacon Press, 1989), p. 56.

7 Michael Silverstein, *The Environmental Economic Revolution: How Business Will Thrive and the Earth Survive in the Years to Come* (New York: St. Martin's Press, 1993), pp. 33, 82, 97.

8 Huey D. Johnson, *Green Plans: Greenprint for Sustainability* (Lincoln: University of Nebraska Press, 1995).

Index

cattle ranching on public lands,
study on, 272, 273
Business initiatives
ecologically sustainable businesses
(*see* Willapa Bay, Washington ecologically sustainable businesses
project)
green plans, involvement in, 394–395
waste reduction and energy conservation, 390
Byproducts
catalytic extraction processing (CEP)
systems, commercial use of liquid
and solid wastes left over by,
152
compostable, 210
consumables, 210
greenhouse sewage treatment, 49–50,
53–54
heat from molten metal slag, of using, 147
service products, 210
steam from electricity generation,
95–96
terminological switch from waste to
products, 209–210
types of, 210
unmarketable, 210

CAA (Clean Air Act)
Chattanooga, Tennessee sustainable
development initiative, effect on,
173
Calcium magnesium acetate (CMA)
salt for road de-icing, as alternative
to, 251
Caliche
bricks, use in making, 28
defined, 28
earthen walls, used in constructing,
23
surfacing material, as, 24
California
colored cotton, laws restricting growing of, 104
Urban Stream Restoration and Flood
Control Act, 362

Camp Fire Girls
Tejas Council facility, 31–32
Capital depreciation, natural, 14, 386
Capital Environmental, 332
Carpet
toxicity of product and process, reducing, 210–211
Carson, Rachel, 395
Carter, President Jimmy, 395
Catalytic extraction processing (CEP)
systems
applicability of, opinion regarding,
157
baghouse waste, 152
Best Demonstrated Available Technology (BDAT) determination, receipt of, 157
byproducts, 152
ceramic manufacturing, 144
chlorinated wastes, use with, 157
closed-loop technology, status as,
151–152
commercial use of liquid and solid
wastes left over by, 152
constant temperature, importance of
establishing, 149
criticisms of, 153, 157
description of, 143–144
determination of equivalency treatment (DET) compared to incinerators, 157
gases, extracting, 143, 144, 157
heat from molten metal slag, developed from idea of using, 147
incineration versus, 151, 157
information about, dissemination of,
153, 156–157
invention of, 147
metals, extracting, 144
percentage of hazardous waste
streams recycled by, 149
priority wastes, 149
radioactive wastes, Quantum CEP
for handling, 154, 155
RCRA regulations, circumvention of,
157
residues left by, handling, 152–153

small-scale facilities, providing technical and information resources for, 84

university environmental programs, 258

urban river pollution, mapping sources of, 366

Zuni tribal land restoration project, finding land management systems applicable to special problems of, 303

Infrastructure, sustainable, 385–397

Insecticides. *See* Pesticides

Institute for Public Representation, Georgetown Law School, 366

Insulation
New Pattonsburg, Missouri flood recovery and rebuilding project, 224
rooftop gardens providing, 166–167

Integrated pest management (IPM) practical farming, 290

International environmental plans
American resource consumption statistics and practices, significance of, 9–11
cooperation, need for, 11–12
Earth Summit, 1, 9
global threats, response to, xi–xii
green plans, national, 391–394
household waste and consumption, minimizing, 346
Montreal Protocol for ozone layer, 15
practical issues in contrast with global talks, 1–2
United Nations Conference on Environment and Development (UNCED), 1

Ion exchange resins
radiation in nuclear power plant cooling towers, soaking up, 154

IPM (integrated pest management) practical farming, 290

Irish potato famine
low diversity, dangers of, 314, 315

Izaak Walton League of America, 362, 367

Jackson, Milton, 191
Jacobs, Lynn, 272
Jaffee, Dan, 249
Jennings, Cynthia R., 322
Jiang Zemin, President of China, 11
Jobs
brownfields redevelopment leading to, 333, 335
Chattanooga, Tennessee sustainable development initiative providing, 185
ecologically sustainable, 231–242
environmental concerns versus, 373
environmental technology's tendency to generate, 3
import-replacement services creating, 231–242
Las Madres del Este de Los Angeles-Santa Isabel, 277–285
loans for ecologically sustainable business enterprises providing, 236–241
local employment, environmental tasks creating, 285
location-efficient mortgages encouraging walk-to-work jobs, 81, 88–89
mass transit access for inner-city workers to suburban employment, 5, 88
Oregon Marketplace project, 231–242
sustainable agriculture, labor required for, 293
toilets, recycling and replacement of older versions with ultra-low-flush (ULFT) models, 5, 277–281
urban river restoration, job creation and training schemes as part of, 368–369
urban rooftop agriculture generating, 165
walk-to-work jobs, encouraging, 81, 88–89, 187
Willapa Bay, Washington ecologically sustainable businesses project, 231, 236

gardeners using, growing numbers
of, 38–39, 42–44
landscaping, use in, 37–45
legal complications of growing,
41–42
low-maintenance advantages of, 40
non-native species, threats from,
233
resource materials on, 37, 44
university environmental programs
encouraging use of, 255–256
weeds, considered as, 37
Natural capital depreciation, 14, 386
Natural gardening movement, 37–45
Natural gas
fuel cell technology, 96
Naturally colored cotton. *See* Cotton,
naturally colored
Natural Resources Defense Council
(NRDC), xiii, 272, 275, 331
NCA (National Cattlemen's
Association)
Environmental Stewardship Award,
269
NEPP (National Environmental Policy
Plan)
Netherlands green plan, 392–393
Netherlands
green plan, 392–393
New Alchemy Institute
founding of, 48
New Jersey
environmental standards, 396
New Pattonsburg, Missouri flood recovery and rebuilding, 213–229
architectural redesign
assembly of team for, 217–219
houses, positioning relative to
street, 228
interstate, orientation to, 223–224
pedestrian-friendly layout, 223
traditional aspects of downtown, re-
taining, 224
visiting relocation site, 221
clothesline regulations, 225
communication of sustainable development idea, 220–221, 227

cost-effectiveness of relocation, 220
cul-de-sac communities, 219, 228
dam, unrealized plans for, 214
energy conservation requirements,
224, 228
flood macho, 213
government funds for relocation
rather than rebuilding, 215, 220
greenways, 223
history of flooding, 214
interstate bypass, economic conse-
quences of, 214
interstate, orientation to, 224
mixed-use development objectives,
228
pedestrian-friendly design, 223,
228
pig waste, methane gas extraction
from, 225–226
planning or visioning sessions,
220–221
public attitude towards relocation
idea, 220
relocation rather than rebuilding, initial idea to use government funds
for, 215
solar power regulations, 224
Soldiers Grove, Wisconsin as earlier
example of flood relocation,
215–216
stormwater drainage, building to encourage, 221–223
street design, 228
sustainable development
communication of idea to townspeo-
ple, 220–221, 227
energy conservation requirements,
224, 228
extent of, 225
solar power regulations, 224
written code requiring, 224–225
Valmeyer, Illinois project, learning
from, 218–219
visioning sessions, 220–221
wetlands construction to control
stormwater runoff, 223
New Road Map Foundation, 7, 67